AF531369

REAL ANALYSIS

REAL ANALYSIS

By

A.K. Sharma

DISCOVERY PUBLISHING HOUSE PVT. LTD.

NEW DELHI-110 002

Reprinted - 2018

First Published - 2008

ISBN: 978-81-8356-340-6

Real Analysis

Published by:

DISCOVERY PUBLISHING HOUSE PVT. LTD.

4383/4B, Ansari Road, Darya Ganj

New Delhi-110 002 (India)

Phone: +91-11-23279245, 43596064-65

Fax: +91-11-23253475

E-mail: discoverypublishinghouse@gmail.com

sales@discoverypublishinggroup.com

web: www.discoverypublishinggroup.com

Printed at:

Infinity Imaging Systems

Delhi

Preface

This book "REAL ANALYSIS" is written according the latest syllabus prescribed by "UGC" for M.Sc. (first) students of all Indian Universities. It is also meant for these students who are appearing in various competitive examinations. Each topic of this book is treated in a systematic and logical manner. The scope and depth of coverage are strictly according to the prescribed syllabus. A large number of solved examples and graded exercise form an integral part of the text. We do not claim any originality of the results but we claim simplicity and lucidity of presentation coupled with comprehensiveness of the material.

Suggestions, constructive criticism for improvement of the text will be highly appreciated and acknowledged.

Author

CONTENTS

1

The Riemann-Stieltjes Integral

INTRODUCTION

We give the definition of the Riemann-Stieltjes Integral (RS-Integral) and some example based on it may be mentioned that the Riemann-Stieltjes Integral is a generalisation of the Riemann Integral. Recall that a real valued functions is monotonically increasing in an interval *l* if

$f(x) \geq f(y)$, when x, $\geq$ y (x,y $\in$ l).

In elementry treatments the process of integration introduces as the inverse of differenatiation. If F ' (x) = f(x) for all x belonging to the domain of the function f, F is called an a German mathematician integral of the given function f.

G.F.B. Riemann was the first mathematician who gave the satisfactory rigrous arithmetic treatment of integration free from geometrical notation.

To formulate an independent theory of integration, the German mathematician, Riemann, gave a purely arithmetic treatment to the subject and Thus, developed the subject entirely free from the intuitive dependence on geometrical concepts. Many refinements and generalisations of the subject followed, the most noteworthy being *Lebesgue theory* of integration.

The present chapter is, Thus, based on a definition of the Riemann integral which depends very explicitly on the order structure of the real line. Accordingly, we begin by discussing the integration of real bounded functions on intervals. Integration over sets other then intervals is beyond the scope of the present discussion.

The function will always be assumed to be bounded unless otherwise stated.

PARTITION OR DIVISION OF AN INTERVAL [A, B]

Definition 1:

Let [a, b] be a given interval. A finite set of points $x_0, x_1, x_2, ..., x_n$ where $a = x_0 \le x_1 \le x_2 < x_3 ...... \le x_n = b$

The closed sub-interval $I_1 = [x_0, x_1]$, $I_2 = [x_1, x_2]$... $I_n = [x_{n-1}, x_n]$ determined by P constitute the segment of the partition.

By changing the set of points the partition can be changed. Thus, there can be an infinite number of partition of the interval [a, b]. The family of all position of [a, b] shall be denoted by P[a, b].

We write $\Delta x_r = x_r - x_{r-1}$ for, i = 1, 2, ... n.

So that Δx_r is the length of the segment $[x_{r-1}, x_r]$. The *Norm* (or *Mash*) of a partition P is the greatest of the legnths of the segments of a partition P and it is denoted by $||P|| = \max (\Delta x_r, r = 1, 2 ..., n)$.

Some time the *norm* of a partition P is also denoted by $\mu(P)$.

Definition 2:

A partition P is called a refinment of another partition P or we say that P* is finer than P iff $P^* \supset P$ i.e., if every point of P is used in the construction of P*.*

If $P^* = P_1 \cup P_2$, then P& is called the common refinement of the given two partitions P_1 and P_2.

Definition 3:

Let f be a bounded function define once bounded interval [a, b]. Also let $P = [\, a = x_0, x_1, ..., x_n = b]$ be any partition of [a, b]. Also let m_r and M_r be the infimum and supremum respectively of the functin f on.

I_r, for r = 1, 2 ..., n i.e.]

$$mr = \inf\{f(x) : x \le x_{r-1} \le x_r\}$$

and $$mr = \sup\{f(x) : x < x_{r-1} \le x_r\}$$

Let us now form two sums $L(P, f) = = \sum_{r=1}^{n} M_r \Delta x_r$

and $U(P, f) = \sum_{r=1}^{n} M_r \Delta x_r$.

Then L (P, f) is called the lower Riemann sum (or lower Darboux's sum of f on (a, b) with respect tothe partition P and U(P, f) is called the upper Riemann sum (or upper Darboux's sum) of f on (a, b) with respect to the partion P obviously $L(P, f) \le U(P, f)$.

Remarks:

1. The numbers a and b are respectively called the lower and upper limits of integration.

   The length of the greatest of all the sub-intervals $[x_{i-1}, x_i]$ of a partition P is called the norm of P and is denoted by $\mu(P)$.

   Thus, $\Delta x_i \leq m(P)$ for i = 1, 2, ... n.

2. If $\int_a^b f dx$ exists, then it should be understood that the function f is bounded and integrable on [a, b].
3. It M, m are the bounds of a bounded function f over [a, b] then, m (b – a) ≤ LL (p, f) ≤ U(p, f) ≤ M(b – a).
4. Since $\sum_{c=1}^{n} m_i \Delta x_i \leq \sum_{i=1}^{n} M_i \Delta x_i$ so

   L(p, f) ≤ U(p, f).
5. If a bounded function f is Riemann integrable on [a, b] then

   $\int_a^b f(x)dx = \overline{\int_a^b} f\,dx = \underline{\int_a^b} f\,dx$.

6. Every bounded function need not be Riemann integrable.

DEFINITION AND EXISTENCE OF THE INTEGRAL

Let *f* and α be bounded functions on [a, b] and α be *monotonic increasing* on [a, b], b ≥ a.

Let P = $\{a = x_0, x_1, \ldots, x_n = b\}$, of [a, b]. Now we write as follow.

$\Delta\alpha_i = a(x_i) - \alpha(x_{i-1})$, i = 1, 2,...., n.

It is clear that $\Delta\alpha_i \leq 0$, Here, we define two sums,

$$U(P, f, \alpha) = \sum_{i=1}^{n} M_i \, \Delta\alpha_i$$

$$L(P, f, \alpha) = \sum_{i=1}^{n} m_i \, \Delta\alpha_i$$

When m_i, M_i are the bounds (infimum and supremum respectively) of *f* in Δx_i. respectively called the *Upper* and the *Lower* sums of *f* corresponding to the partition *P.*

If *m, M* are respectively the lower and the upper bounds of *f* on [a, b], we have.

$$m \leq m_i \leq M_i \leq M$$

$\Rightarrow m\ \Delta\alpha_i \le m_i\ \Delta\alpha_i \le M_i\ \Delta\alpha_i \le M\Delta\alpha_i,\ \Delta\alpha_i \ge 0$

Putting i = 1, 2,...., n and adding all inequalities, we get

$$m\{\alpha_i(b) - \alpha(a)\} \le L(P, f, \alpha) \le U(P, f, \alpha) \le M\{\alpha(b) - \alpha(a) \quad ...(1)$$

We define two integrals, which always exist by a similar reasoning,

$$\overline{\int_a^b} f\,d\alpha = \inf, U(P, f, \alpha) \quad ...(2)$$

$$\underline{\int_a^b} f\,d\alpha = \sup, U(P, f, \alpha).$$

The infimum and supremum being taken over all partitions of [a, b], These are respectively called the *Upper* and the *Lower* integrals of *f* with respect to α.

There two integrals may or may not be equal. In cases these two integrals are equal, i.e.,

$$\overline{\int_a^b} f\,d\alpha = \underline{\int_a^b} f\,d\alpha$$

We say that *f* is integrable with respect to α in the Riemann sense and write $f \in \Re_\alpha$ [a, b] or simply $\Re(\alpha)$. Their common value is denoted by

$$\int_a^b f\,d\alpha$$

or sometimes by $\int_a^b f(x)\,d\alpha(x)$

And is called the *Riemann-Stieltjes integral* (or simply the *Stieltjes integral* or *f* with respect to α, over [a, b].

From (1) and (2), it follows that

$$m\{\alpha(b) - \alpha(a)\} \le L(P, f, \alpha) \le \underline{\int_a^b} f\,d\alpha \le \overline{\int_a^b} f\,d\alpha$$

$$\le U(P, f, \alpha) \le M\{\alpha(b) - \alpha(a)\}.$$

Remark:

The upper and lower integrals always exist for bounded functions but these may not be equal for all bounded functions. Such functions are not integrable. Thus the question of their equality and hence that of the integrability of the function is our main concern. It we take $\alpha(x) = x$ then the *Riemann-Stieltjes Integral* becomes the usual *Riemann-Integral*.

SOME DEDUCTION

(i) If $f \in \Re(\alpha)$, Then ∃ a number λ lying between the bound of *f* such that

$$\int_a^b f d\alpha = \lambda\{\alpha(b) - \alpha(a)\}$$

(ii) If f M continuous on [a, b] Then ∃ a number $\xi \in$ [a, b] such that

$$\int_a^b f d\alpha = f(\xi)[\alpha(b) - \alpha(a)]$$

(iii) If $f \in \Re(\alpha)$ and k is a number such that $|f(x)| \le k$ for all $x \in [a, b]$ then

$$\left|\int_a^b f d\alpha\right| \le k[\alpha(b) - \alpha(a)]$$

(iv) If $f \in \Re(\alpha)$, $g \in \Re(\alpha)$ over [a, b] such that $f(x) \ge g(x)$ Then

$$\int_a^b f d\alpha \ge \int_a^d g d\alpha, \ b \ge a$$

and $$\int_a^b f d\alpha \le \int_a^d g d\alpha, \ b \le a.$$

(v) If $f \in R(\alpha)$ over [a, b] and $f(x) \ge 0$ for all $x \in [a, b]$ then

$$\int_a^b f d\alpha \ge \int_{\le 0\ b \le a}^{\ge 0\ b \ge a}$$

Since $f(a) \ge 0$ the lower bound $m \ge 0$.

Theorem 1:

If f *is monotonic on* [a, b], *and if a is continuous on [a, b], then* $f \in \Re(\alpha)$.

Proof:

Let $\varepsilon > 0$ be a given positive number.

For any positive integer n, choose a partition $P = \{x_0, x_1, ..., x_n\}$ of [a, b] such that

$$\Delta\alpha_i = \frac{\alpha(b) - \alpha(a)}{n}, \ i = 1, 2, ..., n$$

α is continuous and monotonic increasing on the closed interval [a, b] and thus assumes every value between its bounds, $\alpha(a)$ and $\alpha(b)$. Hence it is true.

Let f be monotonic increasing on [a, b], so that its lower and the upper bound, m_i, M_i in Δx_i are given by

$$m_i = f(x_{i-1}), \ M_i = f(x_i), \ i = 1, 2, ..., n$$

$$\therefore U(P, f\ \alpha) - L(P, f, \alpha) \ \sum_{i=1}^{n} (M_i - m_i)\Delta\alpha_i$$

$$\Rightarrow \frac{\alpha(b)-\alpha(a)}{n} \sum_{i=1}^{n}\{f(x_i)-f(x_{i-1})$$

$$\Rightarrow \frac{\alpha(b)-\alpha(a)}{n}\{f(b)-f(a)\}$$

$< \varepsilon$, for large n

$\Rightarrow f \in \Re$ (a) our [a, b], Hence the theorem.

Note: $\Rightarrow f \in \Re(a)$, i.e., $\int f\, d\alpha$ exists when either

(i) f is continuous and α is monotonic, or

(ii) f is monotonic and α is continuous; of course α is still monotonic.

Theorem 2:

If f is RS-integrable on [a, b], then so is |f| and

$$\left|\int_a^b f\, d\alpha\right| \le \int_a^b |f|\, d\alpha.$$

Proof:

Since $\in R(a)$, there exists a partition P of [a, b] such that

$$U(P, f, \alpha) - L(P, f, \alpha) < \varepsilon$$

whereε is an arbitrary positive number

i.e $\left(\sum_{i=1}^{n} M_i \Delta_{\alpha i} - \sum_{i=1}^{n} m_i \Delta_{\alpha i}\right) < \varepsilon$

or $\sum_{i=1}^{n}(M_i - m_i)\, \Delta\alpha_i < \varepsilon$

where M_i = l.u.b. f(x) $(x_{i-1} \le x \le x_i)$

and m_i = g.l.b. f(x) $(x_{i-1} \le x \le x_i)$

and $\Delta_{\alpha i} = \alpha(x_i) - \alpha(x_{i-1})$

Now, let for the same partition P we have

$$M_i' = \text{l.u.b. } |f(x)|\ (x_{i-1} \le x \le x_i)$$

$$m_i' = \text{g.l.b. } |f(x)|\ (x_{i-1} \le x \le x_i)$$

If ξ_1, ξ_2 are any two points in Δx_i, then

$$||f(\xi_2)| - |f(\xi_1)|| \le |f(\xi_2) - f(\xi_1)|$$

$$\therefore M_i' - m_i' \le M_i' - m_i'$$

i.e. $\sum_{i=1}^{n}(M_i - m_i)\,\Delta\alpha_i \le \sum_{i-1}^{n}(M_i - m_i)\,\Delta\alpha_i < \varepsilon$

Hence, $\sum_{i=1}^{n}(M_i - m_i)\,\Delta\alpha_i < \varepsilon$

i.e. $U(P, |f|, \alpha) - L(P, |f|, \alpha) < \varepsilon$

Thus, $|f| \in R(a)$

Again, since M_i can never exceed M_i'

$$\therefore \left|\sum_{i=1}^{n} M_i\,\Delta\alpha_i\right| \le \sum_{i-1}^{n} M'_i\,\Delta\alpha_i$$

$$\therefore \left|\int_a^b f\,d\alpha\right| \le \int_a^b |f|\,dx$$

Hence the theorem.

Theorem 3:

A function f is integrable with respect to α on [a, b] if and only f for every $\varepsilon > 0$ there exists a partition P of [a, b] such that

Proof:

$U(P, f, \alpha) - L(P, f\ \alpha) < \varepsilon$

Condition am Necessary: Let $f \in \Re(a)$ over [a, b]

$$\therefore \int_a^b f\,d\alpha = \overline{\int_a^b} f\,dx = \int_a^b f\,d\alpha$$

Let $\varepsilon \geq 0$ be any number.

Since the upper and the lower integrals are the infimum and the supremum, respectively, of the upper and the lower sums, therefore $\exists$ partitions P_1 and P_2 such that

$$\Rightarrow U(P_1, f, \alpha) > \overline{\int_a^b} f\,d\alpha + \frac{1}{2}\varepsilon = \int_a^b f\,d\alpha + \frac{1}{2}\varepsilon$$

$$\Rightarrow U(P_2, f, \alpha) > \overline{\int_a^b} f\,d\alpha - \frac{1}{2}\varepsilon = \int_a^b f\,d\alpha - \frac{1}{2}\varepsilon$$

Let $P = P_1 \cup P_2$ be the common refinement of P_1 and P_2.

$\therefore U(P\ f, \alpha) \le U(P_1\ f, \alpha)$

$$< \int_a^b f\,d\alpha + \frac{1}{2}\varepsilon < L(P_2, f, \alpha) + \varepsilon$$

$$\leq L(P, f, \alpha) + \varepsilon$$

$$\Rightarrow L(P, f, \alpha) - L(P, f, \alpha) < \varepsilon$$

Condition M Sufficient

For $\varepsilon > 0$, let P be a partition for which

$$U(P, f, \alpha) - L(P, f, \alpha) < \varepsilon$$

For any partition P, we know that

$$L(P, f, \alpha) \leq \underline{\int_a^b} f\, d\alpha \leq \overline{\int_a^b} f\, d\alpha \leq L(P, f, \alpha)$$

$$\therefore \overline{\int} f\, d\alpha - \underline{\int} f\, d\alpha \leq U(P, f, \alpha) - L(P, f, \alpha) < \varepsilon$$

But a non-negative number can be less than every positive number, if it is zero.

$$\therefore \overline{\int} f\, d\alpha - \underline{\int_a^b} f\, d\alpha \quad \text{so that } f \in \Re(a) \text{ over } [a, b]$$

A Definition

(Integral as a Limit Sum)

As an analog to the Riemann sum, we introduce a sum which will lead to a sufficient condition for the existence of a Riemann-Stieltjes Integral.

Definition: Corresponding to a partition P of [a, b] and $t_i \in \Delta x_i$, consider the sum

$$S(P, f, \alpha) = \sum_{i=1}^{n} f(t_i)\Delta\alpha_i$$

we say that S(P, f, α) converges to A as $\mu(P) \to 0$, i.e.,

$$\lim_{\mu(P)\to 0} S(P, f, \alpha) = A$$

if, for every $\varepsilon > 0$ there exists $\delta > 0$ such that $| S(P, f, \alpha) - A | < e$, for every partition $P = \{a = x_0, x_1, x_2, \ldots, x_n = b\}$, of [a, b] with mesh $\mu(P) < \delta$ and every choice of t_i in Δx_i,

Theorem 4:

If lim S(P, f, α) exists as μ(P) → 0, then

$$f \in \Re(\alpha), \text{ and } \lim_{\mu(P)\to 0} S(P, f, \alpha) = \int_a^b f\, d\alpha$$

Let us suppose that lim S(P, f, α) exists as μ(P) → 0 and equal to A.

Therefore, for ε > 0, ∃ > 0 such that for every partition P of [a, b] with mesh μ(P) < δ and every choice of t_1 in Δx_i, we have

$$| S(P, f, \alpha) - A | < \frac{1}{2}\varepsilon$$

$$\text{or } A - \frac{1}{2}\varepsilon < S(P, f, \alpha) < A + \frac{1}{2}\varepsilon. \quad ...(1)$$

Let P be one such partition, If we let the points t_i range over the intervals Δx_i and take the infimum and the supremum of the sums S(P, f, α), Then from (1) we have

$$A - \frac{1}{2}\varepsilon < L(P, f, \alpha) \le U(P, f, \alpha) < A + \frac{1}{2}\varepsilon \quad ...(2)$$

$$\Rightarrow U(P, f, \alpha) - L(P, f, \alpha) < \varepsilon$$

$$\Rightarrow f \in \mathfrak{R}(\alpha) \text{ over } [a, b].$$

Again, since S(P, f, α) and $\int_a^b f\, d\alpha$ lie between U(P, f, α) and L(P, f, α)

$$\therefore \left| S(P, f, \alpha - \int_a^b f\, d\alpha \right| \le U(P, f, \alpha) - L(P, f, \alpha) < \varepsilon$$

$$\Rightarrow \lim_{\mu(P)\to 0} S(P, f, \alpha) = \int_a^b f\, d\alpha \text{ Hence the theorem}$$

Remark:

The theorem asserts that the existence of the limit of S(P, f, α) implies that f ∈ ℜ(α). The existence of the limit is a sufficient condition for f ∈ ℜ(α) but as shown in Example 3 it is not a necessary condition, i.e., functions exist which are integrable but for which limit of S(P, f, α)

(A RELATION BETWEEN RIEMANN INTEGRAL AND RS-INTEGRAL)

Let f be Riemann integrable on [a, b,] and α monotonically increasing on [a, b] such that α' is Riemann integrable on [a, b], then f ∈ ℜ(α) and

$$\int_a^b f\, d\alpha = \int_a^b f\alpha'\, dx.$$

A particular case of the above property is the following:

(IV) *If f is continuous on [a, b] and α has α continuous derivative on [a, b], the*

$$\int_a^b f\, d\alpha = \int_a^b f\alpha'\, dx.$$

Illustration. $\int_0^2 x^2\,dx^2 = \int_0^2 x^2 \,.\, 2x dx = 8.$

Now, we state a few properties of RS-Integrals.

PROPERTIES OF RS-INTEGRALS

The following are the properties of RS-Integrals

1. If $f_1 \in \Re(\alpha)$ and $f_2 \in \Re(\alpha)$ over [a, b], then
   (i) $f_1 + f_2 \in \Re(\alpha)$ and
   $$\int_a^b (f_1 + f_2)\,d\alpha = \int_a^b f_1\,d\alpha + \int_a^b f_2\,d\alpha.$$
   (ii) $f_1 + f_2 \in \Re(\alpha)$ and $\int_a^b cf_1\,d\alpha = c\int_a^b f_1\,d\alpha$, c is a constant.
2. A function f is RS-integrable w.r.t. α if and only if for each $\varepsilon > 0$, there exists a partition P of [a, b] such that $U(P, f, \alpha) - L(P, f, \alpha) < \varepsilon$
3. If $f \in \Re(\alpha_1)$ and $f \in \Re(\alpha_2)$, then $f \in \Re(\alpha_1 + \alpha_2)$ and
   $$\int_a^b f\,d(\alpha_1 + \alpha_2) = \int_a^b f\,d\alpha_1 + \int_a^b f\,d\alpha_2$$
4. *(Integration by Parts)*
   If f is continuous and α monotone on [a, b], then
   $$\int_a^b f\,d\alpha \;= |\; f(x)\,\alpha\,(x)\;|_\alpha^\beta - \int_a^b \alpha\,df$$
5. If $f \in \Re(\alpha)$ over [a, b] and if $a < c < b$, then $f \in \Re(\alpha)$ on [a, c] and [c, b]. Further $\int_a^b f\,d\alpha = \int_a^c f\,d\alpha + \int_c^b f\,d\alpha$.

Theorem 5:

If f is R-S Integrable on [a, b] and if

$f(x) \mid \leq M$ on [a, b]

then, $\left|\int_a^b f\,d\alpha\right| \leq M\,[\alpha\,(b) - \alpha\,(a)].$

Proof:

We have

$$\left|\int_a^b f\,d\alpha\right| \leq \int_a^b |f|\,dx$$

$$\leq \int_a^b M\,d\alpha = M\int_a^b d\alpha = M\;\;[\alpha\,(b) - \alpha\,(a)]$$

Thus $\left|\int_a^b f\,d\alpha\right| \leq M\,[\alpha(b) - \alpha(a)]$

Theorem 6:

If $f \in R(\alpha_1)$ and $f \in R(\alpha_2)$, then

$f \in R(\alpha_1 + \alpha_2)$ and

$$\int_a^b f\,d(\alpha_1+\alpha_2)=\int_a^b f\,d\alpha_1+\int_a^b f\,d\alpha_2 .$$

Proof:

$f \in R(\alpha_1) \Rightarrow \exists$ a partition P_1 s.t.

$U(P_1, f, \alpha_1) - L(P_1, f, \alpha_1) < \varepsilon/2$ ...(1)

$f \in R(\alpha_2) \Rightarrow \exists$ a partition P_2 s.t.

$U(P_2, f, \alpha_2) - L(P_2, f, \alpha_2) < \varepsilon/2$ ...(2)

where ε is an arbitrary positive number.

If the partitions P_1 and P_2 are replaced by their common refinement P, even the inequalities (1) and (2) hold.

thus $U(P, f, \alpha_1) - L(P, f, \alpha_1) < \varepsilon/2$ ...(3)

$U(P, f, \alpha_2) - L(P, f, \alpha_2) < \varepsilon/2$ ...(4)

Since $\sum_{i=1}^{n} M_i\ [\alpha(x_i) - \alpha(x_{i=1})]$

$= \sum_{i=1}^{n} M_i\ [(\alpha_1 + \alpha_2)(x_i) - (\alpha_1 + \alpha_2)(x_{i=1})]$

where $\alpha = \alpha_1 + \alpha_2$

$\sum_{i=1}^{n} M_i\,[(\alpha_1(x_i) + \alpha_2(x_i) - \alpha_1(x_{i=1}) - \alpha_2(x_{i=1})]$

$= \sum_{i=1}^{n} M_i\,[\{\alpha_1(x_i) - \alpha_2(x_{i-1}) + \Sigma\,\{\alpha_2(x_i) - \alpha_2(x_{i-1})\}]$

$= \sum_{i=1}^{n} M_i\,[\alpha_1(x_i) + \alpha_2(x_i)] + \sum_{i=1}^{n} M_i\,[\alpha_1(x_i) - \alpha_2(x_{i-1})]$

Thus, $U(P, f, \alpha) = U(P, f, \alpha_1) + U(P, f, \alpha_2)$. ...(5)

Similarly,

$L(P, f, \alpha) = L(P, f, \alpha_1) + L(P, f, \alpha_2)$. ...(6)

subtracting (6) from (5) we have

$$U(P, f, \alpha) - L(P, f, \alpha) = [U(P, f, \alpha_1) - L(P, f, \alpha_1)] + [U(P, f, \alpha_2) - L(P, f, \alpha_2)]$$

$\in/2 + \in/2 = \in.$

Hence, $f \in R(\alpha)$ where $\alpha = \alpha_1 + \alpha_2$.

Again, from (5) we have

g.*l*.b. $U(P, f, \alpha)$ = g.i.b. $[U(P, f, \alpha_1) + U(P, f, \alpha_2)]$

g.*l*.b. $U(P, f, \alpha) \geq$ g.i.b. $U(P, f, \alpha_1)$ + g.i.b. $U(P, f, \alpha_2)$.

Thus, $\int_a^b f d\alpha \geq \int_a^b f d\alpha_1 + \int_a^b f d\alpha_2$. ...(7)

Similarly,

l.u.b. $L(P, f, \alpha) \leq$ l.u.b. $L(P, f, \alpha_1)$ + l.u.b. $L(P, f, \alpha_2)$

de $\int_a^b f d\alpha \leq \int_\alpha^b f d\alpha_1 + \int_\alpha^b f d\alpha_2$

Now (7) and (8) $\int_a^b d\alpha = \int_\alpha^b f d\alpha_1 + \int_\alpha^b f d\alpha_2$

Theorem 7:

If $f \in \Re[a, b)$ and a is monotone increasing on [a, b] such that $a' \in \Re[a, b,]$, then $f \in \Re(\alpha)$, and

$$\int_a^b f\, d\alpha = \int_a^b f\alpha'\, dx.$$

Proof:

Let $\in > 0$ be any given number.

Since f is bounded, there exists $M > 0$, such that

$|f(x) \leq M, \forall\, x \in [a, b]$

Again since $f, \alpha' \in \Re\,[a, b]$ therefore $f\alpha' \in \Re\,[a, b]$ and consequently $\exists\, \delta_1 > 0, \delta_2 > 0$ such that

$$\left|\sum f(t_i)\, \alpha'(t_i)\, \Delta x_i - \int \alpha'\, dx\right| < \varepsilon/2 \quad ...(1)$$

for $\mu(P) < \delta_1$ and all $t_i \in \Delta x_i$, and

$$\left|\sum \alpha'(t_i) \Delta x_i - \int \alpha'\, dx\right| < \varepsilon/4M \quad ...(2)$$

for $\mu(P) < \delta_1$ and all $t_1 \in \Delta x_i$.

Now, for $\mu(P) < \delta_2$ and all $t_i \in \Delta x_i$, $s_i \in \Delta x_i$, (2) gives

$$\sum |\alpha'(t_i) - \alpha'(s_i)|\, \Delta x_i < 2.\frac{\varepsilon}{4M} = \frac{\varepsilon}{2M}. \quad ...(3)$$

Again Let $\delta = \min(\delta_1, \delta_2)$, and P any partition with $\mu(P) < \delta$.

Then, for all $t_i \in \Delta x_i$, by Lagrange's Mean Value Theorem, there are points $s_i \in \Delta x_i$ such that

$$\Delta\alpha_i = \alpha'(s_i)\,\Delta x_i \qquad ...(4)$$

Thus, we have

$$\left|\Sigma f(t_i)\,\Delta\alpha_i - \int f\alpha'\,dx\right| = \left|\Sigma f(t_i)\alpha'(s_i)\,\Delta x_i - \int f\alpha' dx\right| \,|$$

$$= \left|\Sigma f(t_i)\alpha'(t_i)\,\Delta x_i - \int f\alpha' dx\right.$$

$$\left. + \Sigma f(t_i)[\alpha'(s_i) - \alpha'(t_i)]\,\Delta x_i\right|$$

$$\le \left|\Sigma f(t_i)\alpha'(t_i)\,\Delta x_i - \int f\alpha'\,dx\right|$$

$$+ \Sigma|f(t_i)|\,|\alpha'(s_i) - \alpha'(t_i)\,\Delta x_i$$

$$< \frac{\varepsilon}{2} + M\frac{\varepsilon}{2M} = \varepsilon. \qquad ...(5)$$

Hence for any $\varepsilon > 0$, $\exists\, \delta > 0$ such that for all partitions with $\mu(P) < \delta$, (5) holds

$\Rightarrow \lim_{\mu(P)\to 0} \Sigma f(t_i)\,\Delta\alpha_i$ exists and equals $\int_a^b f\alpha'\,dx$

$\Rightarrow f \in \Re(\alpha)$, and $\int_a^b f\,d\alpha' = \int_a^b f\alpha'\,dx$

Hence the theorem.

Theorem 8:

(First Mean Value Theorem). If a function f is continuous on [a, b] and α is monotonic increasing on [a, b], then their exists a number ξ in [a, b] such that

$$\int_a^b f\,d\alpha = f(\xi)\,\{\alpha(b) - \alpha(a)\}.$$

Proof:

Since f is continuous and α is monotonic, therefore $f \in \Re(\alpha)$.

Let m, M be the infimum and supremum of f in [a, b]. Then and we know that

$$m\{\alpha(b) - \alpha(a)\} \le \int_a^b f\,d\alpha \le M\{\alpha(b) - \alpha(a)\}$$

Hence, there exists a number μ, $m \le \mu \le M$ such that

$$\int_a^b f\, d\alpha = \mu\{\alpha(b) - \alpha(a)\}$$

Again, since f is continuous, there exists a number $\xi \in [a, b]$ such that $f(\xi) = \mu$

$$\therefore \int_a^b f\, d\alpha = f(\xi)\, \{\alpha(b) - \alpha(a)\}$$

Hence the theorem.

Remark: *It is not possible always to choose ξ such that $a < \xi < b$.*

Consider $\alpha(x) = \begin{cases} 0, & x = a \\ 1, & a < x \le b \end{cases}$

For a continuous function f, we have

$$\int_a^b f\, d\alpha = f(a) = f(a)\, \{\alpha(b) - \alpha(a)\}.$$

Theorem 9:

(Change of variable). If

(i) *f is a continuous function on [a, b], and*

(ii) *ϕ is a continuous and strictly monotonic function on $[\alpha, \beta]$ where* $a = \phi(\alpha)$, $b = \phi(\beta)$

then

$$\int_a^b f(x)\, dx = \int_\alpha^\beta f(\phi(y))\, d\phi(y).$$

Proof:

Let ϕ be strictly monotonic increasing function.

Since ϕ is strictly monotonic, it is invertible, i.e.,

$$x = \phi(y) \Rightarrow y = \phi^{-1}(x),\ \forall\, x \in [a, b]$$

so that, we can write

$$\alpha = \phi^{-1}(a),\ \beta = \phi^{-1}(b)$$

Again let

$$P = \{a = x_0, x_1, x_2, \ldots, x_n = b\}$$

be any partition of [a, b], and

$$Q = \{\alpha = y_0, y_1. y_2, \ldots, y_n = \beta\},\ y_i = \phi^{-1}(x_i)$$

be the corresponding partition of $[\alpha, \beta]$, so that

$\Delta x_i = x_i - x_{i-1} = \phi(y_i)\ \phi(y_{i-1}) = \Delta\phi_i$ ...(1)

Again, for any $x_i \in \Delta x_i$, let $\gamma_i \in \Delta y_i$ where

$x_i = \phi(\eta_i)$ ...(2)

Putting $g(y) = \phi(f(y))$, we have

$$S(P, f) = \sum_i f(\xi_i)\,\Delta x_i$$

$$\Rightarrow \sum_i f(\phi(\eta_i))\,\Delta\phi_i \Rightarrow \sum_i g(\eta_i)\,\Delta\phi_i$$

$= S(Q, g, \phi)$ ...(3)

Continuity of f implies that $S(P, f) \to \int_a^b f\,dx$ as $\mu(P) \to 0$. Also, continuity of g implies that $S(Q, g, \phi) \to \int_\alpha^\beta g(y)\,d\phi$ as $\mu(Q) \to 0$.

Since uniform continuity of ϕ on $[\alpha, \beta]$ implies that $\mu(Q) \to 0$ as $\mu(P) \to 0$, therefore letting $\mu(P) \to 0$ in (3), we get

$$\int_a^b f(x)\,dx = \int_\alpha^\beta g(y)\,d\phi = \int_a^b f(\phi(y))\,d\phi(y).$$

Hence the theorem.

INEQUALITY FOR INTERGRALS

We know that for a bounded integrable function f

$$m(b-a) = \int_a^b f dx \le M(b-a), \text{ when } b \ge a.$$

If $b < a$, so that $a > b$ ...(1)

$$m(a-b) \le \int_a^b f dx \le M(a-b), \text{ when } a > b.$$

$$\Rightarrow m(a-b) \ge -\int_a^b f dx \ge M(a-b)$$

$$\Rightarrow m(b-a) \ge \int_a^b f dx \ge M(b-a), \text{ when } b < a\ .$$ ..(2)

We deduce the following result from these two inequality.

1. If f, g are bounded and integrable on [a, b] such that $f \ge g$ then

   $$\int_a^b f dx \ge \int_a^b g dx \quad \text{when } b \ge a$$

   and $\int_a^b f dx \le \int_a^b g dx$ when $b \le a$

2. If f is bounded and integrable on [a, b] and f(x) > 0 for all x ∈ [a, b] then

$$\int_a^b f(x) \quad \begin{matrix} \geq 0 & \text{if } b \geq a \\ \leq 0 & \text{if } b \leq a \end{matrix}$$

3. If f is continuous and integrable on [a, b] then there exist a number c between a and b such that $\int_a^b fdx = (b-a)f(c)$

4. If f is bounded and integrable on [a, b] then there exist a number λ lying between the bounded of f such that

$$\int_a^b fdx = \lambda(b-a).$$

5. If f is bounded and integrable on [a, b] and b is a number such that |f(x)| ≥ k for all n∈(, b) then

$$\left|\int_a^b fdx\right| \leq k(b-a).$$

REFINEMENT OF PARTITIONS

Let P and p* be two partitions of [a, b] if p ∠ p* then p* is called a refinement of p

OR

For any partition p, the length of the largest sub internal is called the norm or mesh of the partition and is denoted as m(p) or simply μ.

$\therefore$ $\mu(p) = \max Dx_i$, $(1 \leq i \in n)$

A partition p^* is said to be a refinement of p if $p^* \geq p$ i.e., every point of p is a point of p^*.

Theorem 10:

If P is a refinement of a partition P, then for a bounded function f,*

(i) L (P*, f) ⊠ L (P, f), and

(ii) U (P*, f) ≤ U (P, f).

Proof:

Suppose first that P* contains just one point more than P, Let this extra point be ξ, and suppose that this point is in Δx_i that is, $x_{i-1} < \xi\ x_i$

As the function is bounded over the entire interval [a, b], it is bounded in every sub-intervals Δx_i (i = 1, 2, ..., n). Let w_1, w_2, m_1 be the infimum (g.l.b.) of f in the intervals $[x_{i-1}\ \xi]$. $[\xi, x_1]$, $[x_{i-1}, x_i]$ respectively.

Clearly $m_i \leq w_1$ $m_i \leq w_2$

$\therefore$ L(P*, f) $= L(p, f)$

$$= w_1(\xi - x_{i-1}) + w_2(x_1 - \xi) - m_i(x_i - x_{i-1})$$

$$= (w_1 - m_i)(\xi - x_{i-1}) + (w_2 - m_i)(x_i - \xi)$$

$$\leq 0 \text{ } (\because \text{ each bracket is positive})$$

If p* contains p points more than p, we repeat the above reasoning p times and arrive at

$$L(p^*, f) \geq L(p, f)$$

Similarly, we can prove that

$$U(p^*, f) \leq U(p, f).$$

Cor. 1:

If M, m are the bounds of an integrable function f over [a, b], then

$$m(b-a) \leq \int_a^b f dx \leq M(b-a)$$

Proof:

We have

$$m(b-a) \leq L(P, f) \leq U(P, f) \leq M(b-a)$$

or $m(b-a) \leq L(P, f) \leq \underline{\int_a^b} f dx \leq \overline{\int_a^b} f dx,$

(By Cor. 3)

$$\leq U(P, f) \leq M(b-a)$$

$$\therefore m(b-a) \leq \underline{\int_a^b} f dx \leq \overline{\int_a^b} f dx \leq M(b-a)$$

Since f is integrable, $\underline{\int_a^b} f\,dx + \overline{\int_a^b} f\,dx = \int_a^b f\,dx$

Hence, $m(b-a) < \int_a^b f\,dx\ M(b-a)$.

Cor. 2:

If f is a bounded function on [a, b], then

$$\underline{\int_a^b} f\,dx \leq \overline{\int_a^b} f\,dx.$$

Proof:

Let P_1 and P_2 be two partitions of [a, b].

(By Cor. 2)

$L(P_1, f) \le U(P_2, f)$.

Keeping P_2 fixed and taking the supremum over all partitions P_1 of [a, b], we obtain

$$\underline{\int_a^b} f\,dx \le U(P_2, f)$$

Taking the infimum over all partitions P_2 of [a, b], we get

$$\underline{\int_a^b} f\,dx \le \overline{\int_a^b} f\,dx.$$

Cor 3:

Let p be a refinement of P containing p points more than P, where P is a partition of [a, b] with norm $\mu(P) < \delta$. Let*

$|f(x)| \le k \ \forall\ x \in [a, b]$.

Then (i) $U(P^*, f) \le U(P, f) - 2pk\delta$,

(ii) $L(P^*, f) \ge L(P, f) + 2pk\delta$.

Proof:

Since $|f(x)| < k \quad \forall\ x \in [a, b]$, therefore

$-k \le f(x) \le k \ \forall\ x \in [a, b]$

$\Rightarrow -k \le Mi' \le Mi \le k$

$\Rightarrow 0 \le m_i - m_i' \le 2k.$ ...(3)

Similarly, $0 \le M_i - M_i \le 2k.$ ...(4)

Using (3) and (4) in (2) above, we obtain

$U(P, f) - U(P^*, f) \le 2k(x_i - x_{i-1}) \le 2k\delta$

$\Rightarrow U(P, f) - 2k\delta \le U(P^*, f)$

$\therefore\ U(P^*, f) \ge U(P, f) - 2pk\delta.$

Taking all the p points of P* one by one, we obtain

$U(P^*, f) \ge U(P, f) - 2pk\delta.$

Similarly we can prove the second part.

Cor 4:

If P_1 and P_2 are any two partitions of [a, b], then

$L(P_1, f) \le U(P_2, f)$.

In other words, an upper sum can never be less than any lower sum.

Proof:

Let $P^* = P_1 \cup P_2$. Then $P^* \supset P_1$ and $P^* \supset P_2$.

By Theorem 1, $L(P_1, f) \leq L(P^*, f)$ and $U(P^*, f) \leq U(P_2, f)$.

Now $L(P_1, f) \leq L(P^*, f) \leq U(P^*, f) \leq U(P_2, f)$.

Hence, $L(P_1, f) \leq U(P_2, f)$.

DARBOUX'S THEOREM

Theorem 11:

If f is a bounded function on [a, b], then to every $\in > 0$, there corresponds $\delta > 0$ such that

$$\text{(i)}\ U(P, f) < \overline{\int_a^b} f\,dx + \in,$$

$$\text{(ii)}\qquad L(P, f) > \underline{\int_a^b} f\,dx + \in$$

for every partition P of [a, b] with norm $\mu(P) < \delta$.

Proof:

(i) As f is bounded, therefore $\exists\ k > 0$, such that

$|f(x)| \leq k, \quad \forall\ x \in [a, b]$

Again, since the upper integral is the infimum (g.l.b.) of the set of upper sums, therefore to every $\in > 0$ there exists a partition $P_1 = \{x_0, x_1, x_2, \ldots, x_p\}$ of [a, b] such that

$$U(P_1, f) < \overline{\int_a^b} f\,dx + \frac{1}{2} \in. \qquad ...(1)$$

The partition P_1 has $p - 1$ points besides x_0 (= a) and x_p(=b). Let δ be a positive number such that

$$2k(p-1)\,\delta = \frac{1}{2} \in. \qquad ...(2)$$

Let P be any partition with norm $\mu(P) < \delta$.

Let, further, P^* be a refinement of P and P_1 so that $P^* = P \cup P_1$.

As P^* is a refinement of P having at the most p–1 more points than P, therefore (using Corollary, Theorem 1) we get

$U(P, f) - 2k(p-1)\delta \leq U(P^*, f)$

$\leq U(P_1, f)$

$$\leq \overline{\int_a^b} f\,dx + \frac{1}{2} \in \text{ (using (1))}$$

$$\Rightarrow U(P, f) < \overline{\int_a^b} f\,dx + \frac{1}{2} \in + \frac{1}{2} \in = \overline{\int_a^b} f\,dx + e \text{ (using (2))}$$

for any partition P with norm $\delta(P) < \delta$.

Similarly we can prove the other part.

Cor. 5:

If f is bounded and integrable on [a, b] such that $|f(x)| \leq k \ \forall x \leq$ (a, b) then

$$\left|\int_a^b f\,dx\right| \leq k\ |b-a|\ (k > 0).$$

Proof:

If a = b then the result is true but M d m be the bounded of f in [a, b]

Since $|f(x)| \leq \forall x \in (a\ b)$ therefore

$$(-k) \leq f(x) \leq k \ \forall x((a, b))$$

Since m = inf f and M = sub f so

$$-k \leq m \leq f(x) \leq M \leq k \ \forall x \in (a, b)$$

Let b > a so that b–a > 0

$$\therefore -k (b-a) \leq m (b-a) \leq \int_a^b f(x)dx$$

$$\leq M(b-a) \leq k(b-a)$$

$$\Rightarrow -k (b-a) \leq \int_a^b f\,dx \leq k(b-a)$$

$$\therefore \left|\int_a^b f\,dx\right| \leq k(b-a) \qquad ...(1)$$

It b < a en a < b from (1)

$$\left|\int_a^b f\,dx\right| \leq k(a-b) \Rightarrow \left|-\int_a^b f\,dx\right| - k(b-a)$$

$$\text{or } \left|\int_a^b f\,dx\right| \leq -k(b-a) \qquad ...(2)$$

from (1) and (2)

$$\left|\int_a^b f\,dx\right| \leq k(b-a).$$

Cor. 6:

If f is bounded and integrable on [a, b] and f(x) ≥ 0 ∀x ∈ [a, b], then

(i) $\int_a^b f\,dx \geq 0$, if b ≥ a and

(ii) $\int_a^b f\,dx \geq 0$, if b ≤ a.

Proof:

f(x) ≥ 0 ∀ x∈ [a, b] ⇒ m = inf f ≥ 0 and M = sup f ≥ 0.

We know $m(b-a) \leq \int_a^b f\,dx$

$\Rightarrow \int_a^b f\,dx \geq m(b-a) \geq 0$, if b ≥ a.

Again $\int_a^b f\,dx \leq M(b-a) \leq 0$, if b ≤ a.

INTEGRAL AS A LIMIT OF SUMS

Let f be bounded real valued function defined on [a, b] and let α be monotonically increasing.

Let p = (a = x_0, x_1, x_2 ..... x_n = b) be any partition of [a, b] choose points $\xi_1, \xi_2, ... \xi_n$ such that

$x_{r-1} \leq \xi_r \leq x_r$. r = [1, 2, 3, n] and form the sum $S(p, f, \alpha) = \sum_{0=1}^{n} f(\xi_r)\Delta x_r$.

It there is a number l for which for every ∈ > 0 there exist a number δ > 0 such that $|l - S(p, f, a)| <$ ∈ whenever μ(P) < δ we say that S(p, f, α) → l as μ (p) → 0 i.e. $\lim_{\mu(p)\to 0} S(p, f, \alpha) = l$.

Theorem 12:

If f is bounded and integrable over [a, b], then to every ε > 0, there corresponds a δ > 0, such that for every partition

P = (a, = x_0, x_1 ...., x_{r-1}, x_r, ... x_n = b) of norm m(P) ≤ d and for every choice of x, ∈ [x_{r-1}, x_r].

$$\left|\sum_{r=1}^{n} f(\xi_r)(x_r - x_{r-1}) - \int_a^b f(x)dx\right| < \varepsilon.$$

Proof:

Since f is a bounded and integrable function, so

$$\overline{\int_a^b} f(x)dx = \underline{\int_a^b} f(x)dx = \int_a^b f(x)dx.$$

Let ε be any positive number. By Darboux's Theorem, these exists a δ > 0 such that for every partition with μ(P) ≤ δ.

$$U(P, f) < \overline{\int_a^b} f(x)dx + \varepsilon = \int_a^b f(x)dx + \varepsilon, \qquad ...(1)$$

and $$L(P, f) > \underline{\int_a^b} f(x)dx - \varepsilon = \int_a^b f(x)dx - \varepsilon \qquad ...(2)$$

If x_r be any point of the interval [xr = 1, x_r] of P, then

$$L(P, f) \le \sum_{r=1}^{n} f(\xi_r)(x_r - x_{r-1}) \le U(P, f). \qquad ...(3)$$

From (1), (2) and (3), we see that for every partition P with μ(P) ≤ δ,

$$\int_a^b f(x)dx - \varepsilon < \sum_{r=1}^{n} f(\xi_r)(x_r - x_{r-1}) < \int_a^b f(x)dx + \varepsilon$$

$$\Rightarrow \left| \sum_{r=1}^{n} f(\xi_r)(x_r - x_{r-1}) - \int_a^b f(x)dx \right| < \varepsilon.$$

Hence the theorem.

Remark:

Let $$S(P,f) = \sum_{r=1}^{n} f(\xi_r)(x_{r-1}) = \sum_{r=1}^{n} f(\xi_r)\Delta x_r,$$

where $\xi_{r-1} \le x_r \le x_r$ (r = 1, 2, ..., n).

The sum S(P, f) is called a *Riemann Sum* of f over [a, b] w.r.t. partition P of [a, b]. The above theorem may be restated as.

If a function f is bounded and integrable over [a, b], then

$$\int_a^b f\,dx = \lim_{\mu(P)\to 0} S(P, f).$$

R-INTEGRABILITY

Definition : Lef f be a bounded function defined on the bouded interval [a, b], then f is called Riemann integrable (or simply R-integrable) on [a, b] iff,

$$\underline{\int_a^b} f = \overline{\int_a^b} f$$

and their common value is called the R-integral of f on [a, b] and is denoted by $\int_a^b f$.

The class of all bounded functions f which ar Riemann integrable on [a, b]. The numbers a and b are called the *lower and upper limits integration* respectively.

If $\int_{-a}^{b} f \neq \int_{a}^{-b} f$ then f is not *Riemann* integrable on [a, b].

Note 1 : The concept of integrability of a function over an integral as introduced here is subjected to very important limitations :

(i) The function is bounded

(ii) The interval of integration is finite.

i.e., neither of the end points is finite.

Note 2: It is not necessary that every bounded function is integrable i.e., There may exist a bounded function f for which.

$$\int_{-a}^{b} f \neq \int_{a}^{-b} f$$

Note 3 : The statement that $\int_{a}^{b} f$ exist indicate that the functions is bounded and integrable over [a, b].

SECOND DEFINITION OF RIEMANN INTEGRAL

We know that a bounded function f on an internal [a, b] is Riemann integrable if

$$\overline{\int_{a}^{b}} f\,dx = \int_{-a}^{b} f\,dx. \qquad \text{(Definition I)}$$

Definition II:

Definition: A function f is said to be integrable over [a, b], if there exists a number I and to every $\in > 0$ corresponds a $\delta > 0$, such that for every partition P : $\{a = x_0, x_1, ..., x_{r-1}, ..., x_n = b\}$ of norm $\leq \delta$ and $\forall\ \xi_r \in [x_{r-1}, x_r]$

$$\left|\sum_{r=1}^{n} f(\xi_r)(x_r - x_{r-1}) - 1\right| < \in.$$

The number I is said to be the integral of the function f over the interval [a, b].

i.e., $\quad I = \int_{a}^{b} f(x)dx.$

Theorem 13 : Definition 2 ⇒ Definition 1

Proof:

Let f be integrable by second Def. Now we show that f is bounded in [a, b].

By Def. 2, for $\varepsilon = 1$, there exists a partition P of [a, b] such that for every choice of $x_r \varepsilon [x_{r-1}, x_r]$,

$$|\Sigma f(\xi_r)\delta_r - I| < 1 \Rightarrow |\Sigma f(\xi_r)\delta_r| < |I| + I, [d_r = x_r - x_{r-1}]$$

Since f is not bounded in [a, b], it must also be so in at lest one sub-interval say $[x_{m-1}, x_m]$.

We take $\xi_r = x_r$ when $r \neq m$ so that every number ξ_r (except ξ_m) is fixed. Accordingly, every term of the sum $\Sigma f(\xi_r) \delta_r$ (except the term $f(\xi_m)\delta_m$) is also fixed. Since f is not bounded in $[x_{m-1}, x_m]$, we choose a point x_m in $[x_{m-1}, x_m]$, such that

$$|\Sigma f(\xi_r)\delta_r| > |I| + 1.$$

and Thus, we arrive at a contradiction. Hence, f is bounded in [a, b].

Now, let ε by any point positive number. There exists a $\delta > 0$ such that for every partition P whose norm is $\leq\delta$,

$$|\Sigma f(\xi_r)\delta_r - I| < \frac{1}{2}\varepsilon, \quad \text{(Def. 2)}$$

$$\Rightarrow I - \frac{1}{2} \leq < \Sigma f(\delta_r)\delta_r < I + \frac{1}{2}\varepsilon \quad ...(1)$$

for every choice of point x_r in $[x_{r-1}, x_r]$

Let $M_r = \sup f$, $m_r = \inf f$ in $[x_{r-1}, x_r]$. Then there exist points α_r, β_r in $[x_{r-1}, x_r]$ such that

$$f(\alpha_r) > M_r - \frac{1}{2}\varepsilon\,(b-a),\ f(\beta_r) < m_r + \frac{1}{2}\varepsilon\,(b-a).$$

From these relations, we obtain

$$\Sigma f(\alpha_r)\delta_r > U(P, f) - \frac{1}{2}\varepsilon \Leftrightarrow U(P, f)$$

$$< \Sigma f(\alpha_r)\delta_r + \frac{1}{2}\varepsilon, \quad ...(2)$$

$$\Sigma f(\beta_r)\delta_r < U(P, f) + \frac{1}{2}\varepsilon \Leftrightarrow L(P, f)$$

$$> \Sigma f(\beta_r)\delta_r - \frac{1}{2}\varepsilon, \quad ...(3)$$

Taking $x_r = a_r$ and b_r separately in (1) and using (2) and (3);

$$I - \varepsilon \le L(P, f) \le U(P, f) < I + \varepsilon \quad ...(4)$$

for every partition P whose norm is $\le \delta$.

We know that

$$L(P, f) \le \underline{\int_a^b} f(x)dx \le \overline{\int_a^b} f(x)dx \le U(P, f), \quad ...(5)$$

From (4) and (5), we have

$$I - \varepsilon < \underline{\int_a^b} f(x)dx \le \overline{\int_a^b} f(x)dx < I + \alpha \quad ...(6)$$

$$\Rightarrow 0 < \overline{\int_a^b} f(x)\,dx - \underline{\int_a^b} f(x)dx < 2\varepsilon \quad ...(7)$$

Since ε is arbitrary, so (7) gives

$$\overline{\int_a^b} f(x)\,dx - \underline{\int_a^b} f(x)dx = 0$$

$$\Rightarrow \overline{\int_a^b} f(x)\,dx = \underline{\int_a^b} f(x)dx \quad ...(8)$$

so that the function f is integrable according to he first definition,

From (6) and (8), we have

$$I - \varepsilon < \int_a^b f(x)dx < I + \varepsilon .$$

Since ε is arbitrary, this gives

$I = \int_a^b f(x)dx$ Hence Def. 2 $\Rightarrow$ Def. 1.

CONDITIONS OF INTEGRABILITY

Theorem 14:

(First Form) A necessary and sufficient condition for the integrability of a bounded function f is that to every $\varepsilon > 0$, there corresponds a $\delta > 0$ such that for every partition P of [a, b] with norm $\mu(P) < \delta$.

$$U(P, f) - L(P, f) < \varepsilon$$

or the oscillatory sum $w(P, f) < \varepsilon$.

Proof:

(The Condition is Necessary)

Let f be integrable. Then we have

$$\int_a^b f\,dx = \overline{\int_a^b} fdx = \underline{\int_a^b} fdx \quad ...(1)$$

Now let ε > 0 be given, ∃ some δ > 0 such that for every partition P with norm m(P) < δ,

$$\therefore\ U(P,f) < \overline{\int_a^b} f dx + \frac{1}{2}\varepsilon = \int_a^b f\,dx + \frac{1}{2}\varepsilon, \quad ...(2)$$

$$\text{and}\quad L(P, f) > \underline{\int_a^b} f\,dx - \frac{1}{2}\varepsilon = \int_a^b f\,dx - \frac{1}{2}\varepsilon \quad \text{[Using (1)]}$$

$$\text{or}\quad -L(P, f) < -\int_a^b f\,dx + \frac{1}{2}\varepsilon, \quad ...(3)$$

From (2) and (3), we get

$U(P, f) - L(P, f) < \varepsilon$,

for every partition P with norm μ(P) < δ.

(The Condition is Sufficient)

Let $U(P, f) - L(P, f) < \varepsilon$, ...(4)

for every partition P of [a, b] with norm μ(P) < δ.

We have

$$L(P, f) \le \underline{\int_a^b} f\,dx \le \overline{\int_a^b} f\,dx \le U(P,f)$$

$$\Rightarrow \overline{\int_a^b} fdx + \underline{\int_a^b} f\,dx < U(P, f) - L(P, f) < \varepsilon, \text{ by (4)}$$

$$\therefore\ \overline{\int_a^b} fdx - \underline{\int_a^b} f\,dx < \varepsilon \quad ...(5)$$

Since (5) is true ∀ ε > 0, so

$$\overline{\int_a^b} f\,dx - \underline{\int_a^b} f\,dx = 0 \Rightarrow \overline{\int_a^b} f\,dx = \underline{\int_a^b} f\,dx.$$

Hence, f is integrable.

Hence, the theorem.

Theorem 15:

(Second Form) A bounded function f is integrable on [a, b] iff for every ε > 0, there exists a partition P of [a, b] such that

$U(P, f) - L(P, f) < \varepsilon$ or $w(P, f) < \varepsilon$.

Proof:

(N. Condition) Let f be integrable.

$$\Rightarrow \underline{\int_a^b} f\,dx = \overline{\int_a^b} f\,dx = \int_a^b f\,dx. \quad ...(1)$$

Let $\varepsilon > 0$ be given. Since the upper integral is the infimum of the upper sums, so there exists a partition P_1 of [a, b] such that

$$U(P_1, f) < \overline{\int_a^b} f\,dx - \frac{1}{2}\varepsilon. \qquad ...(2)$$

Similarly, there exists a partition P_2 of [a, b] such that

$$L(P_2, f) > \underline{\int_a^b} f\,dx - \frac{1}{2}\varepsilon. \qquad ...(3)$$

Let $P = P_1 \cup P_2$ so that P is a refinement of P_1 and P_2.

We have

$$\left.\begin{aligned} U(P, f) &\leq U(P_1, f), \\ L(P, f) &\geq L(P_2, f). \end{aligned}\right] \qquad ...(4)$$

Now, $U(P, f) \leq U(P_1, f) < \overline{\int_a^b} f\,dx + \frac{1}{2}\varepsilon$, by (2) and (4)

Using (1)m $U(P, f) < \int_a^b f\,dx + \frac{1}{2}\varepsilon.$

Again $L(P, f) \geq L(P_2, f) > \underline{\int_a^b} f\,dx - \frac{1}{2}\varepsilon.$, by (3) and (4)

$\therefore L(P, f) > \int_a^b f\,dx + \frac{1}{2}\varepsilon,$ using (1)

or $-L(P, f) < -\int_a^b f\,dx + \frac{1}{2}\varepsilon, \qquad ...(6)$

From (5) and (6), we get $U(P f) - L(P, f) < \varepsilon$.

(S. Condition)

Let $\varepsilon > 0$ be given. Let P be a partition for which

$U(P, f) - L(P, f) < \varepsilon \qquad ...(7)$

We know that

$$L(P, f) \leq \underline{\int_a^b} f\,dx \leq \overline{\int_a^b} f\,dx \leq U(P, f)$$

$$\Rightarrow \overline{\int_a^b} f\,dx - \underline{\int_a^b} f dx \leq U(P, f) - L(P, f) < \varepsilon \quad \text{by (7)}$$

$$\Rightarrow \overline{\int_a^b} f\,dx - \underline{\int_a^b} f dx < \varepsilon \qquad ...(8)$$

Since (8) is true $\forall\ \varepsilon > 0$, so

$$\overline{\int_a^b} f\,dx - \underline{\int_a^b} f dx = 0$$

$$\Rightarrow \overline{\int_a^b} f\,dx - \underline{\int_a^b} f dx.$$

Hence, f is integrable.

Theorem 16:

If a function f is monotonic on [a, b], then it is integrable on [a, b].

Proof:

Since f is monotonic on [a, b] so f is bounded on [a, b]. Suppose f is monotonically increasing. So $f(b) \geq f(a)$.

Let $\varepsilon > 0$ be given. We choose $\eta > 0$ such that

$$\eta < \frac{\varepsilon}{f(b) - f(a) + 1}$$

Let $P = \{a = x_0, x_1, \ldots.., x_n = b)$ be a partition of [a, b] with $\mu(P) < \eta$.

Let M_i, m_i be the bounds of f in $[x_{i-1}, x_i]$.

Since f is monotonically increasing, so $f(x_{i-1}) \leq f(x_i)$.

Thus, $M_i = f(x_i)$, $m_i = f(x_{i-1})$ for $i = 1, 2, \ldots, n$.

Now, $U(P, f) - L(P, f) = \sum_{i=1}^{n}(M_i - m_i)(x_i - x_{i-1})$

$$= \sum_{i=1}^{n}\{f(x_i) - f(x_{i-1})\}(x_i - x_{i-1})$$

$$< \eta \sum_{i=1}^{n}\{f(x_i) - f(x_{i-1})\}(\because \mu(P) < \eta)$$

$$< \frac{\varepsilon}{f(b) - f(a) + 1}.\{f(b) - f(a)\}$$

$\therefore \quad U(P, f) - L(P, f) < \varepsilon \; [\; \because \; f(b) - f(a) < f(a) + 1]$

Hence, f is integrable on [a, b].

Theorem 17:

Every continuous function is integrable.

Proof:

We now prove that f is continuous and integrable on [a, b]

let $\in > 0$ be given

Let us choose a positive number η, such that

$$\eta(b - a) < \in$$

Since f is continuous on the closed interval [a, b], therefore, it is bounded and is *uniformly continuous* on [a, b], which implies that there exists $\delta > 0$, such that

$| f(t_1) - f(t_2) | < \eta$, if $| t_1 - t_2 | < \delta$,

and $t_1, t_2 \in [a, b]$ ...(1)

We, now, choose a partition P with norm $\mu(P) < \delta$.

Then by (1), we have

$M_i - m_i \le \eta$ (i = 1, 2, ... n)

$$\therefore U(P, f) - L(P, f) = \sum_{i=1}^{n} (M_i - m_i)\,\Delta x_i$$

$$\le \eta \sum_{i=1}^{n} \Delta x_i$$

$$= \eta\,(b - a) < \epsilon \quad \text{....(2)}$$

Thus, f is integrable [4 Theorem 3].

Corollary: *If a function f is continuous, then to every* $\epsilon > 0$ *there corresponds* $\delta > 0$ *such that*

$$\left| \sum_{i=1}^{n} f(t_i)\,\Delta x_i - \int_a^b f\,dx \right| < \epsilon$$

for every partition $P = \{x_0, x_1, x_2, ..., x_n\}$ *of [, b] with* $\mu(P) < \delta$, *and for every choice of points* t_i *in* $[x_{i-1}, x_i]$.

Follows from (2) above, since the two numbers $\sum f(t_1)\,\Delta x_i$ and $\int f\,dx$ lie between U(P, f) and L(P, f).

Remark:

For continuous function $\lim_{\mu(P)\to 0} \Sigma f(t_i)\,\Delta x_i$ exists and equals $\int_a^b f dx$.

Theorem 18:

If function f is bounded in [a, b] and the set of its points of discontinuity has a finite number of limit points, then f is integrable in [a, b]

Proof:

Let $a_1, a_2, a_3, ..., a_p$

be p limit points of the set of points of discontinuity of f in [a, b]. We enclose them in, p, non-overlapping intervals

$[a_1', a_1''], [a_2', a_2''], \ldots\ldots, [a_p', a_p'']$

such that the sum of their lengths is $< \varepsilon/2(M - m)$; M, m being the bounds of f in [a, b]. The contribution to the oscillatory sum by these intervals is $<[\varepsilon/2\ (M-m)], (M - m) = \varepsilon/2$.

It is clear that only a *finite* number of points of discontinuity of f can lie in each of the (p + 1) intervals

$[a, a_1'], [a_2'', a_1'], \ldots\ldots, [a_p'', b]$.

By Theorem 8, the contribution to the oscillatory sum arising from each of these (p + 1) intervals is separately

$< \varepsilon/2\ (p + 1)$.

Thus, there exists a partition of [a, b] such that the corresponding oscillatory sum is

$$< \frac{\varepsilon}{2} + \frac{\varepsilon}{2(p+1)}(p+1) = \varepsilon$$

Hence, f is integrable in [a, b].

Theorem 19:

A bounded function f which has only a finite number of points of discontinuity in [a, b] is integrable in [a, b].

Proof:

Let M, m be the bounds of f in [a, b]

Let $a_1, a_2, a_3, \ldots, a_p$

be p points of discontinuity of f in [a, b]

Let ε be a positive number. We enclose the points $a_1, a_2, \ldots a_p$, in p non-overlapping intervals

$[a_1', a_1''], [a_2', a_2''], \ldots, [a_p', a_p'']$

such that the sum of their lengths is $< \varepsilon/2\ (M - m)$. The oscillation of f in each of these intervals in $\leq (M - m)$ and accordingly the contribution to the oscillatory sum from these is

$<[\varepsilon/2(M-m)], (M - m) = \varepsilon/2$. ...(1)

Now, f is continuous in the (p + 1) sub-intervals

$[a, a_1'], [a_1'', a_2'], \ldots [a_p'', b]$

and so the contribution to the oscillatory sum from each of these (p + 1) sub-intervals is

$<\varepsilon/2\ (p+1)$. ...(2)

Thus, there exists a partition of [a, b] such that the corresponding oscillatory sum is

$$< \frac{\varepsilon}{2} + \frac{\varepsilon}{2(p+1)}(p+1) = \varepsilon. \qquad \text{[From (1) and (2)]}$$

Hence, the function f is integrable in [a, b].

Theorem 20:

It a bounded function f is integrable on [0, 1] then it is also integrable on [a, c] where c is a point of [a, b] (ii) conversely, if f is bounded and integrable on [a, c], [a, b] then it is also integrable on [a, b] (iii) Also in either case

$$\int_a^b f dx = \int_a^c f dx + \int_c^b f dx \quad a \le c \le b.$$

Proof:

(i) Since $f \in R\,[a, b]$ therefore for $\in > 0 \ \exists$ a partition p such that

$$U(P, f) - L(P, f) < \in$$

Let P* be a refinement of P such that

$$P^* = P \cup \{c\}$$

$$\therefore \ L(P, f) \le L(P^*, f) \le U(P^*, f) \le U(P, f) \qquad ...(1)$$

$$\Rightarrow U(P^*, f) - L(P^*, f) \le U(P, f) - L(P, f)$$

$$< \in \qquad ...(2)$$

Let P_1, P_2 is the sets of points of P* between [a, c], [c, b] respectively. Clearly P_1, P_2 are partitions of [a, c], [c, b] respectively and $P^* = P_1 \cup P_2$. Also

We have

$$U(P^*, f) = U(P_1, f) + U(P_2, f) \qquad ...(3)$$

and

$$L(P^*, f) = L(P_1, f) + U(P_2, f) \qquad ...(4)$$

$$\therefore \ \{U(P_1, f) - L(P_1, f)] + \{U(P_2, f) - L(P_2, f)\} = U(P^*, f) - L(P^*, f)$$

$$< \in$$

Since each bracket on the left is non-negative, it follows that partitions P_1, P_2 exist such that

$$U(P_1, f) - L(P_1, f) < \in/2$$

$\Rightarrow U(P_2, f) - L(P_2, f) < \in/2$

$\Rightarrow$f is integrable on [a, c] and [c, b]

Hence, the theorems

(ii) Let $f \in R$ over [a, c] and [c, b].

Therefore for $\varepsilon < 0$ we can find partitions P_1, P_2 of [a, c], [c, b] respectively such that

$$U(P_1, f) - L(P_1, f) < \frac{1}{2} \in \quad ...(5)$$

$$U(P_2, f) - L(P_2, f) < \frac{1}{2} \in \quad ...(6)$$

Let $\qquad P^* = P_1 \cup P_2.$

Clearly P^* is a partition of [a, b]

$\therefore U(P^*, f) - l(P^*, f) = U(P_1, f) + U(P_2, f) - L(P_1, f) - L(P_2, f)$

$$< \frac{1}{2} \in + \frac{1}{2} \in = \in$$

Thus, for $\in > 0$, $\exists$ a partition P^* of [a, b] such that

$U(P^*, f) - L(P^*, f) < \in$

$\Rightarrow f \in R$ over [a, b]

Hence the theorem.

(iii) We know that for any functions S and T, if W = S + T, then inf W $\geq$ inf S + inf T.

Now for any partitions P_1, P_2 of [a, c], [c, b] respectively, if $P^* = P_1 \cup P_2$, then

$U(P^*, f) = U(P_1, f) + U(P_2, f)$

Hence, the theorem.

$$\int_a^b fdx \geq \bar{\int_a^c} fdx + \bar{\int_c^b} fdx$$

but since f is integrable on [a, c], [ab], [a,b]

$$\therefore \int_a^b fdx \geq \bar{\int_a^c} fdx + \int_c^b fdx \quad ...(7)$$

proceeding with (f) in place of f, we have

$$\int_a^b fdx \leq \int_a^c fdx + \int_c^b fdx \quad ...(8)$$

(7) and (8) imply that

$$\int_a^b fdx = \int_a^c fdx + \int_c^b fdx.$$ **Hence Proved.**

Theorem 21:

If f is bounded and integrable on [a, c], [c, b]; then it is integrable [a, b]. Further

$$\int_a^b fdx = \int_a^c fdx + \int_c^b fdx.$$

Proof:

Since f is bounded and integrable on [a, c] and [c, b], so for $\varepsilon < 0$, there exist partitions P_1 and P_2 of [a, c], [c, –b] respectively, such that

$$U(P_1, f] - L(P_1, f) < \frac{1}{2}\varepsilon,$$

$$U(P_2, f] - L(P_2, f) < \frac{1}{2}\varepsilon.$$

Let $P^* = P_1 \cup P_2$ so that P^* is partition of [a, b] and

$$U(P^*, f) - L(P^*, f) = \{P_1, f) + U(P_2, f)\} - \{L(P_1, f) + L(P_2, f)\}$$

$$= \{U(P_1, f) - L(P_2, f)\} + \{U(P_2, f) - L(P_2, f)\}$$

$$< \frac{1}{2}\varepsilon + \frac{1}{2}\varepsilon = \varepsilon.$$

$\therefore U(P^*, f) - L(P^*, f) < \varepsilon$, where P^* is a partition of [a, b].

Hence, f is integrable on [a, b]. The proof of part (ii) is similar to that of Theorem 11.

Theorem 22:

The oscillation of a bounded function f on an internal [a, b] is the supremum of the set $(\,|f(x_1) - f(x_2)| : x_1, x_2\ \varepsilon(a, b)$ *of numbers.*

Proof:

Let M, m be the bounds of f, on [a, b]

Now

$m \le f(x_1)\ f(x_2) \le M\ \forall\ x_1, x_2\ \varepsilon\ [a, b]$

$\Rightarrow |f(x_1) - f(x_2)| \le M - m\ \forall\ x_1, x_2\ \varepsilon\ [a, b]$ ...(1)

$\Rightarrow$ M – m is an upper bund of the set.

Let $\varepsilon > 0$ be any given number sine M is supremum of f, therefore

$\exists$ $x' \varepsilon$ (a, b) such that

$$f(x') < - \frac{1}{2} \varepsilon \qquad ...(2)$$

Similarly $\exists$ $x'' \varepsilon$ [a, b] such that

$$f(x'') < m + \frac{1}{2} \varepsilon. \qquad ...(3)$$

from (2) and (3)

$\exists$ x', x'' ε (a, b) such that

$\Rightarrow f(x') - f(x'') > M - m - \varepsilon$

$\Rightarrow f(x') - f(x'') \mid > M - m - \varepsilon$ ...(4)

(1) and (4) imply that M – m is an upper bound of the set in question.

$\therefore M - m = \sup (f(x_1) - f(x_2) \mid : x_1, x_2 \ \varepsilon \ [a, b].$

Theorem 23:

(i) If f_1 and f_2 are two bounded and integrable functions on [a, b], then $f_1 + f_2$ is bounded and integrable on [a, b], and

$$\int_a^b (f_1 + f_2)dx = \int_a^b f_1 dx + \int_c^b f_2 dx.$$

(ii) Also $f_1 - f_2$ is bounded and integrable on [a, b], and

$$\int_a^b (f_1 - f_2)dx = \int_a^b f_1 dx - \int_c^b f_2 dx.$$

Proof:

Let $f = f_1 + f_2$ so that $f(x) = f_1(x) + f_2(x) \ \forall \ x \in [a, b]$

$\Rightarrow \mid f(x) \mid \le \mid f_1(x) \mid + \mid f_2(x) \mid \ \forall \ x \in [a, b],$ ...(1)

Since f_1 and f_2 are bounded on [a, b], so there exist two positive constants A and B such that

$\mid f_1(x) \mid \le A, \quad \mid f_2(x) \mid \le B \ \forall \ x \in [a, b].$ ...(2)

Thus, $\mid f(x) \mid \le A + B \ \forall \ x \in [a, b]$ from (1) and (2) we have and so $f = f_1 + f_1$ is bounded on [a, b].

Let $P = \{a = x_0, x_1, x_2 \ \ldots\ldots, x_n = b\}$ be any partition of [a, b].

Let M_i', $\mid m_i'$; M_i'', m_i''; M_i, m_i be the suprema and infima of f_1, f_2 and f respectively in $[x_{i-1}, x_i]$. Then

$$m_i' + m_i'' \le m_i \le M_i \le M_i' + M_i''.$$

or (mi' + mi")(xi – xi – 1) ≤ mi(xi – xi – 1) ≤ $M_i(x_i - x_{i-1}) \le$ (Mi' + Mi") × $(x_i - x_{i-1})$

Putting i = 1, 2, ... n and adding, we get

$L(P, f_1) + L(P, f_2) \le L(P, f) \le U(P, f) \le U(P, f_1) + U(P, f_2).$ ...(3)

Let ε > 0 be given. Since f_1, f_2 are integrable, so ∃ δ > 0 such that for every partition P with μ(P) < δ, we have

$$U(P, f_1) - L(P, f_1) < \frac{1}{2}\varepsilon, \quad ...(4)$$

$$U(P, f_2) - L(P, f_2) < \frac{1}{2}\varepsilon, \quad ...(5)$$

From equ. (3), for any partition P with μ(P) < δ, we have

$$U(P, f) - L(P, f) \le \{U(P, f_1) - L(P, f_1)\} + \{U(P, f_2) - L(P, f_2)\}$$

$$< \frac{1}{2}\varepsilon + \frac{1}{2}\varepsilon, \text{ by (4) and (5)}$$

∴ U(P, f) – L(P, f) < ε,

for all partitions P of [a, b] with μ(P) < δ.

Hence, f is integrable.

(ii) Since f_1 and f_2 are integrable, so by Darboux's Theorem, for any ε > 0, ∃ some δ > 0 such that for all partitions P with μ(P) < δ,

$$U(P, f_1) < \int_a^b f_1 dx + \frac{1}{2}\varepsilon \text{ and } U(P, f_2) < \int_a^b f_2 dx + \frac{1}{2}\varepsilon. \quad ...(6)$$

Now $\overline{\int_a^b} f dx \le U(P, f) \le U(P, f_1) + U(P, f_2)$, by (3)

or $\int_a^b f dx < \int_a^b f_1 dx + \int_a^b f_2 dx + \varepsilon,$

using (6) and the fact that f is integrable.

Since ε is arbitrary, therefore

$$\int_a^b f dx \le \int_a^b f_1 dx + \int_a^b f_2 dx. \quad ...(7)$$

Taking $(-f_1)$, $(-f_2)$ for f_1, f_2 in (7), we get

$$\int_a^b (-f) dx \le \int_a^b (-f_1) dx + \int_a^b (-f_2) dx$$

or $\int_a^b f dx \ge \int_a^b f_1 dx + \int_a^b f_2 dx.$ ...(8)

Hence, $\int_a^b f dx = \int_a^b f_1 dx + \int_a^b f_2 dx$, by (7) and (8).

Similarly we can prove the second part of the theorem.

Cor. :

If f, g are bounded and integrable on [a, b] such that f ≥ g, then

(i) $\int_a^b fdx \geq \int_a^b gdx$, if $b \geq a$

(ii) $\int_a^b fdx \leq \int_a^b gdx$, if $b \leq a$.

Proof:

$f \geq g \Rightarrow f(x) \geq g(x) \Rightarrow f(x) - g(x) \geq 0 \ \forall \ x \in [a, b]$, where $f - g$ is bounded and integrable on [a, b].

$$\int_a^b [f(x) - g(x)] \geq 0 \text{ or } \leq o \text{ according as } b \geq a \text{ or } b \leq a$$

$$\Rightarrow \int_a^b f(x) - \int_a^b g(x)dx] \geq 0 \text{ or } \leq o \text{ according as } b \geq a \text{ or } b \leq a$$

Hence, $\int_a^b f(x)dx \geq \int_a^b g(x)dx$, if $b \geq a$

and $\int_a^b f(x)dx \leq \int_a^b g(x)dx$, if $b \leq a$.

Theorem 24:

If f_1 and f_2 are two bounded and integrable functions on [a, b] and there exists a number $\lambda > 0$ such that $| f_2(x) | \geq \lambda$ for all x in [a, b], then f_1/f_2 is bounded and integrable on [a, b].

Proof:

Since f_1 and f_2 are bounded on [a, b], so there exists a positive number k such that

$$|f_1(x)| \leq k, \ |f_2(x)| \leq k \ \forall \ x \in [a, b] \quad ...(1)$$

$$\text{Also } |f_2(x)| \geq \lambda \Rightarrow \frac{1}{|f_2(x)|} \leq \frac{1}{\lambda} \forall x \in [a, b] \quad ...(2)$$

Now we have $\left|\frac{f_1}{f_2}(x)\right| = \frac{|f_1(x)|}{|f_2(x)|} \leq \frac{k}{\lambda} \ \forall \ x \in [a, b]$

$\Rightarrow f_1/f_2$ is bounded on [a, b].

Let $P = \{a = x_0, x_1, ..., x_n = b\}$ be a partition of [a,b].

Let M_i', m_i'; M_i'', m_i'', M_i, m_i be the bounds of f_1, f_2, f_1/f_2 respectively in $[x_{i-1}, x_i]$

We know that

$$\left|(f_1/f_2)(x_2)-(f_1/f_2)(x_2)\right| = \left|\frac{f_1(x_2)}{f_2(x_2)} - \frac{f_1(x_1)}{f_2(x_1)}\right|$$

$$= \left|\frac{f_2(x_1)[f_1(x_2)-f_1(x_1)] - f_1(x_1)[f_2(x_2)-f_2(x_1)]}{f_2(x_2)f_2(x_1)}\right|$$

$\leq (k/\lambda^2) \mid f_1(x_2) - f_1(x_1) \mid + (k/\lambda^2) \mid f_2(x_2) - f_2(x_1)$ by (1) and (2)

$\leq (k/\lambda^2)\ (M_i'-m_i') + (k/\lambda^2)\ (M_i''-m_i'')$.

$\therefore\ M_i-m_i \leq (k/\lambda^2)\ (M_i'-m_i') + (k/\lambda^2)\ (M_i''-m_i'')$. ...(3)

Let $\varepsilon > 0$ be given.

Since f_1, f_2 are integrable, so $\exists$ a $\delta > 0$ such that for any partition P with norm $\mu(P) < \delta$,

$$\left.\begin{aligned} U(P,f_1)-L(P,f_1)<\varepsilon\lambda^2/2k,\\ U(P,f_2)-L(P,f_2)<\varepsilon\lambda^2/2k,\end{aligned}\right\} \quad ...(4)$$

From (3), for any partition P with $\mu(P) < \delta$, we have

$U(P, f_1/f_2) - L(P, f_1/f_2) \leq (k/\lambda^2)\ [U(P, f_1) - L(P, f_1)]$

$\qquad + (k/\lambda^2)\ [U(P, f_2) - L(P, f_2)]$

$\leq (k/\lambda^2)\ (\varepsilon\lambda^2/2k) + (k/\lambda^2)\ (\varepsilon\lambda^2/2k)$, by (4)

$= \varepsilon$.

Hence, f_1/f_2 is integrable on [a, b]. **Hence Proved.**

Theorem 25:

If f is bounded and integrable on [a, b], then | f | is also bounded and integrable on [a, b] and

$$\left|\int_a^b f dx\right| \leq \int_a^b |f|\, dx.$$

Proof:

Since f is bounded, so $\exists$ $k > 0$ such that

$| f(x) | \leq k\ \forall\ x \in [a, b]$

So he function | f | is bounded.

Since f is integrable, so for any $\varepsilon > 0$, there exists a partition P of [a, b] such that

$U(P, f) - L (P, f) < \varepsilon$. ...(1)

Let M_i, m_i; M_i', m_i' be the bounds of f and | f | in $[x_{i-1}, x_i]$.

Let x_i, $x_2 \in [x_{i-1}, x_i]$. Then

$|\,|f(x_2)| - |f(x_1)|\,| \le |f(x_2) - f(x_1)| \le M_i - m_i$ (Theorem 12)

$\Rightarrow M_i' - m_i' \le M_i - m_i$

$\Rightarrow (M_i' - mi')(xi - xi - 1) \le (M_i - mi)(xi - xi - 1)$

Taking i = 1, 2, 3, ... n and adding, we get

$U(P, |f|) - L(P, |f|) \le U(P, f) - L(P, f) < \varepsilon$, by (1)

$\therefore U(P, |f|) - L(P, |f|) < \varepsilon$

Hence, | f | is integrable.

Since $f(x) \le |f(x)| = |f|(x)\ \forall\ x \in [a, b]$

and f, | f | are integrable on [a, b], so $\int_a^b f dx \le \int_a^b |f| dx$. ...(2)

Similarly $-f(x) \le |f|\ f(x)| = |f|(x)\ \forall\ x \in [a, b]$

$\Rightarrow \int_a^b -f dx \le \int_a^b |f| dx.$

$\Rightarrow -\int_a^b f dx \le \int_a^b |f| dx.$ ...(3)

From (2) and (3), $\left|\int_a^b f dx\right| \le \int_a^b |f| dx.$

Remark:

The converse of the above theorem is not true i.e., if | f | is integrable, then f need not be integrable.

Proof:

Consider the function

$$f(x) = \begin{cases} 1 & \text{when } x \text{ is rational,} \\ -1 & \text{when } x \text{ is irrational.} \end{cases}$$

$\therefore |f|(x) = |f(x)| = 1\ \forall\ x \in R$

$\Rightarrow \int_a^b |f|\, dx = (b - a).$

$\therefore$ | f | is integrable.

Now $U(P, f) = \sum_{i=1}^{n} M_i\ (x_i - x_{i-1}) = \sum_{i=1}^{n} 1(x_i - x_{i-1}) = (b - a)$

$\therefore \overline{\int_a^b} fdx = b - a$

Again $L(P, f) = \sum_{i=1}^{n} m_i (x_i - x_{i-1}) = \sum_{i=1}^{n} (-1)(x_i - x_{i-1}) = -(b-a) = (b-a)$

$\therefore \underline{\int_a^b} fdx = a - b$. Since $\overline{\int_a^b} fdx \neq \underline{\int_a^b} fdx$, f is not integrable.

Theorem 26:

If f_1 and f_2 are two bounded and integrable functions on [a, b], then their product f_1f_2 is also bounded and integrable on [a, b].

Proof:

Since f_1, f_2 are bounded therefore $\exists$ k > 0, such that, for all $x \in [a, b]$,

$| f_1 (x) | \leq k, | f_1 (x) | \leq k$

$\Rightarrow | (f_1 f_2) (x) | = | f_1 (x) f_2(x) | \leq k^2, \forall x \in [a, b]$

$\Rightarrow f_1.f_2$ is bounded

Let $P = (a = x_0, x_1, x_2, ..., x_n = b)$ be any partition of [a, b].

Let $M_i', m_i'; M_i'', m_i''; M_i, m_i$ be the bounds of f_1, f_2 and f_1, f_2 respectively in Δx_i,

We have for all $x_1, x_2 \in \Delta x_i$,

$$(f_1f_2)(x_2) - (f_1f_2)(x_2) = f_1(x_2)f_2(x_2) - f_1(x_1)f_2(x_1)$$

$$= f_2(x_2)[f_1(x_2) - f_1(x_1)] + f_1(x_1)[f_2(x_2) - f_2(x_1)]$$

$$\Rightarrow | (f_1f_2)(x_2) - (f_1f_2)(x_1) | \leq | f_2(x_2) |.|f_1(x_2) - f_1(x_1) |$$

$$+ |f_1(x_1) |.|f_2(x_2) - f_2(x_1)| \leq k(M_i'-m_i') + k(M_i''-m_i'')$$

$$\Rightarrow M_i - m_i \leq k(M_i'-m_i') + k(M_i'' - m_i'') \quad ...(1)$$

Now let $\in > 0$ be a given number.

Since f_1, f_2 are integrable, therefore $\exists\ \delta > 0$ such that for any partition P with norm $\mu(P) < \delta$,

$$U(P, f_1) - L(P, f_1) < \in/2k, \quad U(P, f_2) - L(P, f_2) < \in/2k$$

Hence, from (1), by multiplying by δx_i and adding all such inequalities, we have for any partition P with norm $\mu(P) < \delta$,

$$U(P, f_1f_2) - L(P, f_1f_2) \leq k(U(P, f_1) - L(P, f_1) + k[U(P, f_2) - L(P, f_2)]$$

$$< K(\in/2k) + k(\in/2k) = \in.$$

Implying that f_1f_2 is integrable on [a, b].

Hence the theorem.

Theorem 27:

If f is integrable on [a, b], then f^2 is also integrable on [a, b].

Proof:

Since f is bounded on [a, b], so $\exists$ M > 0 such that

$$| f(x) | \le M \ \forall \ x \in [a, b]. \qquad ...(1)$$

Since f is integrable on [a,b], | f | is integrable on [a, b] and so for any $\varepsilon > 0$, there exists a partition P of [a, b] such that

$$U(P, | f |) - L(P, | f |) < \frac{\varepsilon}{2M} \qquad ...(2)$$

Since $| f^2(x) | = | f(x) |^2 \le M^2$, so f^2 is bounded on [a, b].

Let M_i, m_i be the bounds of | f | in $[x_{i-1}, x_i]$. ...(3)

Then M_i^2, m_i^2 are the bounds of f^2 in $[x_{i-1}, x_i]$, $[\because f^2(x) = |f(x)|^2)$

$$\text{Now } U(P, f^2) - L(P, f^2) = \sum_{i=1}^{n} (M_i^2 - m_i^2)\, \Delta x_i$$

$$= \sum_{i=1}^{n} (M_i - m_i)(M_i + m_i) \Delta x_i$$

$$\le 2M \left\{ \sum_{i=1}^{n} (M_i - m_i) \Delta x_i \right\}, \text{ using (1) and (3)}$$

$$= 2M \{U(P, | f |) - L(P, | f |)\}$$

$$< 2M . \frac{\varepsilon}{2M} = \varepsilon, \text{ by (2)}$$

Hence, f^2 is integrable on [a, b]

Hence Proved.

Definition (Primitive)

A differentiable function ϕ defined on [a, b] such that its derivetive ϕ' equals a given function f on [a, b] is called a primitive or anti derivative of f on [a, b].

If $\phi(x)$ is a primitive of f then f + c will also be a primitive of f where c is any constant Hence, the primitive of a function is not unique.

Every continuous function f possesses a primitive F.

where $F(x) = \int_a^x f(t)dt$.

Remark 1:

The converse of the above statement may not be true i.e., a function which possesses a primitive may not be continuous.

Consider the function f defined on [0, 1] as follows:

$$f(x) = 2x \sin(1/x) - \cos(1/x),\ x \neq 0,$$
$$= 0 \qquad\qquad x = 0.$$

Let
$$F(x) = \begin{cases} x^2 \sin(1/x), & x \neq 0, \\ 0 & x = 0. \end{cases}$$

It is clear that

$$F'(x) = f(x)\ \forall\ x \in [0, 1].$$

Thus, f has a primitive F. Now we show that f is not continuous at x = 0. We have

$$\lim_{x\to 0} (fx) = 2 \lim_{x\to 0} x \sin(1/x) - \lim_{x\to 0} \cos(1/x)$$

$$= 2 \times 0 - \lim_{x\to 0} \cos(1/x) = -\lim_{x\to 0} \cos(1/x)$$

which does not exist.

Hence, f is not continuous on [0, 1], but f has a primitive.

Remark 2:

The integral function of an integrable function is continuous.

Remark 3:

(A Useful Formula)

Theorem 28:

If a function f is bounded and integrable on [a, b], them the function F defined as

$$F(x) = \int_a^x f(t)dt \quad (a \le x \le b).$$

is continuous on [a, b]. Further, if f is continuous in [a, b], then F' = f.

Proof:

Since f is bounded on [a, b], so ∃ k > 0 such that

$$| f(x) | \le k\ \forall\ x \in [a, b]. \qquad ...(1)$$

Let e > 0 be given and c be any point of [a, b].

$$\therefore \qquad F(c + h) = \int_a^{c+h} f(t)dt,\ F(c) = \int_a^c f(t)dt$$

and $$|F(c+h) - F(c)| = \left| \int_a^{c+h} f(t)dt - \int_a^c f(t)dt \right|$$

$$= \left| \int_a^{c+h} f(t)dt \right| \le |f(t)||c+h-c|$$

$$\le |h| k, \text{ using (1)}$$

$$\le e, \text{ if } |h| < \varepsilon/k.$$

Choosing $\delta = \varepsilon/k < 0$, we see that

$| F(c + h) - F(c) | < \varepsilon$ if $| h | < \delta$.

Thus, F is continuous at any point $c \in [a, b]$ and so F is continuous on [a, b].

Second part of this theorem is as under.

We have $F(c + h) - F(c) = \int_a^{c+h} f(x)dx$.

Since f is continuous and integrable in [a, b], so

$$\int_a^{c+h} f(x)dx = (c+h-c)\ f(c+\theta h),\ 0 \le \theta \le 1\ [\text{Cor. 6, p.64}]$$

$$\therefore\ F(c+h) - F(c) = hf(c+\theta h)$$

$$\Rightarrow \lim_{h\to 0} \frac{F(c+h)-F(c)}{h} = \lim_{h\to 0} f(c+\theta h)$$

$$= f(c), \text{ since f is continuous at c.}$$

$\therefore F'(c) = f(c)\ \forall\ c \in [a, b]$. Hence, $F' = f$.

Hence the theorem.

GENERALIZED FIRST MEAN VALUE THEOREM

Theorem 29:

If f and g are integrable on [a, b] and g keeps the same sign over [a, b], then there exists a number μ lying between the bounds of f such that

$$\int_a^b fg\,dx = \mu \int_a^b g\,dx.$$

Proof:

Since g keeps the same sign over [a, b], we may take

$g(x) \ge 0\ \forall\ x \in [a, b]$.

Let m = inf. f and M = sup. f be the bound of f in [a, b] so that

$m \le f(x) \le M\ \forall\ x \in [a, b]$

$\Rightarrow mg(x) \le f(x) \le g(x) \le Mg(x)$ $[\because g(x) \ge 0 \ \forall \ x \in [a, b]$

Since f, g and f g are integrable in [a, b], so

$$m\int_a^b g\,dx \le \int_a^b fg\,dx \le M\int_a^b g\,dx$$

Hence, $\int_a^b fg\,dx = \mu \int_a^b g\,dx,$ ...(1)

where μ lies between m and M. Hence the theorem.

Cor.

If in the Generalized Mean Value Theorem, f is continuous on [a,. b], then there exists a number $\xi \in [a, b]$ such that

$$\int_a^b fg\,dx = f(\xi)\int_a^b g\,dx.$$

Proof:

Since f is continuous on [a, b], so f assume every value lying between the bounds m and M of f.

Since $m \le \mu < M$, there exists some $\xi \in [a, b]$ such that f(x) = m. Putting this value in (1), we get the desired result.

The following result will be used in proving the *Second Mean Value Theorem.*

BONNET'S THEOREM

Theorem 4:

If $\int_a^b \phi\,dx$ and $\int_a^b g\,dx$ *both exist and ϕ is positive and monotonically decreasing on [a, b], then there exists a point $\xi \in [a, b]$ such that*

$$\int_a^b \phi g\,dx = \phi(a)\int_a^\xi g\,dx.$$

Proof:

Let $P = \{a = x_0, x_1, \ldots, x_n = b\}$ be a partition of [a, b] and M_i, m_i the bounds of g in $[x_{i-1}, x_i]$.

Let $t_1 = a$ and t_i $(i \ne 1)$ be any point of $[x_{i-1}, x_i]$.

We know, $m_i\,\Delta x_i \le \int_{x_{i-1}}^{x_i} g\,dx \le M_i\,\Delta x_i,$ ...(1)

Since $m_i \le g(t_i) \le M_i$, therefore

$m_i \Delta x_i \leq g(t_i)\, \Delta x_i \leq M_i\, \Delta x_i$, ...(2)

Putting i = 1, 2, 3, ..., p (p ≤ n) in (1) and (2) and adding vertically, we respectively obtain, then

$$\sum_{i=1}^{p} m_i\, \Delta x_i \leq \int_a^{x_i} g\, dx \leq \sum_{i=1}^{p} m_i\, \Delta x_i \quad ...(3)$$

$$\sum_{i=1}^{p} m_i\, \Delta x_i \leq \sum_{i=1}^{p} g(t_i)\, \Delta x_i \leq \sum_{i=1}^{p} m_i\, \Delta x_i. \quad ...(4)$$

From (3) and (4), we obtain

$$\left| \int_a^{x_p} g\, dx - \sum_{i=1}^{p} g(t_i)\, \Delta x_i \right| \leq \sum_{i=1}^{p} (M_i - m_i)\, \Delta x_i \leq \omega(P, g)$$

$$\Rightarrow \int_a^{x_p} g\, dx - \omega\,(P, g) \sum_{i=1}^{p} g(t_i)\, \Delta x_i \leq \int_a^{x_p} g\, dx + \omega\,(P, g) \quad ...(5)$$

where $\omega(P, g) = \sum_{i=1}^{n} (M_i - m_i)\, \Delta x_i = U(P, g) - L(P, g).$

Since $\int_a^t g\, dx$ is continuous in [a, b], so it is bounded.

Let A and B be its bounds so that

$$B \leq \int_a^t g\, dx \leq A \quad \forall\, t \in [a, b]. \quad ...(6)$$

From (5) and (6), we obtain

$$B - \omega\,(P, g) \leq \sum_{i=1}^{p} g(t_i)\, \Delta x_i \leq A + \omega\,(P, g).$$

Using the Able's Lemma, where

$a_i = \phi(t_i)$, $v_i = g(t_i)\, \Delta x_i$,

$k = B - \omega(P, g)$, $K = A + \omega(P, g)$; we obtain

$$a_i k \leq \sum_{i=1}^{p} a_i v_i \leq a_i K \quad (a_1 = \phi(t_i) = \phi(a))$$

$$\text{or } \phi(a)\, [B - \omega(P, g)] \leq \sum_{i=1}^{n} \phi(t_i) g(t_i)\, \Delta x_i \leq \phi(a)\, [A + \omega(P, g)]$$

Since $\omega(P, g) \to 0$ as $m(P) \to 0$, so

$$f(a) \leq \int_a^b \phi g\, dx \leq A\phi(a).$$

$\therefore \int_a^b \phi g\,dx = \mu\,\phi(a).$, where $b \le m \le A$. ...(7)

The function $\int_a^t g\,dx$ being continuous in [a, b], assumes every value between its bounds B and A. Thus, $\exists$ some $\xi \in [a, b]$ such that $\int_a^{\xi} g\,dx = \mu$. ....(8)

From (7) and (8), we obtain

$$\int_a^b \phi g\,dx = \phi(a) \int_a^{\xi} g\,dx.$$

Theorem 30:

(Abel's Lemma)

If (i) $a_1, a_2, \ldots\ldots\ldots, a_n$ *is a monotorically decreasing set of n numbers,*

(ii) $v_1, v_2, \ldots\ldots\ldots, v_n$ *is a set of any n numbers,*

(iii) k, K *are two numbers such that* $k < v_1 + v_2 + \ldots\ldots\ldots + v_p < K,\ 1 \le p \le n,$

then $a_1 k < \sum_{r=1}^{n} a_r v_r < a_1 K.$

Proof:

Let $S_1 = v_1,\ S_2 = v_1 + v_2,\ S_3 = v_1 + v_2 + v_3,$

$S_n = v_1 + v_2 + \ldots\ldots + v_n.$

$$\therefore \sum_{r=1}^{n} a_r v_r = a_1 v_1 + \ldots\ldots + a_n v_n$$

$$= a_1 S_1 + a_2(S_2 - S_1) + \ldots\ldots + = a_{n-1}(S_{n-1} - S_{n-2}) + a_n(S_2 - S_{n-1})$$

$$= (a_1 - a_2)\,S_1 + (a_2 - a_3)\,S_2 + \ldots\ldots + (a_{n-1} - a_n)\,S_{n-1} + a_n S_n. \qquad \text{...(iv)}$$

By (i), $a_{i+1} \le a_i$ for $i = 1, 2, \ldots\ldots, n$.

$\therefore a_1 - a_{i+1} \ge 0$ for $i = 1, 2, \ldots., n$. ...(v)

From (ii), $k < S_p < K$, for $p = 1, 2, \ldots., n$. ...(vi)

Using (v) and (vi) in (iv), we obtain

$$\sum_{r=1}^{n} a_r v_r < (a_1 - a_2)K + (a_2 - a_3)K + \ldots. + (a_{n-1} - a_n)K + a_n K = a_1 K,$$

and $$\sum_{r=1}^{n} a_r v_r > a_1 K$$

Hence, $a_i k < \sum_{r=1}^{n} a_r v_r < a_1 K.$

SECOND MEAN VALUE THEOREM

Theorem 31:

If $\int_a^b f\,dx$ and $\int_a^b g\,dx$ *both exist and f is monotonic on [a, b], then there exists some* $\xi \in [a, b]$ *such that*

$$\int_a^b fg\,dx = f(a)\int_a^\xi g\,dx + f(b)\int_\xi^b g\,dx.$$

Proof:

Suppose f is monotonically decreasing on [a, b],

$\therefore f(x) \geq f(b)\ \forall\ x \in [a, b],$

Thus, $\phi(x) = f(x) - f(b)\ \forall\ x \in [a, b]$, is positive and monotonically decreasing on [a, b].

By Bonnet's Theorem, there exists some $\xi \in [\alpha, b]$ such that

$$\int_a^b \phi g\,dx = \phi(a)\int_a^\xi g\,dx$$

or $$\int_a^b \{f(x)-f(b)\}g(x)\,dx = \{f(a)-f(b)\}\int_a^\xi g\,dx$$

or $$\int_a^b fg\,dx - f(b)\int_a^b g\,dx = f(a)\int_a^\xi g\,dx - f(b)\int_a^\xi g\,dx$$

or $$\int_a^b fg\,dx = f(a)\int_a^\xi g\,dx + f(b)\left[\int_a^b g\,dx - \int_a^\xi g\,dx\right]$$

Hence, $$\int_a^b fg\,dx = f(a)\int_a^\xi g\,dx + f(b)\int_a^b g\,dx. \qquad ...(1)$$

Suppose Now, f is monotonically increasing. Then (–f) is monotonically decreasing and so by (1)

$$\int_a^b (-f)g\,dx = -f(a)\int_a^\xi g\,dx - f(b)\int_a^b g\,dx.$$

Hence, $$\int_a^b fg\,dx = f(a)\int_a^\xi g\,dx + f(b)\int_\xi^b g\,dx.$$

Hence the theorem.

MEAN VALUE THEOREMS OF INTEGRAL CALCULUS

We have two Mean Value Theorems for derivatives. In the same way, we have two mean value theorems for integrals as well.

FIRST MEAN VALUE THEOREM

Theorem 32:

If a function f is continuous on [a, b], then there exists a number e in [a, b] such that

$$\int_a^b f\,dx = f(\xi)\,(b-a)$$

f is continuous, therefore $f \in R\,[a, b]$.

Let m, M be the infimum and supremum of f in [a, b]. We have

$$m(b-a) \le \int_a^b f\,dx \le M(b-a)$$

Hence, there is a number $\mu \in [m, M]$, such that

$$\int_a^b f\,dx = \mu(b-a)$$

Since f is continuous on [a, b], it attains every value between its bounds m, M. Therefore, there exists a number $\xi \in [a, b]$ such that $f(\xi) = \mu$.

$$\therefore \quad \int_a^b f\,dx = f(\xi)(b-a).$$

Remark:

The preceding theorem says that the condition of continuity is necessary for the function to assume its mean value in the given interval, for example the function f(x) defined on [2, 5] as follows:

$$f(x) = \begin{cases} 1 & \text{if } 2 \le x \le 3 \\ 3 & \text{if } 3 \le x \le 5 \end{cases}$$

Now, $\int_2^5 f\,dx = \int_2^3 1\,dx + \int_3^5 3\,dx = 7$

and so the mean value of the function is

$$\frac{1}{5-2}\int_2^5 f\,dx = \frac{7}{3}$$

Which the function fails to assume on the interval.

FUNDAMENTAL THEOREM OF INTEGRAL CALCULUS

Theorem 33:

If f is bounded and R-integrable on [a, b] and if there is a differentiable function ϕ on [a, b] such that $\phi' = f$, then

$$\int_0^1 f(x)\,dx = \phi(b) - \phi(a).$$

Proof:

Let ε be an arbitrary positive number. Since $\phi' = f$ is bounded and R-integrable on [a, b], there exists a partition P of [a, b]

$$P = \{a = x_0, x_1, x_2, ..., x_n = b\}$$

Since ϕ is a real continuous function on $[x_{i-1}, x_i]$ and is differentiable on $[x_{i-1}, x_i]$ therefore by the mean value theorem we have

$$\phi(x_i) - \phi(x_{i-1}) = (x_i - x_{i-1})\, f(t_i)$$

where $t_i \leq [x_{i-1}, x_i]$.

Thus,

$$\phi(b) - \phi(a) = \sum_{i=1}^{n}\left[\phi(x_i) - \phi(x_{i-1})\right]$$

$$= \sum_{i=1}^{n}(x_i - x_{i-1})\, f(t_i)$$

$$= \sum_{i=1}^{n} f(t_i)\Delta_i$$

The R.H.S. tends to $\int_a^b f(x)\,dx$ as $\mu(P) \to 0$.

Therefore $\phi(b) - \phi(a) = \int_a^b f(x)\,dx$.

SOLVED EXAMPLES

Example 1:

Let f(x) = k be a constant function on [a, b] and α monotonically increasing function on [a, b]. Prove that

$$\int_a^b f\,d\alpha = k[\alpha(b) - \alpha(a)].$$

Solution:

Let $P = \{a = x_0, x_1, ..., xi-1, x_i, ... x_n = b\}$ be any partition of [a, b]. Then

$$\sum_{i=1}^{n} f(t_i)\,\Delta\alpha_i = \sum_{i=1}^{n} k[\alpha(x_i) - \alpha(x_{i-1})]$$

$$= k[\{\alpha(x_1) - \alpha(x_0)\} - \{\alpha(x_2) + \{\alpha(x_2) - \alpha(x_1)\} + ...$$
$$+ \{a(x_n) - a(x_{n-1})\}]$$

$$= k[\alpha(x_n) - \alpha(x_0)] = k[\alpha(b) - \alpha(a)].$$

Thus, $\int_a^b k\, d\alpha = \lim_{i=1} t(t_i)\Delta\alpha_i = k[\alpha(b)-\alpha(a)]$ . . . (1)

Note: If k = 1, then $\int_a^b d\alpha = \alpha(b)-\alpha(a)$.

$\therefore\ k\int_a^b d\alpha = k[\alpha(b)-\alpha(a)]$.

From (1) and (2), we obtain . . . (2)

$\int_a^b k\, d\alpha = k\int_a^b d\alpha$. **Ans.**

Example 2:

Suppose a function α increases on [a, b], $a \le x_0 \le b$, α is continuous at x_0. Let f be a function defined in [a, b] such that $f(x_0) = 1$ and $f(x) = 0$ for $x \ne x_0$. Prove that $f \in R(\alpha)$ on [a, b] and $\int_a^b f d\alpha = 0$.

Solution:

Since α is continuous at x_0, so for any $\varepsilon > 0$, there exists scme $\delta > 0$ such that

$| \alpha(x) - \alpha(x_0) | < \varepsilon/3$ for $| x - x_0 | < \delta$ or $x_0 - \delta < x < x_0 + \delta$.

For $y \in]x_0 - \delta, x_0 + \delta[$, $| \alpha(y) - \alpha(x_0)| < \varepsilon/3$. Thus

$| \alpha(x) - \alpha(y) | = | \{\alpha(x) - \alpha(x_0)\} - \{\alpha(y) - \alpha(x_0)\}[$

$\le | \alpha(x) - \alpha(x_0) | + | \alpha(y) - \alpha(x_0) | < \varepsilon/3 + \varepsilon/3$

$\therefore | \alpha(x) - \alpha(y) < 2\,\varepsilon/3$ for all $x, y \in]x_0 - \delta, x_0 + \delta[$. ...(1)

We are given that $f(x_0) = 1$ and $f(x) = 0$ for $x \ne x_0$. ...(2)

Consider a partition P of [a, b,] as follows:

$P = \{a, x_0 - \frac{1}{2}\delta, x_0 + \frac{1}{2}\delta, b\}$.

From (2), it follows that

sup f = 0 in $]a\, x_0 - \frac{1}{2}\delta$ [and] $]x_0 + \frac{1}{2}\delta, b[$, and

sup f = 1 in $]x_0 - \frac{1}{2}\delta, x_0 + \frac{1}{2}\delta_0[$ $(\because f(x_0) = 1)$

Obviously inf f = 0 in each of the three sub-intervals stated above.

Now,

$U(P, f, \alpha) = 0.\left[\alpha\left(x_0 - \frac{1}{2}\delta\right) - \alpha(a)\right] + 1.\left[\alpha\left(x_0 + \frac{1}{2}\delta\right) - \alpha\left(x_0 - \frac{1}{2}\delta\right)\right]$

$$+\left[0.\alpha(b)-\alpha\left(x_0+\frac{1}{2}\delta\right)\right] \qquad ...(3)$$

$$=\alpha\left(x_0+\frac{1}{2}\delta\right)-\alpha\left(x_0-\frac{1}{2}\delta\right)<\frac{2\varepsilon}{3}, \text{ using (1).}$$

$$\therefore\ U(P, f, \alpha)<\frac{2\varepsilon}{3}<\varepsilon. \qquad ...(4)$$

As argued in (3), we have

$L(P, f, \alpha) = 0\ [....] + 0\ [....] = 0.$

$\therefore U(P, f, \alpha) - L(P, f, \alpha) < \varepsilon.$

Hence, $f \in R_\alpha [a, b]$.

It follows that

$$\int_a^b f d\alpha = \inf \{U(P, f, \alpha)\} < \varepsilon, \text{ using (4)}$$

Now, $f \leq 0$ on $[a, b] \Rightarrow \int_a^b f d\alpha \geq 0$. Also $\int_a^b f d\alpha \leq \varepsilon$.

Hence, $\int_a^b f d\alpha = 0$, since ε is arbitrary. **Ans.**

Example 3:

Let $\alpha(x) = 0$ *for* $x \leq 0$; $\alpha(x) = 1$ *for* $x > 0$. . . . (1)

Let f be a bounded function on $[-1, 1]$. *Prove that*

$f \in \mathfrak{R}(\alpha)$ *iff* $f(0+) = f(0)$ *and then* $\int_{-1}^{1} f\, d\alpha = f(0)$.

Solution:

We have

$$S(P, f, \alpha) = \sum_{j=1}^{n} f(t_j)\ \{\alpha(x_j)-\alpha(x_{j-1})\}$$

$$= 0+f(t_{i-1})\{\alpha(x_{i-1}) - \alpha(x_{i-2})\} + f(t_i)\ \{\alpha(x_i) - \alpha(x_{i-1})\}+0$$

$$= 0 + 0 + f(t_i)\ (1-0) + 0 = f(t_i), \text{ by (1)}$$

$\therefore S(P, f, \alpha) = f(0)$, when $t_i = 0$.

Using variable x instead of t_i, we find that $x \to 0$ from the right, when the norm m(P) tends to zero. Thus lim $S(P, f, \alpha)$ exists iff $f(0+)$ exists and equals f(0). Hence $f \in \mathfrak{R}(\alpha)$ iff $f(0+) = f(0)$ and

$$\int_{-1}^{1} f d\alpha = \lim S(P, f, \alpha) = f(0).$$ **Ans.**

Example 4:

Prove that $\int_0^3 x^2 d([x]-x) = 5$.

Solution:

Integrating by parts we obtain

$$\int_0^3 x^2\, d([x]-x) = \left| x^2([x]-x) \right|_0^3 - \int_0^3 2x([x]-x)dx$$

$$= \{9(|3|-3) - 0\} - 2\int_0^3 x[x]\,dx + 2\int_0^3 x^2\,dx$$

$$= 0 - 2\left[\int_0^1 x.0\,dx + \int_1^2 x.1\,dx + \int_2^3 x.2dx\right] + 2\left|\frac{x^3}{3}\right|_0^3$$

$$= -2\left[0 + \left|\frac{x^2}{2}\right|_1^2 + \left|x^2\right|_2^3\right] + 18$$

$$= -2(2 - \frac{1}{2} + 9 - 4) + 18 = -13 + 18 = 5.$$ **Ans.**

Example 5:

Show $\int_{-1}^{1} x d|x| = 1$, *by definition of RS-integral.*

Solution:

Let f (x) = x and α (x) = | x |. Then

$\alpha(x) = -x, - \leq x < 0$ and $\alpha(x) = x, 0 \leq x \leq 1$.

Let $P_n = \{-1 = x_0, x_1, ..., x_n = 0 = y_0, y_1, ... y_n = 1\}$

be a partition of [–1, 1], where

$$x_r = -1 + \frac{r}{n},\ y_r = \frac{r}{n};\ r = 1, 2, ..., n.$$

Now, $M_r = \sup f(x)$ in $[x_{r-1}, x_r] = x_r = -1 + \frac{r}{n}$,

$M'_r = \sup f(x)$ in $[y_{r-1}, y_r] = y_r = \frac{r}{n}$. $(\therefore f(x) = x)$

The corresponding infima are

$$m_r = -1+\frac{r-1}{n},\ m'_r = \frac{r-1}{n}.$$

We have

$$\Delta\,\alpha_r = \alpha\,(x_r) - \alpha\,(x_{r-1}) = \begin{cases} -x_r + x_{r-1}, & \text{if } 1 \le r < n \\ \alpha(0) - \alpha(x_{n-1}), & \text{if } r = n \end{cases}$$

$$= \begin{cases} 1-\dfrac{r}{n}-1+\dfrac{r-1}{n} = -\dfrac{1}{n}, & \text{if } 1 \le r < n \\ 0+\left(-1+\dfrac{n-1}{n}\right) = -\dfrac{1}{n}, & \text{if } r = n \end{cases}$$

and $$\Delta\,\alpha_r' = \alpha\,(y_r) - \alpha\,(y_{r-1}) = \begin{cases} y_r - y_{r-1}, & \text{if } 1 \le r < n \\ 1 - y_n, & \text{if } r = n \end{cases}$$

$$= \begin{cases} \dfrac{r}{n}-\dfrac{r-1}{n} = \dfrac{1}{n}, & \text{if } 1 \le r < n \\ 1-\dfrac{n-1}{n} = \dfrac{1}{n}, & \text{if } r = n. \end{cases}$$

$$\therefore\ U(P_n, f, \alpha) = \sum_{r=1}^{n} M_r\ \Delta\alpha_r + \sum_{r=1}^{n} M'_r\ \Delta\alpha'_r$$

$$= \sum_{r=1}^{n}\left(-1+\frac{r}{n}\right)\left(-\frac{1}{n}\right) + \sum_{r=1}^{n}\left(\frac{r}{n}\right)\left(\frac{1}{n}\right)$$

$$= \frac{1}{n}\sum_{r=1}^{n} 1 - \frac{1}{n^2}\sum_{r=1}^{n} r + \frac{1}{n^2}\sum_{r=1}^{n} r = 1 \to 1 \text{ as } n \to \infty,$$

and $$L(P_n, f, \alpha) = \sum_{r=1}^{n}\left(-1+\frac{r-1}{n}\right)\left(-\frac{1}{n}\right) + \sum_{r=1}^{n}\left(\frac{r-1}{n}\right)\frac{1}{n}$$

$$= \frac{1}{n}\sum_{r=1}^{n} 1 - \frac{1}{n^2}\sum_{r=1}^{n}(r-1) + \frac{1}{n^2}\sum_{r=1}^{n}(r-1)$$

$$= \frac{1}{n}.n = 1 \to 1 \text{ as } n \to \infty.$$

Hence, $\int_{-1}^{1} f d\alpha = \int_{-1}^{1} x d|x| = 1$.

Example 6:

Compute $\int_{-1}^{1} f\,dx$, *where f(x) = |x|.*

Solution:

The function f(x) = | x | i bounded and continuous o [–1, 1]. By definition,

$$f(x) = \begin{cases} x \text{ if } x \geq 0, \\ -x \text{ if } x \leq 0. \end{cases}$$

Consider a partition P of [–1, 1] into 2n equal sub-intervals each of length 1/n as follows:

$$P = \left\{-1 = x_0, x_1 = -1 + \frac{1}{n}, x_2 = -1 + \frac{2}{n}, \ldots, x_n = 0 = y_0, \right.$$
$$\left. = y_0, y_1 = \frac{1}{n}, y_2 = \frac{2}{n}, \ldots, y_4 = \frac{i}{n}, \ldots, y_n = 1\right\}$$

Let $\xi_i = x_i \in [x_{i-1}, x_i]$ and $x_i' = y_i \in [y_{i-1}, y_i]$, (i = 1, 2, ..., n)

$\therefore$ $x_i = -1 + i/n$, $y_i = i/n$ for each i.

$$\text{Now } S(P, f) = \sum_{i=1}^{n} f(\xi_1)\,\Delta x_i + \sum_{i=1}^{n} f(\xi_i')\,\Delta y_i$$

$$= \Sigma(-x_i)\,\Delta x_i + \Sigma\, f(y_i)\,\Delta y_i$$

$$= \Sigma\,(-x_i)\,\Delta x_i + \Sigma\, y_i\,\Delta y_i$$

[$\because$ $x_i \leq 0$ and $y_i \leq 0$ for i = 1, 2, ... n]

$$= \sum -\left(-1 + \frac{i}{n}\right)\Delta x_i + \sum \frac{i}{n}\Delta y_i$$

$$= \sum -\left(1 - \frac{i}{n}\right)\frac{1}{n} + \sum \frac{i}{n}.\frac{1}{n}$$

$$\left(\because \Delta x_i = \Delta y_i = \frac{1}{n}\right)$$

$$= \sum_{i=1}^{n} \frac{1}{n} = \frac{n}{n} = 1$$

$$\therefore \lim_{\mu(P)\to 0} S(P, f) = 1.$$

$$\text{Hence } \int_{-1}^{1} |x|\,dx = \lim_{\mu(P)\to 0} S(P, f) = 1.$$

Example 7:

Show that $\int_1^2 f\,dx = \frac{11}{2}$ *where* $f(x) = 3x + 1$.

Let $P = \{1 = x_0, x_1, x_2, ..., x_n = 2\}$ *be a partition which divides [1, 2] into n equal sub-intervals, each of length* $\frac{2-1}{n} = \frac{1}{n}$, *so that*

$$\mu(P) = \frac{1}{n} \to 0 \text{ as } n \to \infty$$

$$x_i = 1 + \frac{i}{n}, i = 1,2, ...n$$

$$\Delta x_i = \frac{i}{n}, i = 1,2, ...n$$

$$\sum_{i=1}^{n} \Delta x_i = n, \frac{1}{n} = 1$$

Let $t_1 = x_i$, when $i = 1, 2, ..., n.$

$$\therefore S(P, f) = \sum_{i=1}^{n} f(t_1)\,\Delta x_i = \sum_{i=1}^{n} f(x_i)\Delta x_i = \sum_{i=1}^{n} (3x_i + 1)\Delta x_i$$

$$= \sum_{i=1}^{n} \left\{3\left(1 \div \frac{i}{n}\right) + 1\right\} \Delta x_i = 4\sum_{i=1}^{n} \Delta x_i + \frac{3}{n^2} \sum_{i=1}^{n} i$$

$$= 4 + \frac{3}{n^2} \frac{n(n+1)}{2} = \frac{11}{2} + \frac{3}{2n}$$

Proceeding to limits when $\mu(P) \to 0$.

$$\lim_{\mu(P) \to 0} S(P, t) = \frac{11}{2}$$

Since the limit exists the function is integrable

and $s_1^{\,2}$ f dx = lim s(P, f) $= \frac{11}{2}$).

Example 8:

A function f is defined for the interval (0, 1) by

$$f(x) = \frac{1}{2^n} \text{ for } \frac{1}{2^{n+1}} < x \le \frac{1}{2^n},$$

n= 0, 1, 2, 3, ...; $f(0) = 0$. *Prove that f is integrable and calculate*

$\int_0^x f(x)\,dx$ where x lies between $\frac{1}{2^m}$ and $\frac{1}{2^{m-1}}$

Solution:

Obviously f(x) is bounded and monotonically increasing in (0, 1) [f(0)=0 and f(1) = 1 and the upper and lower bounds of f(x) are 1 and 0 in (0, 1)].

Hence, f is integral on (0, 1).

We can prove the problem as follows:

Here f(x) has ordinary discontinuities only at the set of points 0, 1/2, $1/2^2, 1/2^3, \ldots, 1/2^n \ldots$ which form an enumerable set with 0 as the only limiting point. So the function f is R-integrable.

Note that $f\left(\frac{1}{2^n}+0\right) = \frac{1}{2^{n-1}}, f\left(\frac{1}{2^n}-0\right) = \frac{1}{2^n}$

And $f\left(\frac{1}{2^n}\right) = \frac{1}{2^n}$ so that f is discontinuous at $x = \frac{1}{2^n}$,

$n = 1, 2, 3, \ldots, \infty$

Again the required integral $= \int_0^{\infty} f(x)\,dx$

$$= \int_{1/2^m}^{x} f(x)\,dx + \int_{1/2^{m+1}}^{1/2^m} f(x)\,dx + \int_{1/2^{m+2}}^{1/2^{m+1}} f(x)\,dx + \ldots \quad \ldots(1)$$

But in any interval $\left(\frac{1}{2^{r+1}}, \frac{1}{2^r}\right)$ the functional value is $\frac{1}{2^r}$ at all points except $x = \frac{1}{2^{r+1}} \frac{1}{2^m}$ where it is $\frac{1}{2^{r+1}}$.

Similarly in the interval $\left(\frac{1}{2^m}, x\right)$ the functional value is $\frac{1}{2^{m-1}}$ at all points except at the point $x = \frac{1}{2^m}$ where it is $\frac{1}{2^m}$.

In view of the above, the integral $\int_0^x f(x)\,dx$

$$= \frac{1}{2^{m-1}}\left(x - \frac{1}{2^m}\right) + \frac{1}{2^m}\left(\frac{1}{2^m} - \frac{1}{2^{m+1}}\right) + \frac{1}{2^{m+1}}\left(\frac{1}{2^{m+1}} - \frac{1}{2^{m+2}}\right) + \ldots$$

$$= \frac{x}{2^{m-1}} - \frac{1}{2^{2m-1}} + \frac{1}{2^{2m}}$$

$$- \frac{1}{2^{2m+1}} + \frac{1}{2^{2m+2}} - \frac{1}{2^{2m+3}}$$

$$= \frac{x}{2^{m-1}} - \frac{1}{2^{m-1}} + \left\{\frac{1}{1-(-\frac{1}{2})}\right\}$$ summing up the above infinite G.P.

$$= \frac{x}{2^{m-1}} - \frac{1}{3-2^{2m-2}}$$

Example 9:

If f(x) $= \begin{cases} 0 \text{ when x is an integer} \\ 1 \text{ when x is not an integer} \end{cases}$.

Show that f is integrable on any interval.

Solution:

Consider any internal (o, p) where p is a positive integer.

This function is obviously continuous at all points x = 1, 2, 3, ... p since $f(\alpha) = 0$ if a is any integer and $f(\alpha - 0) = 1 = f(\alpha + 0)$

Thus, the function f has only a finite number of discontinuities in a given interval (0, p). Hence, if each of these p points of discontinuity be enclosed in an interval of length < ϵ/p, all these points will be enclosed in a finite number of non-overlapping intervals whose total length (p ϵ/p) = ϵ. where ϵ is an arbitrarily chosen small positive number.

Hence, f is integrable in (0, p).

Example 10:

Find the upper and lower R-integrals for the function f defined on [0, 1] as follows:

$$f(x) = \begin{cases} \sqrt{1-x^2} & \text{if x is rational} \\ 1-x & \text{if x is irrational} \end{cases}$$

Solution:

Here we have

$$f(x) = \begin{cases} \sqrt{1-x^2} & \text{if x is rational} \\ 1-x & \text{if x is irrational} \end{cases}$$

$(1 - x^2) - (1 - x^2) = 2x - 2xb = 2x (1 - x) > x \ \forall x \in (0, 1)$ $\sqrt{1-x^2}$ $> (1 - x) \ \forall x \in (0, 1)$

with usefule notation we have for all values of r

$Mr = \sqrt{1-x^2}$ and $mr = (1 - x)$

Hence, $\underline{\int_0^1} f = \int_0^1 (1-x)du = \left[x - \frac{x^2}{2}\right]_0^1 = 1 - \frac{1}{2} = \frac{1}{2}$

and $\overline{\int_0^1} f = \int_0^1 \sqrt{(1-x^2)}du = = \left[\frac{1}{2}x\sqrt{1-x^2} + \frac{1}{2}\sin^{-1} x\right]_0^1$

$= \frac{1}{2} \cdot \frac{\pi}{2} = \frac{\pi}{2}$

Thus, $\underline{\int_0^1} f \neq \overline{\int_0^1} f$ it follows that $f \in (0, 1)$.

Example 11:

Find the upper and lower R-integrals for the function defined in the interval (0, 1) as follows:

$f(x) = \sqrt{(1-x^2)}$ *when x is rational*

$f(x) = (1 - x^2)$ *when x is irrational*

and show that $f(x)$ *is not R-integrable in (0, 1).*

Solution:

Here we have $[\sqrt{(1-x^2)}\,^2] - (1 - x^2)^2$

$= 1 - x^2 - (1 - 2x + x^2)$

$= 2x - 2x^2 = 2x(1 - x) > 0$ for all values of x in the open interval (0, 1)

$\therefore [\sqrt{(1-x^2)}\,^2] > (1 - x)^2$

or $\sqrt{(1-x^2)} > (1 - x)$ for $0 < x < 1$ ...(1)

Thus, if we suppose that $x_0 = 0, x_1, x_2 ...., x_{n-1}, x_n = 1$ is any mode of sub-division *D* of (0, 1) then in any interval $\delta_r \equiv (x_r - x_{r-1})$ however small it may be, the values of $f(x)$ range between $(1 - x)$ and $\sqrt{(1-x^2)}$

$$\left.\begin{array}{ll}\text{So} & M_r \text{ The Supremum } f(x) \text{ in } \delta_r = \sqrt{(1-x^2)} \\ \text{and} & m_r \text{ The Infimum of } f(x) \text{ in } \delta_r = (1-x)\end{array}\right\} \text{ for all r} \quad ...(2)$$

Hence, the upper integral = J = $\overline{\int_0^1} f(x)\,dx$

$\lim_{\Delta\to 0}\sum_{r=1}^{n} M_r\delta_r$ where Δ is the norm of the division *D*

$$\int_0^1 \sqrt{(1-x^2)}\,dx = \left[\frac{x\sqrt{(1-x^2)}}{2}+\frac{1}{2}\sin^{-1}x\right]_0^1 \quad \text{Integrating by parts.}$$

$$= \frac{1}{2}\,[\sin^{-1}(1)] = \frac{1}{2}\,\pi/2 = \pi/4$$

Similarly the lower integral = *I* = $\underline{\int_0^1} f(x)\,dx$

$\lim_{\Delta\to 0}\sum_{r=1}^{n} m_r\,\delta_r$

$$\int_0^1 (1-x)\,dx = \left[x-\frac{x^2}{2}\right]$$

$= 1 - \frac{1}{2} - \frac{1}{2}$ Since X ≠ J

Hence, the function l is not R-integrable in [0, 1].

Example 12:

Show that (3x + 1) is integrable on [1, 2] and

$$\int_1^2 (3x+1)\,dx = \frac{11}{2}.$$

Solution:

Let f (x) = 3x + 1. Consider the following partition of [1, 2] :

$$P=\left\{1,1+\frac{1}{n},1+\frac{2}{n},......,1+\frac{i-1}{n},\frac{1+i}{n},.....,1+\frac{n}{n}=2\right\}$$

Then $\Delta x_i = \frac{1}{n}$ for i = 1, 2,......n. In fact, μ (P) = $\frac{1}{n}$

Thus, m (P) → 0 iff n → ∞

Clearly $M_i = 3\left(1+\frac{i}{n}\right) + 1$, $m_i = 3\left(1+\frac{i-1}{n}\right)+1$

in the *i*th sub-interval of P, i = 1, 2.....n.

$\therefore M_i = 4 + \frac{3i}{n}$, $m_i = 4 + \frac{3(i-1)}{n}$

$$\text{Now } U(P, f) = \sum_{i-1}^{n} M_i \Delta x_i = \sum_{l-1}^{n}\left(4+\frac{3i}{n}\right).\frac{1}{n}$$

$$= \frac{1}{n^2}\sum_{i-1}^{n}(4x+3i) = \frac{1}{n^2}[4n^2 + 3)(1+ 2 +,....+ n)]$$

$$= 4 + \frac{3}{n^2}\frac{n(n+1)}{2} = 4+\frac{3}{2}\left(1+\frac{1}{n}\right)$$

$$\to 4 + \frac{3}{2} \text{ as } n\to \infty$$

$$\text{We know } \quad \overline{\int_a^b} f\,dx = \lim_{n\to\infty} \{U(P, f)\} = 4 + \frac{3}{2} = \frac{11}{2}$$

$$\text{Again } L(P, f) = \sum_{i=1}^{n} m_i \Delta x_i = \sum_{i=1}^{n}\left[4+\frac{3(i-1)}{n}\right].\frac{1}{n}$$

$$= \frac{1}{n^2}\sum_{i=1}^{n}[4n^2 + 3(i-1)]$$

$$= \frac{1}{n^2}[4n^2 + 3(1 + 2+.......+ n - 1)]$$

$$= 4\ \frac{3}{n^2}(n - 1).\ \frac{n}{2}$$

$$= 4 + \frac{3}{2}\left(1-\frac{1}{n}\right) \to 4 + \frac{3}{2} \text{ as } n \to \infty$$

$$\text{We know that } \quad \underline{\int_a^b} f\,dx = \lim_{n\to\infty} [L(P, f)] = 4 + \frac{3}{2} = \frac{11}{2}$$

Hence, $\quad f(x) = 3x + 1$ is integrable on $\{1, 2\}$

and $\quad \int_1^2 (3x+1)\,dx = \dfrac{11}{2}$ **Hence Proved.**

Example 13:

Show that x^2 is integrable on any interval [0, k).

Solution:

Let us consider the partition P of [0, k] obtained by dividing the interval into n equal parts. Thus, [0, k/n, 2k/n....., nk/n] is the partition P. $[(i - 1)(k/n)]^2$ and $[i(k/n)]^2$ are the lower and upper bounds of the function in Δx_i and the length of each such interval is k/n.

$$\therefore\ U(P, x^2) = \frac{k^2}{n^3}(1^2 + 2^2 + 3^2 + ... + n^2)$$

$$= \frac{k^3}{n^3} \cdot \frac{n}{6}(n + 1)(2n + 1)$$

$$= \frac{k^3}{6}\left(1 + \frac{1}{n}\right)\left(2 + \frac{1}{n}\right)$$

and

$$L(P, x^2)\ \frac{k^3}{n^3}\{0 + 1^2 + 2^2 + ... + (n - 1)^2\}$$

$$= \frac{k^3}{6}\left(1 - \frac{1}{n}\right)\left(2 - \frac{1}{n}\right)$$

$$\therefore\ \inf U(P, x^2) = \frac{k^3}{3} - \sup L(P, x^2)$$

Hence, the function x^2 is integrable and

$$\int_0^k x^2\, dx\ k^3/3.$$

Example 14:

$$\textit{If } f(x) = \begin{cases} 0, \text{when x is rational,} \\ 1, \text{when x is irrational,} \end{cases}$$

show that f is not integrable on any interval.

Solution:

Let $P = \{a = x_o, x_1, ... , x_{i-1}\ .....\ x_n = b$

be any partition of an interval [a, b]. $(a \neq b)$

Then $\sum_{i=i}^{n} \Delta x_i\ \sum_{i=1}^{n}(x_i - x_{i-1}) = (b - a)$

In any sub-interval $[x_{i-1,}\ x_i]$, it is clear that

$$M_i = 1\ m_i = 0\ (i = 1, 2,.....,n).$$

$$\therefore\ U(P, f) = \sum_{i=1}^{n} M_i\, \Delta x_i = \sum_{i=1}^{n} 1.\ \Delta x_i = (b - a)$$

$$L(P, f) = \sum_{i=1}^{n} m_i\, \Delta x_i = \sum_{i=1}^{n} 0.\ \Delta x_i = 0$$

Thus, $\overline{\int_0^b} f\,dx = \inf\{U(p, f)\} = b - a,$

$\underline{\int_a^b} f\,dx = \sup\{L(p, f)\} = 0,$

$\therefore \overline{\int_a^b} f\,dx \neq \underline{\int_a^b} f dx$ and so f is not integrable on [a, b].

Note that f is a bounded function on [a, b].

Example 15:

Show that any constant function is integrable.

Solution:

Let f(x) = k on [a, b] (k being a constant).

Then for any partition

$P = [a = x_0, x_1 \ldots\ldots x_i ,,,,, x_n = b)$

$M_i = m_i = k$ for i = 1, 2, ... n

$$U(P, f) = \sum_{c=1}^{n} M_i \Delta x_i = k\sum_{i=1}^{n} \Delta x_i = k(b-a)$$

$$L(P, f) = \sum_{i=1}^{n} m_i \Delta x_i = k\sum_{i=1}^{n} \Delta x_i = k(b-a)$$

Thus,

$$\overline{\int_0^b} f\,dx = \inf\left[U(P,f)\right] = k(b-a)$$

$$\underline{\int_a^b} f\,dx = \sup\left[L(P,f)\right] = k(b-a)$$

Hence, $\int_a^b f dx = k(b-a)$

Hence, the function x^2 is integrable and $\int_0^k x^2 dk\, k\frac{2}{3}$.

Example 16:

If $f(x) = x$ and $\alpha(x) = \sin(\pi x/2)$, find $U(P, f, \alpha)$ and $L(P, f, \alpha)$, when $P[0, 4] = \{0, 2, 4\}$.

Solution:

The partition of inf [0, 4] is given as P = {0, 2, 4}.

$\therefore m_1 = \inf$ in $[0, 2] = 0$; $m_2 = \inf f$ in $[2, 4] = 2$;

M_1 = sup f in [0, 2] = 2; M_2 = sup f in [2, 4] = 4.

$U(P, f, \alpha) = M_1 \{\alpha(2) - \alpha(0)\} + M_2 \{\alpha(4) - \alpha(2) = 2(0 - 0)$
$+ 4(0 - 0) = 0.$

$[\therefore \alpha(2) = \sin \pi = 0, \alpha(4) = \sin 2\pi = 0]$

Similarly $L(P, f, \alpha) = m_1 \{\alpha(2) - \alpha(0)\} + m_2 \{\alpha(4) - \alpha(2)\} = 0.$

Example 17:

Let $\alpha(x) = f(x) = \begin{Bmatrix} 0 \text{ if } 0 \leq x < 1 \\ 1 \text{ if } \leq x \leq 2 \end{Bmatrix}$ *and* $g(x) = \begin{Bmatrix} 0 \text{ if } 0 \leq x \leq 1 \\ 1 \text{ if } < x \leq 2 \end{Bmatrix}$

(i) *Is f (α)? If so, compute* $\int_0^2 f d\alpha$.

(ii) *Is* $g \in \mathfrak{R}(\alpha)$*? If so, compute* $\int_0^2 g d\alpha$.

Solution:

Consider any portion of [0, 2] as follows:

$P = \{0 = x_0, x_1 ..., x_{1-1}, x_1 = 1, x_1 = 1, x_{i+1}, ..., x_m = 2\}.$

Let M_k = sup f and m_k = inf f over $[x_{k-1}, x_k]$. Then

$m_k = M_k = 0$ for k = 0, l, ..., l – 1;

$m_k = M_k = 1$ for k = l,........, m.

Now, we have

$$U(P, f, \alpha) = \sum_{k=1}^{l-1} M_k \Delta\alpha_k + M_l \Delta\alpha_l + \sum_{k=l+1}^{m} M_k \Delta\alpha_k$$

$$= 0 + 1. \{\alpha(x_{1-1}) - \alpha(x_{1-1})\} + \sum_{k=l+1}^{m} 1.\{\alpha(x_k) - a(x_{k-1})\}$$

$$= 1(1 - 0) + \sum_{k=l+1}^{m} 1(1-1) = 1.$$

Similarly, L(P, f, a) = 1.

$$\therefore \int_a^b f d\alpha = \inf\{U(P,f,\alpha)\} = 1 \text{ and } \int_a^b f d\alpha = \sup\{L(P,f,\alpha)\} = 1$$

Hence, $f \in \mathfrak{R}(a)$ and $\int_1^2 f d\alpha = 1$.

(ii) Let m_k = inf g and M_k = sup g over $[x_{x-1}, x_k]$.

$\therefore m_k = M_k = 0$ for k = 0, 1....., l

$m_k = M_k = 1$ for k = l + 1...., m

$$\text{Now, } U(P, g, \alpha) = \sum_{k=0}^{l} M_k \Delta\alpha_k + \sum_{k=l+1}^{m} M_k \Delta\alpha_k$$

$$= 0 + \sum_{k=l+1}^{m} 1.\{\alpha(x_k) - \alpha(x_{k-1})\} = \sum_{k=l+1}^{m} 1\{1-1\} = 0$$

Similarly, L(P, g, a) = 0.

As argued above, $g \in \Re(\alpha)$ and $\int_0^2 g d\alpha = 0$.

Example 18:

Let f (x) = [x] ∀ x ∈ [– 1, 2] and

$$\alpha(x) = \begin{cases} 1, x < 0 \\ 2, 0 \le x < 1 \\ 3, 1 \le x \le 2. \end{cases} \quad \text{Is } f \in \Re(\alpha).$$

Solution:

Let $P = \{-1 = x_0, x_1, ..., x_l = 0, x_{l+1}, ...x_m = 1, x_{m+1}, ..., x_n = 2\}$ be any partition of [– 1, 2].

Let M_k = sup f over $[x_{k-1}, x_k]$ and m_k = inf f over $[x_{k-1}, x_k]$.

$\therefore m_k = -1$ and $M_k = 0$ for $k = 0, 1, ..., l$

$m_k = 0$ and $M_k = 1$ for $k = l + 1..., m$

$m_k = 1$ and $M_k = 2$ for $k = m + 1...., n$

We have

$$\Delta\alpha_k = \alpha(x_k) - a(x_{k-1}) = \begin{cases} 1-1=0 & k=1,2,...,l-1 \\ 2-1=1 & k=l \\ 2\text{-}2=0 & k=l+1.....,m-1 \\ 3\text{-}2=1 & k=m \\ 3\text{-}3=0 & k=m+1.....,n. \end{cases}$$

$$\text{Now } U(P, f, a) = \sum_{k=1}^{1} M_k \Delta\alpha_k + \sum_{k=l+1}^{m-1} M_k \Delta\alpha_k + M_m \Delta\alpha_m + \sum_{k=m+1}^{n} M_k \Delta\alpha_k$$

$$= 0 + 1.0 + 1.1 + 2.0 = 1.$$

$$\therefore \quad \overline{\int_{-1}^{2}} f d\alpha = \inf \{U(P, f, \alpha)\} = .$$

$$\text{Now } L(P, f, \alpha) \therefore \sum_{k=1}^{l-1} m_k \Delta\alpha_k + m_l \Delta\alpha_l + \sum_{k=l+1}^{m} m_k \Delta\alpha_k + \sum_{k=m+1}^{n} m_k \Delta\alpha_k$$

$$= -1.0 + (-1).1 + 0 + 1.0 = -1.$$

$$\therefore \quad \underline{\int_{-1}^{2}} f d\alpha = \sup \{L(P, f, \alpha)\} = -1.$$

Since $\overline{\int_{-1}^{2}} f d\alpha \neq \underline{\int_{-1}^{2}} f d\alpha$, $f \notin \Re(\alpha)$.

Hence f is not RS-integrable.

Example 19:

Evaluate $\int_0^3 x\, d[x]$.

Solution:

The function $\alpha(x) = (x)$ is monotonically increasing on [0, 3].

For $0 \leq x < 1$, $\alpha(x) = 0$,

For $1 \leq x < 2$, $\alpha(x) = 1$,

For $2 \leq x < 3$, $\alpha(x) = 2$ and $\alpha(3) = [3] = 3$.

Consider any partition P of [0, 3] :

$P = \{0 = x_0, x_1, ..., x_{l-1}, x_l = 1, x_{l+1}, ..., x_{m-1}, x_m = 2, x_{m+1}, ..., x_{n-l}, x_n = 3\}$

$$\therefore\ U(P, f, \alpha) = \sum_{i=1}^{l-1} M_i\, \Delta\alpha_i + M_l\,(x_l - x_{l-1}) + \sum_{i=l+1}^{m-1} M_i . \Delta\alpha_i$$

$$+ M_m(x_m - x_{m-1}) + \sum_{i=m+1}^{n-1} M_i\, \Delta\alpha_i + M_n(x_n - x_{n-1}) \quad ...(1)$$

$$= \sum_{i=1}^{l-1} M_i\,(0-0) + 1(1-0) + \sum_{i=l+1}^{m-1} M_i(1-1)$$

$$+ 2(2-1) + \sum_{i=m+1}^{n-1} M_i(2-2) + 3(3-2) = 6.$$

$$\therefore\ \overline{\int_0^3} x d[x] = \inf\{U(P, f, \alpha)\} = 6.$$

Similarly, $\underline{\int_0^3} x d[x] = \sup\{L(P, f, \alpha)\} = 6$.

Hence $\int_0^3 x d[x] = 6$. **Ans.**

Example 20:

Evaluate $\int_0^3 x d([x] - x)$.

Solution:

By R.S.-Integral property.

We have $\int_0^3 x d([x] - x) = \int_0^3 x d[x] - \int_0^3 x dx = 6 - \frac{9}{2} = \frac{3}{2}$ **Ans.**

Example 21:

Give an example of a function f and integrator α such that $f \notin R[a, b]$, but $f \in R_\alpha [a, b]$.

Solution:

We have to give an example of a function f and integrator α such that f is RS-integrable w.r.t. α but f is not Riemann integrable. Consider a function

$$f(x) = \begin{cases} 0, & \text{if } x \text{ is rational} \\ 1, & \text{if } x \text{ is irrational.} \end{cases}$$

We know f is not Riemann integrable over any interval [a, b] i.e. $f \notin \mathfrak{R}[a, b]$. Obviously f is bounded over [a, b].

Let us take $\alpha(x) = c$ (constant) for all $x \in [a, b]$.

Then $\int_a^b f(x)\, d\alpha = \int_a^b f(x)\, dc = 0.$

Hence $f \in \mathfrak{R}_\alpha [a, b]$ but $f \notin \mathfrak{R}[a, b]$.

Example 22:

Evaluate $\int_0^1 dx^2$ from the definition of RS-integral.

Solution:

Consider a partition of [0, 1] as follows:

$$P = P = (0, \frac{1}{n}, \frac{2}{n}, \ldots, \frac{k-1}{n}, \ldots, \frac{n}{n} = 1\}.$$

We take $t_k = \frac{k}{n} \in \left[\frac{k-1}{n}, \frac{k}{n}\right]$. We have $f(x) = x$ and $\alpha(x) = x^2$.

Now $\int_0^1 x\, dx^2 = \lim_{n\to\infty} S(P, f, \alpha)$ (by definition)

$$\Rightarrow \lim_{n\to\infty} \sum_{k=1}^{n} f(t_k)\, \{\alpha(x_k) - \alpha(x_{k-1})\}$$

$$\Rightarrow \lim_{n\to\infty} \sum_{k=1}^{n} f\left(\frac{k}{n}\right) \left\{\alpha\left(\frac{k}{n}\right) - \alpha\left(\frac{k-1}{n}\right)\right\} \qquad \ldots(1)$$

$$\Rightarrow \lim_{n\to\infty} \sum_{k=1}^{n} \frac{k}{n} \left\{\left(\frac{k}{n}\right) - \left(\frac{k-1}{n}\right)^2\right\}$$

$$\Rightarrow \lim_{n\to\infty} \sum_{k=1}^{n} \frac{k}{n}\left(\frac{2k-1}{n^2}\right) = \lim_{n\to\infty} \sum_{k=1}^{n} \frac{1}{n^3}(2k^2-k) \qquad ...(2)$$

We know $\sum_{k=1}^{n} k = \frac{n(n+1)}{2}, \sum_{k=1}^{n} k^2 = \frac{n(n+1)(2n+1)}{6}$. ...(3)

From (2) and (3), we get

$$\int_0^1 x\,dx^2 = \lim_{n\to\infty}\left[\frac{n(n+1)(4n-1)}{6n^3}\right], \text{ after simplication}$$

We have $= \frac{1}{6}\lim_{n\to\infty}\left(1+\frac{1}{n}\right)\left(4-\frac{1}{n}\right) = \frac{2}{3}$. **Ans.**

Example 23:

Show that $\int_{-1}^{1} x\,d|x| = 1$.

Solution:

Let $f(x) = x$, $\alpha(x) = |x|$.

Integrating by parts, we get

$$\int_{-1}^{1} x\,d|x| = \{x|x|\}_{-1}^{1} - \int_{-1}^{1} |x|.1dx = 2 - 2\int_0^1 xdx = 2 - 1 = 1.$$

Example 24:

Evaluate: $\int_0^3 x^2 d[x]$.

Solution:

We know that

$$U(P,f,a) = \sum_{i=1}^{l-1} M_i(0-0) + 1^2(1-0) + \sum_{i=l+1}^{m-1} M_i(1-1)$$

$$+2^2(2-1) + \sum_{i=m+1}^{n-1} M_i(2-2) + 3^2(3-2) = 14.$$

$\therefore \overline{\int_0^3} x^2 d[x] = \inf\{U(P,f,\alpha) = 14$.

Similarly, $\underline{\int_0^3} x^2 d[x] = 14$. Hence $\int_0^3 x^2 d[x] = 14$. **Ans.**

Example 25:

Let f be bounded on [–1, 1] and the function α be defined as α(x)

$= 0$ *if* $x < 0$; $\alpha(0) = \frac{1}{2}$; $\alpha(x) = 1$ *if* $x > 0$. ...(1)

Prove that is RS-integrable w.r.t. α *iff f is continuous at* $x = 0$ *and then*

$$\int_{-1}^{1} f\, d\alpha = f(0).$$

Solution:

Let $P = \{-1 = x_0, x_1, ..., x_{i-2}, x_{i-1} = 0, x_i, ..., x_n + 1\}$ be a partition of $[-1, 1]$. Let $t_i \in \Delta x_i = [x_{i-1}, x_i]$, $x_{i-1} = 0$.

Now know that $S(P, f, \alpha) = \sum_{j=1}^{n} f(t_j)\{\alpha(x_j) - \alpha(x_{j-1})\}$

$= 0 + f(t_{i-1})\{\alpha(x_{i-1}) - \alpha(x_{i-2})\} + f(t_i)\{\alpha(x_i) - \alpha(x_{i-1})\} + 0$

$= f(t_{i-1})\left(\frac{1}{2} - 0\right) + f(t_i)\left(1 - \frac{1}{2}\right)$, using (1)

$= \frac{1}{2}[f(t_{i-1}) + f(t_i)] = f(0)$, when $t_{i-1} = 0 = t_i$.

Using variable x instead of t_i and t_{i-1}, we find that $x \to 0$ from the left as well as from the right, when the norm $\mu(P)$ tends to zero. Thus

$\lim S(P, f, \alpha)$ exists iff $f(0-)$ and $f(0+)$ both exist and equal $f(0)$.

Hence f is RS-integrable w.r.t. α if f is continuous at $x = 0$.

We know that if f is continuous, then $f \in \Re(\alpha)$.

Hence f is RS-integrable iff f is continuous at $x = 0$ and

$\int_{-1}^{1} f d\alpha = \lim S(P, f, \alpha) = f(0)$. **Ans.**

Example 26:

Evaluate: $\int_0^2 [x]\, dx^2$, *using definition of integral.*

Solution:

It is clear that $\alpha(x) = x^2$ is monotonically increasing in $[0, 2]$. Also $[x] = 0$, $0 \le x < 1$; $[x] = 1$, $1 \le x < 2$ and $[x] = 2$ for $x = 2$.

Take $P_n = \left\{0, \frac{1}{n}, \frac{2}{n}, ... \frac{2n}{n} = 2\right\}$ as partition of $[0, 2]$.

$\therefore \Delta\alpha_r = \alpha\left(\frac{r}{n}\right) - \alpha\left(\frac{r-1}{n}\right) = \left(\frac{r}{n}\right)^2 - \left(\frac{r-1}{n}\right)^2 = \frac{2r-1}{n^2}$ ...(1)

Now

$$M_r = \text{suo } [x] \text{ in } \left[\frac{r-1}{n}, \frac{r}{n}\right] = \begin{cases} 0, 1 \le r < n \text{ or } 1 \le r \le n-1 \\ 1, r = n \\ 1, n < 2n \text{ or } n+1 \le r \le 2\,n-1 \\ 2, r = 2n \end{cases} \quad ...(2)$$

and

$$m_r = \inf [x] \text{ in } \left[\frac{r-1}{n}, \frac{r}{n}\right] = \begin{cases} 0, 1 \le r < n \text{ or } 1 \le r \le n-1 \\ 0, r = n \\ 1, n < 2n \text{ or } n+1 \le r \le 2\,n-1 \\ 1, r = 2n \end{cases} \quad ...(3)$$

$$\therefore\ U(P_n, f, \alpha) = \sum_{r=1}^{2n} M_r\, \Delta\alpha_r = \sum_{r=1}^{2n} M_r\left(\frac{2r-1}{n^2}\right), \text{by (1)}$$

Using (2), we get

$$\therefore\ U(P_n, f, \alpha) = \frac{1}{n^2}\sum_{r=1}^{n-1} 0, (2r-1) + \frac{1}{n^2}(2n-1)$$

$$+ \frac{1}{n^2}\sum_{r=n+1}^{2n-1} 1, (2r-1) + \frac{1}{n^2}, 2\,(4n-1)$$

$$= \left(\frac{2}{n} - \frac{1}{n^2}\right) + \frac{1}{n^2}\left[2\sum_{r=n+1}^{n+(n-1)} r - \sum_{r=n+1}^{n+(n-1)} 1\right] + 2\left(\frac{4}{n} - \frac{1}{n^2}\right)$$

$$= \frac{2}{n} - \frac{1}{n^2} + \frac{1}{n^2}\left[2\left\{(n-1)\,n + \frac{(n-1)n}{2}\right\} - (n-1)\right] + \frac{8}{n} - \frac{2}{n^2}$$

$$= \frac{10}{n} - \frac{3}{n^2} + \frac{3(n-1)}{n} - \frac{1}{n} + \frac{1}{n^2}$$

$$= \frac{6}{n} - \frac{2}{n^2} + 3 \to 3, \text{ as } n \to \infty.$$

$$\text{Similarly, } L(P_m, f, \alpha) = 0 + 0 + \frac{1}{n^2}\sum_{r=n+1}^{2n-1} 1.\,(2r-1) + \frac{1}{n^2}(4n-1)$$

$$= \frac{4}{n} - \frac{1}{n^2} + \frac{3(n-1)}{n} - \frac{1}{n} + \frac{1}{n^2} = \frac{3}{n} + 3 - \frac{3}{n} = 3$$

$$\to 3, \text{ as } n \to \infty.$$

$$\therefore\ \overline{\int_0^2} f\, d\alpha = 3 = \underline{\int_0^2} f\, d\alpha. \text{ Hence } \int_0^2 [x]\, dx^2 = 3.$$ **Ans.**

Example 27:

Using only the definition of the integral, show that:

$$\int_0^3 x^3 d[x] = 1.$$

Solution:

Use know that

$$\int_0^1 x^3 d[x] = \lim_{n\to\infty} \sum_{k=1}^{n} \left(\frac{k}{n}\right)^3 \left\{\left[\frac{k}{n}\right] - \left[\frac{k-1}{n}\right]\right\}$$

By definition, $\alpha(x) = [x] = 0$ for $0 \le x < 1$ and $\alpha(1) = 1$.

$$\int_0^1 x^3 d[x] = \lim_{n\to\infty} \sum_{k=1}^{n-1} \left(\frac{k}{n}\right)^3 \left\{\left[\frac{k}{n}\right] - \left[\frac{k-1}{n}\right]\right\} + \lim_{n\to\infty} \left(\left\{\left[\frac{n}{n}\right] - \left[\frac{n-1}{n}\right]\right\}\right.$$

$$= \lim_{n\to\infty}\left[\sum_{k=1}^{n-1} \left(\frac{k}{n}\right)^3 (0-0) + 1(1-0)\right] = 1.$$ **Ans.**

Example 28:

Give an example of a function f and integrator α such that $f \notin R$ [a, b], but $f \in R_\alpha[a, b]$.

Solution:

We solve this problem by choosing f and α in a different manner.

We choose

$$f(x) = \begin{cases} 0, & x = 0 \\ 1/x, & 0 < x < 1 \\ 1, & 1 \le x \le 2 \end{cases} \text{ and } a(x) = \begin{cases} 1, & 0 \le x \le 1 \\ x, & 1 \le x \le 2 \end{cases}.$$

It is easy to verify that the functions f and α are both continuous at x = We see that

$$\int_0^2 f dx = \int_0^1 f dx + \int_1^2 f dx = \int_0^1 \frac{1}{x} dx + \int_1^2 1.dx = \infty + 1 = \infty$$

(Notice that $\int_0^1 \frac{1}{x} dx$ is divergent.)

Thus, $f \notin R\,[0, 2]$.

Now, $\int_0^2 f d\alpha = \int_0^f f(x) d\alpha(x) + \int_1^2 f(x)\, d\alpha\,(x)$

$$= \int_0^1 \frac{1}{x}.0dx + \int_1^2 1.1\,dx = 0+1=1.$$

($\because$ $\alpha(x) = 1$ for $0 \le x \le 1 \Rightarrow d\,\alpha(x) = 0$ etc.)

Hence, $f \in R_\alpha[0, 2]$, but $f \notin R[0, 2]$. **Ans.**

Example 29:

Evaluate: $\int_{-1}^{2} x^5 d(|x|^3)$

Solution:

$$\int_{-1}^{2} x^5 d(|x|^3) = \{x^5\,|x|^3\}_{-1}^{2} - \int_{-1}^{2} |x|^3 .5x^4\,dx$$

$$= (32\times 8+1) - 5\left[\int_{-1}^{2} |x| x^6 dx\right] \quad (\because |x|^2 = x^2)$$

$$= 257 - 5\left[-\int_{-1}^{0} x^7\,dx + \int_0^1 x^7\,dx + \int_1^2 x^7\,dx\right]$$

$$= 257 - 5\left[\frac{1}{8} + \frac{1}{8} + \frac{255}{8}\right] = 257 \times \frac{3}{8} = \frac{771}{8}.$$ **Ans.**

Example 30(a):

Evaluate: $\int_0^2 [x]\,dx^2$

Solution:

Let $f(x) = [x]$ and $\alpha(x) = x^2$

By the relation between Riemann Integral and RS-Integral,

$$\int_0^2 [x]\,dx^2 = [x]\,2xdx$$

$$= 2[\int_0^1 [x]xdx + \int_1^2 [x]\,x\,dx]$$

$$= 2\left[\int_0^1 0.xdx + \int_1^2 1.xdx\right] = 3.$$ **Ans.**

Example 30(b):

Using definition of RS-integral, evaluate $\int_0^3 x\,d([x]-x)$.

Solution:

Let $f(x) = x$ and $\alpha(x) = [x] - x$. Then

$$\alpha(x) = \begin{cases} 0-x, 0 \leq x < 1 \\ 1-x, 1 \leq x < 2 \\ 2-x, 2 \leq x < 3 \\ 0, x = 3. \end{cases} \qquad ...(1)$$

We observe that a (x) is piecewise monotonically decreasing function in [0, 1), [1, 2) and [2, 3) respectively.

Let $P_n = \left\{0, \frac{1}{n}, \frac{2}{n}, ... \frac{r-1}{n}, \frac{r}{n}, ... \frac{3n}{3} = 3\right\}$

be a partition of [0, 3]. If M_r and m_r are the sup, and inf. of f (x) = x in [r – 1/n, r/n] respectively. Then

$$M_r = \frac{r}{n} \text{ and } m_r = \frac{r-1}{n}. \qquad ...(2)$$

Using (1), we obtain

$$\Delta\alpha_r = \alpha\left(\frac{r}{n}\right) - \alpha\left(\frac{r-1}{n}\right) = \begin{cases} -\frac{r}{n} + \frac{r-1}{n} = -\frac{1}{n} & 1 \leq r < n \\ 0 + \frac{n-1}{n} = \left(1 - \frac{1}{n}\right), & r = n \\ \left(1 - \frac{r}{n}\right) - 1 - \frac{r-1}{n} = -\frac{1}{n}, & n < r < 2n \\ 0 - \left(1 - \frac{2n-1}{n}\right) = 1 - \frac{1}{n}, & r = 2n \\ \left(2 - \frac{r}{n}\right) - \left(2 - \frac{r-1}{n}\right) = -\frac{1}{n}, & 2n < r < 3n \\ 0 - \left(2 - \frac{3n-1}{n}\right) = 1 - \frac{1}{n}, & r = 3n. \end{cases}$$

$\therefore\ \Delta\alpha_r = -\frac{1}{n}$, except at r = n, 2n, 3n

and $\Delta\,\alpha_r = 1 - \frac{1}{n}$ at r = n, 2n, 3n.

Now

$$U(P_n, f, \alpha) = \sum_{r=1}^{3n} M_r\,\Delta\alpha_r = \sum_{r=1}^{n-1} \frac{r}{n}\,\Delta\alpha_r + \frac{n}{n}\,\Delta\alpha_n + \sum_{r=n+1}^{2n-1} \frac{r}{n}\Delta\alpha_r$$

$$+ \frac{2n}{n}\Delta\alpha_{2n} + \sum_{r=2n+1}^{3n-1} \frac{r}{n}\,\Delta\alpha_r + \frac{3n}{n}\,\Delta\alpha_{3n}$$

$$= -\frac{1}{n^2}\sum_{r=1}^{n-1} r + \left(1-\frac{1}{n}\right) - \frac{1}{n^2}\sum_{r=n+1}^{2n-1} r + 2 - \frac{2}{n} - \frac{1}{n^2}\sum_{r=2n+1}^{3n-1} r + 3 - \frac{3}{n}$$

$$= 6 - \frac{6}{n} - \frac{1}{n^2}\cdot\frac{(n-1)n}{2} - \frac{1}{n^2}\left[(n-1)n + \frac{(n-1)n}{2}\right]$$

$$-\frac{1}{n^2}\left[(n-1)2n + \frac{(n-1)n}{2}\right]$$

$$= 6 - \frac{6}{n} - \frac{1}{2n^2}\{n^2 - n + 3n^2 - 3n + 5n^2 - 5n\}$$

$$= 6 - \frac{6}{n} - \frac{1}{2n^2}\{9n^2 - 9n\} = 6 - \frac{6}{n} - \frac{9}{2} + \frac{9}{2n}$$

$$= \frac{3}{2} - \frac{3}{2n} \to \frac{3}{2}, \text{ as } n \to \infty.$$

$\therefore \overline{\int_0^3} f\,d\alpha = \frac{3}{2}$. Similarly, $\underline{\int_0^3} f\,d\alpha = \frac{3}{2}$. Hence $\int_0^3 f\,d\alpha = \frac{3}{2}$. **Ans.**

Example 30(c):

Using only the definition of RS-Integral, evaluate

$$\int_0^2 x^2\, d[x].$$

Solution:

It is clear that f (x) = x^2 is bounded in [0, 2 and

$$\alpha(x) = [x] = \begin{cases} 0, & 0 \le x < 1 \\ 1, & 1 \le x < 2 \\ 2, & x = 2 \end{cases}$$

is monotonically increasing in [0, 2].

Let $P_n = \left\{0, \frac{1}{n}, \frac{2}{n}, \ldots, \frac{r-1}{n}, \frac{r}{n}, \ldots, \frac{2n}{n} = 2\right\}$ be a partition of [0, 2], for any positive integer n.

For each sub-interval [r – 1/n, r/n] (r = 1, 2, ..., n), we have

$$\Delta\alpha_r = \alpha\left(\frac{r}{n}\right) - \alpha\left(\frac{r-1}{n}\right) = \begin{cases} 0 - 0 = 0, & \text{if } 1 \le r < n \\ 1 - 0 = 1, & \text{if } r = n \\ 1 - 1 = 0, & \text{if } n < r < 2n \\ 1 - 0 = 1, & \text{if } r = 2n. \end{cases}$$

$\therefore\ \Delta\alpha_n = \Delta\alpha_{2n} = 1$ and $\Delta\alpha_r = 0$ if $r \neq n, 2n$. ...(1)

Now $M_r = \sup f$ in $[r - 1/n, r/n] = (r/n)^2$, ...(2)

$m_r = \inf f$ in $[r - 1/n, r/n] = (r - 1/n)^2$. ...(3)

$$\therefore U(P_n, f, \alpha) = \sum_{r=1}^{n} M_r\ \Delta\alpha_r = 0 + M_n\ \Delta\alpha_n + M_{2n}, \Delta\alpha_{2n}, \quad \text{using (1)}$$

Now $\overline{\int_0^2} f\,d\alpha = \lim_{n\to\infty} U(P_n, f, \alpha) = 5.$

Similarly we have, $L(P_n, f, \alpha) = 0 + m_n\ \Delta\ \alpha_n + m_{2n}\ \Delta\ \alpha_{2n}$

$$= \left(\frac{n-1}{n}\right)^2 + \left(\frac{2n-1}{n}\right)^2, \text{ by (1) and (3)}$$

$$= \left(1 - \frac{1}{n}\right)^2 + \left(2 - \frac{1}{n}\right)^2 \to 5 \text{ as } n \to \infty$$

$\therefore\ \underline{\int_0^2} f d\alpha = \lim_{n\to\infty} L(P_n, f, \alpha) = 5$. Hence $\int_0^2 f d\alpha = 5.$

Example 31:

If $f(x) = x$ and $\alpha(x) = \sin(\pi x/2)$, find $U(P, f, \alpha)$ and $L(P, f, \alpha)$, where $P[0, 2, 4]$. Prove that, refinements of P need not decrease upper sums or increase Lower sums.

Solution:

We have

$$\alpha(x) = \sin(\pi x/2) = \begin{cases} \sin(\pi\ x/2), 0 \leq x \leq 2 \text{ or } 0 \leq \pi\ x/2 \leq \pi \\ \sin(\pi\ x/2), 2 \leq x \leq 4 \text{ or } \pi \leq \pi\ x/2 \leq 2\pi \end{cases}$$

It follows that α (x) is not monotonic in the subintervals [0, 2] and [2, 4]. Also $\alpha(0) = \alpha(2) = 4$.

$U(P, f, \alpha) = L(P, f, \alpha) = 0.$

We consider a refinement of partition P = [0. 2. 4] as follows:

P^* [0, 4] = {0, 1, 2, 3, 4}. Then $\alpha(1) = 1$, $\alpha(3) = -1$.

Now $U(P^*, f, \alpha) = M_1 \{\alpha(1) - \alpha(0)\} + M_2 \{\alpha(2) - \alpha(1)\}$

$+ M_3 \{\alpha(3) - \alpha(2) + M_4 \{\alpha(4) - \alpha(3)\}$; where

M_1 = sub f in [0, 1] = 1, M_2 = sup f in [1, 2] = etc.

$\therefore U(P^*, f, \alpha) = 1(1 - 0) + 2(0 - 1) + 3(-1 - 0) + 4\ \{0 - (-1)$

$= 1 - 2 - 3 + 4 = 0 = U(P, f, \alpha).$

Similarly, $L(P^*, f, \alpha) = 0(1-0) + 1(0-1) + 2(-1-0) + 3\{0-(-1)\}$

$$= -1 - 2 + 3 = 0 = L(P, f, \alpha).$$

Hence, the refinements of P need not decrease upper sums or increase lower sums.

Example 32:

Prove that $\int_0^1 x\, d(e^{2x}) = \frac{1}{2}(\text{H } e^2)$.

Solution:

Let $f(x) = x$, $\alpha(x) = e^{2x}$. Integrating by parts, we obtain

$$\int_0^1 x\, d(e^{2x}) = \left|xe^{2x}\right|_0^1 \int_0^1 e^{2x}.1dx$$

$$= e^2 - \frac{1}{2}\left|e^{2x}\right|_0^1 = e^2 - \frac{1}{2}(e^2 - 1) = \frac{1}{2}(1 + e^2).$$ **Ans.**

Example 33(a):

A function f is defined in the interval (0, 1) as

$$f(x) = \frac{1}{n}, \text{ for } \frac{1}{n} > x > \frac{1}{n+1}$$

and $\qquad f(0) = 0.$

Prove that f(x) is R-integrable in (0, 1) and evaluate $\int_0^1 f(x)\, dx$.

Solution:

Here

$$f\left(\frac{1}{n}\right) = \frac{1}{n}$$

$$f\left(\frac{1}{n} + 0\right) = \frac{1}{n-1}$$

$$f\left(\frac{1}{n} - 0\right) = \frac{1}{n}.$$

Thus, there is a discontinuity of the first kind at each point

$$x = \frac{1}{n}, \; n = 1, 2, 3, \ldots\ldots$$

also at $x = 0$, $f(0)$, $f(0 + 0) = 0$, so $f(x)$ is continuous on the right of $x = 0$.

Now the set of points 1, $\frac{1}{2}, \frac{1}{3}, \frac{1}{4}$....., the points of discontinuity of f, has only one limit point. Hence, f is R-integrable.

Again,

$$\int_0^1 f(x)\,dx = \lim_{n\to\infty} \sum_1^n \int_{1/(n+1)}^{1/n} \frac{1}{n} dx$$

$$= \sum_1^{\infty} \frac{1}{n}\left(\frac{1}{n} - \frac{1}{n+1}\right)$$

$$= \left(1-\frac{1}{2}\right) + \frac{1}{2}\left(\frac{1}{2} - \frac{1}{3}\right) + \ldots$$

$$= \left(1+\frac{1}{2^2} + \frac{1}{3^3} + \ldots\right) - \left(\frac{1}{2} + \frac{1}{2.3} + \frac{1}{3.4} + \ldots\right)$$

$$= \frac{\pi^2}{6} - \left(\frac{1}{2} + \frac{1}{2} - \frac{1}{3} + \frac{1}{3} - \frac{1}{4} + \ldots.\right)$$

$$= \frac{\pi^2}{6} - 1$$ **Ans.**

Example 33(b):

The function f is defined in the interval (0, 1) by the condition that, (f r is a positive integer),

$$f(x) = 2rx \text{ when } \frac{1}{r+1} < x \le \frac{1}{r}$$

show that f is integrable in (0, 1) and that

$$\int_0^1 f(x)\,dx = \frac{\pi^2}{6}.$$

Solution:

Obviously, the function is not defined for

$$x = \frac{1}{r}, \quad r = 1, 2, 3, \ldots.$$

Also $f\left(\frac{1}{r}+0\right) = \lim_{h\to 0} 2\,(r-1)\left(\frac{1}{r}+h\right)$

$$= 2 - \frac{2}{r}$$

$$\Rightarrow f\left(\frac{1}{r}-0\right) = \lim_{h\to 0} 2r\left(\frac{1}{r}-h\right) = 2$$

$$\therefore \Rightarrow f\left(\frac{1}{r}+0\right) \neq f\left(\frac{1}{r}-0\right)$$

Also $\quad f(1-0) = \lim_{h\to 0} 2(1-h) = 2$

and $\quad f(1) = 2$

Thus, $\frac{1}{r}$ is a point of discontinuity of f when r = 2, 3, ..., i.e. the points $\frac{1}{2}, \frac{1}{3}, \frac{1}{4}$, ..., are the points of discontinuity. This is an enumerable set.

So the function f is R–integrable.

Let $\frac{1}{2}, \frac{1}{3}, \frac{1}{4}$, ..., be the points of division of (0, 1) then

$$\int_0^1 f(x)\,dx = \sum_{r=1}^{\infty} \int_{1/(r+1)}^{1/r} 2rx\,dx = \sum_1^{\infty}\left[rx^2\right]_{1/(r+1)}^{1/r}$$

$$= \sum_1^{\infty} r\left[\frac{1}{r^2} - \frac{r}{(r+1)^2}\right]$$

$$= \sum_1^{\infty}\left(\frac{r}{r^2} - \frac{r}{(r+1)^2}\right)$$

$$= \left(\frac{1}{1^2} - \frac{1}{2^2}\right) + \left(\frac{2}{2^2} - \frac{2}{3^2}\right) + \ldots.$$

$$= \frac{1}{12} + \frac{1}{22} + \frac{1}{32} + \frac{1}{42} + \ldots.$$

$$= \frac{\pi^2}{6}.$$ **Ans.**

Example 34:

If $f(x) = 0\ \forall\ x \in [0, 1]$ except at the set of points

$x_1, x_2, \ldots, s_n, \ldots, n \in N$ and $f(x_n) = 1/_n$.

Show that f is integrable in [0, 1].

Solution:

We know, that the function has countable number of discontinuities in [0, 1] viz. x_1, x_2, ..., x_3, ...

Let $\varepsilon > 0$ be given

Consider the interval $\left]x_n - \frac{\varepsilon}{2^n}, x_n + \frac{\varepsilon}{2^n}\right[$ for n = 1, 2, 3, ...

Consider a partition P of [0, 1] as follows :

$$0 < x_1 - \frac{\varepsilon}{2} < x_1 < x_1 + \frac{\varepsilon}{2} < x_2 - \frac{\varepsilon}{2^2} < x_2 < x_2 + \frac{\varepsilon}{2^2} < ..... <_1$$

$$\therefore U(P,f) - L(P,f) < \varepsilon\left(\frac{1}{2} + \frac{1}{2^2} + ...\right)$$

$$= \frac{\varepsilon}{2}\left(1 + \frac{1}{2} + \frac{1}{2^2} + ...\right)$$

$$= \frac{\varepsilon}{2}\frac{1}{(1-\frac{1}{2})} = \varepsilon.$$

Hence, f is integrable over [0, 1]. Hence the problem.

Example 35:

Show that $\int_0^1 \sin x \, dx = 1 - \cos t \ (0 \le x \le t \le \frac{1}{2}\pi)$.

Solution:

Let the partition P of [0, t] as follows:

$$P = \left\{0, \frac{t}{n}, \frac{2t}{n}, ..., \frac{(n-1)t}{n}, \frac{nt}{n}\right\}, \Delta x_i = \frac{t}{n}.$$

Since sin x is an increasing function in $0 \le t \le \frac{1}{2}\pi$, so

$$M_i = \sin\left(\frac{it}{n}\right), M_i = \sin\left(\frac{i-1}{n}t\right). \text{ Now,}$$

$$U(P, f) = \sum_{i=1}^{n} M_i . \Delta x_i = \frac{t}{n}\sum_{i=1}^{n} \sin\left(\frac{it}{n}\right)$$

$$= \frac{t}{n}\left[\sin\left(\frac{t}{n}\right) + \sin\left(\frac{2t}{n}\right) + ... + \sin\left(\frac{nt}{n}\right)\right] = \frac{tS}{n},$$

where $S = \sin\frac{t}{n}\sin\frac{t}{n} + \sin\left(\frac{2t}{n}\right) + ... + \sin\left(\frac{nt}{n}\right)$

$\Rightarrow 2 \sin (t/2n) S = 2 \sin (t/2n) \sin (t/n) + 2 \sin (t/2n) \sin (2t/n) + ... + 2 \sin (t/2n) \sin (nt/n)$

$= \{\cos (t/2n) - \cos (3t/2n)\} + \{\cos(3t/2n) - \cos (5t/2n)\} + ...$

$$\therefore \quad S = \frac{\cos(t/2n) - \cos\{(n+1)t/n\}}{2 \sin(t/2n)}$$

$$\text{Thus, } U(P, f) = \frac{tS}{n} = \frac{t}{n} \frac{\cos(t/2n) - \cos\{(n+1)t/n\}}{2 \sin(t/2n)}$$

$$\text{Again } L(P, f) = \sum_{i=1}^{n} m_i \, \Delta x_i = \frac{t}{n} \sum_{i=1}^{n} \sin\left(\frac{i-1}{n} t\right)$$

$$\text{or} \quad L(P, f) = \frac{t}{n} \frac{\cos (t/2n) - \cos(nt/n)}{2 \sin(t/2n)}$$

$$\therefore \; w (P, f) = U(P, f) - L(P, f) = \frac{t/n}{2 \sin(t/2n)}$$

$$X[\cos (nt/n) - \cos \{(n+1)t/n\}]$$

$$= \frac{t/n}{2 \sin(t/2n)} [2 \sin (t+t/2n) \sin (t/2n)]$$

$$= \frac{t}{n} \sin\left(2 + \frac{t}{2n}\right).$$

Thus, $w(P, f) \to 0$ as $n \to \infty$ and so $f(x) = \sin x$ is integrable over $[0, t]$.

$$\int_0^t \sin x \, dx = \lim_{n \to \infty} L(P, f)$$

$$= \lim_{n \to \infty} [\cos(t/2n) - \cos(nt/n)] . \frac{(t/2n)}{\sin(t/2n)}$$

$$= (\cos 0 - \cot t). \, 1 = (1 - \cos t).$$

Example 36:

Integrate on [0, 2] the function f(x) = x [x], where [x] denotes the greatest integer not greater than x.

Solution:

We have $x[x] = x, \; 0 = 0, \;$ if $0 \le x < 1$

$= x, \; 1 = x,$

$$\text{if } 1 \le x < 2$$

$$= 2x, \text{ if } x = 2$$

It follows that the function x[x] is bounded and has only two points of discontinuity at 1, 2.

Thus, x[x] is integrable over [0, 2] and so

$$\int_0^2 x[x]\,dx = \int_0^1 x[x]\,dx + \int_1^2 x[x]\,dx$$

$$= \int_0^1 0\,dx + \int_1^2 x\,dx = \frac{3}{2}.$$

Example 37:

Prove that if the functions f and ϕ are bounded and integrable in [a, b], then

$$\left[\int_a^b f(x)\,\phi(x)\,dx\right]^2 \le \int_a^b [f(x)]^2\,dx.\int_a^b [\phi(x)]^2\,dx.$$

Under what conditions does the sign of equality hold?

Solution:

Since f and ϕ are integrable in [a, b], so f^2, ϕ^2 and $f\phi$ are bounded and integrable in [a, b]. Thus, for any partition

$$P = \{a = x_0, x_1, ..., x_{r-1}, x_r, ..., x_n = b\}$$

of [a, b] and for any $\xi_r \in [x_{r-1}, x_r]$, we have

$$\left[\int_a^b f(x)\,\phi(x)\,dx\right]^2 = \left[\lim \sum (x_r - x_{r-1})\, f(\xi_r)\,\phi(\xi_r)\right]^2,$$

$$\int_a^b [f(x)]^2\,dx = \lim \left[\sum\left(\sqrt{x_r - x_{r-1}}\right) f(\xi_r)\right]^2,$$

$$\int_a^b [\phi(x)]^2\,dx = \lim \left[\sum\left(\sqrt{x_r - x_{r-1}}\right) \phi(\xi_r)\right]^2,$$

Putting $a_r = \sqrt{(x_r - x_{r-1})}\, f(\xi_r)$, $b_r = \sqrt{(x_r - x_{r-1})}\, \phi(\xi_r)$

in the Cauchy's inequality

$$(\Sigma a_r b_r)^2 \le \Sigma a_r^2\, \Sigma b_r^2,$$

we get the required result.

The sign of equality holds if

$$\frac{a_1}{b_1} = \frac{a_2}{b_2} = ...$$

i.e., if $\frac{f(\xi_1)}{\phi(\xi_1)} = \frac{f(\xi_2)}{\phi(\xi_2)}$ ......

i.e., if f and f are both constant functions.

Example 38(a):

If f(y, x) = 1 + 2x for y rational and f(y, x) = 0 for y irrational, calculate

$$F(y) = \int_0^1 f(y, x)\,dx.$$

Solution:

If y irrational, f(y, x) = 0 and so f is continuous and consequently integrable on [0. 1].

$$\therefore \quad F(y) = \int_0^1 f(y, x)\,dx = 0.$$

If y rational, f(y, x) = 1 + 2x is again continuous on [0, 1] and consequently integrable in [0, 1].

Now, consider the partition of [0, 1] as follows:

$$P = \left\{0, \frac{1}{n}, \frac{2}{n}, \ldots, \frac{n}{n} = 1\right\}, \Delta x_r = \frac{1}{n} \text{ for } r = 1, 2, \ldots, n.$$

$$\int_0^1 f dx = \lim_{n\to\infty} \sum_{r=1}^{n} f\left(\frac{r}{n}\right).\frac{1}{n} = \lim_{n\to\infty} \sum_{r=1}^{n} \frac{1}{n}\left(1+2\frac{r}{n}\right)$$

$$= \lim_{n\to\infty} \left[\frac{1}{n}\left\{\left(1+\frac{2}{n}\right)+\left(1+\frac{4}{n}\right)+..+\left(1+\frac{2n}{n}\right)\right\}\right]$$

$$= \lim_{n\to\infty} \left[\frac{1}{n}\left\{(1+1+\ldots+1)++\frac{2}{n}(1+2+\ldots+n)\right\}\right]$$

$$= \lim_{n\to\infty} \left[\frac{1}{n}\left\{n+\frac{2}{n}.\frac{n(n+1)}{2}\right\}\right] = \lim_{n\to\infty} \frac{2n+1}{n} = 2.$$

Hence, F(y) = 2. **Ans.**

Example 38(b):

A function f is defined on [0, 1] by

$f(x) = 1/n$ for $1/n \geq x > 1/(n+1)$, n = 1, 2, 3, ...

Prove that f is integrable over [0, 1] and evaluate $\int_0^1 f(x)\,dx$.

Solution:

The points of discontinuity of f are $\left\{1, \frac{1}{2}, \frac{1}{3}, \frac{1}{4}, \ldots\right\}$. The above set has only one limit point viz., 0. Thus, f is integrable over [0, 1]. Now,

$$\int_0^1 f(x)\,dx = \lim_{n\to\infty}\sum_{r=1}^{n}\int_{1/(r+1)}^{1/r}\frac{1}{r}\,dx = \lim_{n\to\infty}\sum_{r=1}^{n}\frac{1}{r}\left(\frac{1}{r}-\frac{1}{r+1}\right)$$

$$= \lim_{n\to\infty}\left[\left(1-\frac{1}{2}\right)+\frac{1}{2}\left(\frac{1}{2}-\frac{1}{3}\right)+\ldots+\frac{1}{n}\left(\frac{1}{n}-\frac{1}{n+1}\right)\right]$$

$$= \lim_{n\to\infty}\left[\left(1+\frac{1}{2^2}+\ldots+\frac{1}{n^2}\right)-\left\{\frac{1}{2}+\frac{1}{2.3}+\frac{1}{3.4}+\ldots+\frac{1}{n(n+1)}\right\}\right]$$

$$= \lim_{n\to\infty}\left[\left(1+\frac{1}{2^2}+\ldots+\frac{1}{n^2}\right)-\left\{\frac{1}{2}+\left(\frac{1}{2}-\frac{1}{3}\right)+\left(\frac{1}{3}-\frac{1}{4}\right)+\ldots+\left(\frac{1}{n}-\frac{1}{n+1}\right)\right\}\right]$$

$$\lim_{n\to\infty}\left[\left(1+\frac{1}{2^2}+\ldots+\frac{1}{n^2}\right)-\left(1-\frac{1}{n+1}\right)\right] = \frac{\pi^2}{6}-1.$$ **Ans.**

Example 38(c):

A function f is defined on [0, 1] as follows:

$$f(x) = \frac{1}{a^{r-1}}, \text{ when } \frac{1}{a^r} < x \le \frac{1}{a^{r-1}} \quad (r = 1, 2, 3, \ldots..)$$

f(0) = 0, where a is an integer than 2.

Show that $\int_0^1 f dx$ *exists and is equal to* $\frac{a}{a+1}$.

Solution:

Now, we have

$$f(x) = 1, \quad \frac{1}{a} < x \le 1$$

$$\Rightarrow \frac{1}{a}, \quad \frac{1}{a^2} < x \le \frac{1}{a}$$

$$\Rightarrow \frac{1}{a^2}, \quad \frac{1}{a^3} < x \le \frac{1}{a^2}$$

....

Clearly $1/a$, $1/a^2$, $1/a^3$, ... are the points of discontinuity of f with 0 as their only limit point. Thus, f is integrable on [0, 1]

Now, $\int_{1/a^n}^{1} f dx = \int_{1/a^n}^{1/a^{n-1}} \frac{1}{a^{n-1}} dx + \dots + \int_{1/a^2}^{1/a^2} \frac{1}{a} dx + \int_{1/a}^{1} 1 dx$

$$= \frac{1}{a^{n-1}}\left(\frac{1}{a^{n-1}} - \frac{a}{a^n}\right) + \dots + \frac{1}{a}\left(\frac{1}{a} - \frac{1}{a^2}\right) + \left(1 - \frac{1}{a}\right)$$

$$= \left(\frac{1}{a^{2(n-1)}} + \frac{1}{a^{2(n-2)}} + \dots + \frac{1}{a^2} + 1\right)\left(1 - \frac{1}{a}\right)$$

$$= \left(1 - \frac{1}{a}\right)\frac{\left[1 - \left(\frac{1}{a^2}\right)^n\right]}{(1 - 1/a^2)} = \frac{1}{\left(1 + \frac{1}{a}\right)}\left[1 - \frac{1}{a^{2n}}\right]$$

$$\therefore \int_0^1 f dx = \lim_{n\to\infty} \int_{1/a^n}^{1} f dx = \frac{1}{\left(1 + \frac{1}{a}\right)} = \frac{a}{a+1} \qquad (1)$$

Hence Proved.

$$\left(\because \; a > 2 \Rightarrow 0 < \frac{1}{a^2} < \frac{1}{4} \Rightarrow \frac{1}{a^{2n}} \to 0 \text{ as } n \to \infty.\right)$$

Example 39(a):

Show that the function f defined on [0, 1] by the conditions

$$f(x) = 2rx \quad \text{when } \frac{1}{r+1} < x < \frac{1}{r}, \; r = 1, 2, 3, \dots\dots$$

is integrable over [0, 1] and that

$$\int_0^1 f(x)\, dx = \frac{\pi^2}{6}.$$

Solution:

The given function f is not defined at the points

$$\left\{0, 1, \frac{1}{2}, \frac{1}{3}, \dots\dots\dots 1/r, \dots\dots\right\}.$$

The above set of points of discontinuity of f has only one limit point viz. 0 and consequently, f is integrable over [0, 1].

Now, $\int_0^1 f(x)\, dx = \lim_{n\to\infty} \int_{1/n}^{1} f(x)\, dx.$

We have

$$\int_{1/n}^{1} f(x)\,dx = \int_{1/2}^{1} f + \int_{1/2}^{1} f + ...+ \int_{1/(r+1)}^{1/r} f + ...+ \int_{1/n}^{1/(n-1)} f.$$

Now, $\int_{1/(r+1)}^{1/r} f(x)\,dx = \int_{1/(r+1)}^{1/r} 2rx dx = r\left[\frac{1}{r^2} - \left(\frac{1}{(r+1)^2}\right)\right] = \frac{2r+1}{r(r+1)^2}.$

Therefore, $\int_{1/n}^{1} f(x)\,dx = \sum_{r=1}^{n-1} \frac{2r+1}{r(r+1)^2} = \sum_{r=1}^{n-1}\left[\frac{1}{r} - \frac{1}{r+1} + \frac{1}{(r+1)^2}\right]$

$$= \sum_{r=1}^{n-1}\left(\frac{1}{r} - \frac{1}{r+1}\right) + \sum_{r=1}^{n-1} \frac{1}{(1+r)^2} = \left(1 - \frac{1}{n}\right) + \sum_{r=1}^{n-1} \frac{1}{(1+r)^2}.$$

We know, that the series

$$\frac{1}{1^2} + \frac{1}{2^2} + \frac{1}{3^2} + \frac{1}{4^2} + ...$$

is convergent and has the sum $\pi^2/6$. Hence,

$$\int_0^1 f(x)\,dx = \lim_{n\to\infty} \int_{1/n}^{1} f(x)\,dx = 1 - 0 + \left(\frac{1}{2^2} + \frac{1}{3^2} + \frac{1}{4^2} + ...\right.$$

$$= 1 + \frac{\pi^2}{6} - 1 = \frac{\pi^2}{6}.$$ **Ans.**

Example 39(b):

If f be defined in [0, 1] by the condition
$f(x) = (-1)^{r-1}$, when $1/(r+1) < x < 1/r$, $(r = 1, 2, 3, ...)$, $f(0) = 0$.

Show that $\int_0^1 f\,dx = \log 4 - 1$.

Solution:

We have

$f(x) = 1, \quad \frac{1}{2} < x \le 1$

$= -1, \quad \frac{1}{3} < x \le \frac{1}{2}$

$= 1, \quad \frac{1}{4} < x \le \frac{1}{3}$ and so on.

Thus, we have the given function f is discontinuous at the points

$0, \frac{1}{2}, \frac{1}{3}, \frac{1}{4}, ...$

Further, the above set of points of discontinuity of f has only one limit point viz., 0. Hence, the function f is integrable on [0, 1].

$$\therefore \int_0^1 f\,dx = \lim_{n\to\infty} \sum_{r=1}^{n} \int_{1/r+1}^{1/r} f\,dx.$$

$$\text{Now, } \sum_{r=1}^{n} \int_{1/r+1}^{1/r} f dx = \int_{1/2}^{1} 1.\,dx + \int_{1/3}^{1/2} (-1)\,dx + \int_{1/4}^{1/3} 1.\,dx + \ldots + \int_{1/n+1}^{1/n} (-1)^{n-1}\,dx$$

$$\Rightarrow \left(1-\frac{1}{2}\right) - \left(\frac{1}{2}-\frac{1}{3}\right) + \left(\frac{1}{3}-\frac{1}{4}\right) + \ldots + (-1)^{n-1}\left(\frac{1}{n}-\frac{1}{n+1}\right)$$

$$\Rightarrow \left\{\frac{1}{3}-\frac{1}{4} + \ldots + (-1)^{n-1}.\frac{1}{n}\right\} - \frac{(-1)^{n-1}}{n+1}$$

$$\Rightarrow -1+2\left\{1-\frac{1}{2}+\frac{1}{3}-\frac{1}{4}+\ldots+(-1)^{n-1}.\frac{1}{n}\right\} - \frac{(-1)^{n-1}}{n+1}$$

$$\therefore \int_0^1 f\,dx = \lim_{n\to\infty} \left[-1+2\left\{1-\frac{1}{2}+\frac{1}{3}- \ldots + (-1)^{n}.\frac{1}{n}\right\} - \frac{(-1)^{n-1}}{n+1}\right]$$

$$= -1 + 2\log(1+1) - 0 = -1 + 2\log 2 = \log 2 = \log 4 - 1.$$

Hence proved.

Example 40:

Let f and g be bounded integrable function on [a, b] and p, q be any constants. Then pf + qg is integrable and

$$\int_a^b (pf+qg)\,dx = p\int_a^b f\,dx + q\int_a^b g\,dx.$$

Solution:

Let $h(x) = pf(x) + q\,g(x)\ \forall\ x \in [a, b]$.

For any partition P of [a, b], we have

$S(P, h) = \Sigma\, h(t_i)\,\Delta x_i = p\,\Sigma\, f(t_i)\,\Delta x_i + q\,\Sigma g(t_i)\,\Delta x_i$

or $S(P, h) = p\,S(P, f) + q\,S(P, g)$. ...(1)

Since f and g are integrable, for any $\varepsilon > 0$, there exists some $\delta > 0$ such that for all partitions P with $\mu(P) < \delta$.

$$\left|S(P,f) - \int_a^b f\,dx\right| < \frac{\varepsilon}{2|p|},\quad \left|S(P,g) - \int_a^b g\,dx\right| < \frac{\varepsilon}{2|q|}, \qquad ...(2)$$

$$\text{Now, } \left|S(P,h) - p\int_a^b f\,dx - q\int_a^b g\,dx\right|$$

$$= \left| p\left[S(P, f) - \int_a^b f\,dx \right] + q\left[S(P, g) - \int_a^b g\,dx \right] \right|, \text{ using (1)}$$

$$\leq |p| \left| S(P, f) - \int_a^b f\,dx \right| + |q| \left| S(P, g) - \int_a^b g\,dx \right| < \varepsilon, \text{ by (2)}$$

Thus, $\lim\limits_{\mu(P)\to 0} S(P, h)$ exists and equals

$$p \int_a^b f\,dx + q \int_a^b g\,dx.$$

Hence, h = pf + qg is integrable and

$$\int_a^b (pf + qg)\,dx = p \int_a^b f\,dx + q \int_a^b g\,dx.$$

Example 41:

Let f be bounded and integrable in [a, b]. Then

$$\int_a^b [f(x)^2\,dx = 0 \Leftrightarrow f(c)\ 0$$

at every point, c, of continuity of f.

Solution:

Let $\int_a^b [f(x)^2\,dx = 0.$ ...(1)

Let f be continuous at $c \in [a, b]$. We have to show f(c) = 0. Suppose f(c) > 0. Some $\exists\ \delta > 0$ such that

$f(x) > f(c)\ \forall x \in]c - \delta, c + \delta[$.

$$\text{Now, } \int_a^b f^2 dx = \int_a^{c-3} f^2 dx + \int_{c-3}^{c+3} f^2 dx + \int_{a+3}^b f^2 dx$$

$$\geq \int_{c-3}^{c+3} f^2\,dx \left[\because f^2(x) \geq 0\ \forall x \in [a,b]\right]$$

$$> \frac{1}{2} f^2(c) \int_{c-3}^{c+3} dx = \delta f^2(c) > 0, \text{ by (2)}$$

Thus, $\int_a^b f^2\,dx > 0$, which is contradiction to (1).

Hence, f(c) = 0 at every point, c, of continuity of f.

Conversely, let f(c) = 0 at every point of continuity of f.

Let $P = \{a = x_0, x_1, ..., x_{i-1}, x_o, ..., x_n = b\}$

be any partition of [a, b]. Since f is integrable over [a, b], so f is integrable over $[x_{i-1}, x_i]$ for each i. Thus, there exists some $y \in [x_{i-1}, x_i]$ such that f is continuous at y and by the given hypothesis,

$f(y) = 0 \quad \Leftrightarrow \quad \{f(y)\}^2 = 0.$

Let m_i = inf. $\{f(x)\}^2$ over $[x_{i-1}, x_i]$.

Then $m_i = 0$ and consequently

$$L(P, f^2) = \sum_{i=1}^{n} m_i (x_i - x_{i-1}) = 0, \text{ for each partition P.}$$

$$\therefore \int_a^b f^2(x)\, dx = \text{sup. } \{L(P, f^2)\} = 0.$$

Hence, $\int_a^b f^2(x)\, dx = 0$, since f^2 is integrable over [a, b].

Remark:

It can be proved that : *A bounded function f is Riemann integrable over [a, b] if and only if f is continuous almost everywhere on [a, b].*

Example 42(a):

Let f be integrable on [a, b]. For each x∈[a, b], let F(x) = $\int_a^x f(t)\, dt$. *Show that F is uniformly continuous on [a, b].*

Solution:

Let $x_1, x_2 \in [a, b]$. Then

$$| F(x_2) - F(x_1) | \leq \int_{x_1}^{x_2} |f(t)|\, dt, \text{ if } x_1 \leq x_2. \qquad ...(1)$$

Since f is integrable on [a, b], so f is bounded on [a, b]. Consequently, there exists some M > 0 such that

$$| f(x) | \leq M \ \forall \ x \in [a, b]. \qquad ...(2)$$

Let e > 0 be given. From (1) and (2), we obtain

$| F(x_2) - F(x_1) | \leq M | x_2 - x_1 |.$

$\therefore | F(x_2) - F(x_1) < \varepsilon$ if $M | x_3 - x_1 | < \varepsilon$

$| F(x_2) - F(x_1) < \varepsilon$ if $| x_2 - x_1 | < \delta = \varepsilon/M, \ \forall \ x_1, x_2 \in [a, b].$

Hence, F is uniformly continuous on [a, b].

Example 42(b):

Evaluate $\int_0^2 f(x)\, dx$, *where*

f(x) = 0, when x = n/(n+1), (n+1)/n (n = 1, 2, 3,...)

f(x) = 1, elsewhere.

Is f integrable in [0, 2]?

Examine for continuity the function f so defined at x = 1.

Example 42(c):

A function f is bounded in [a, b]; show that

(i) $\overline{\int_a^b} kf\,dx = k\overline{\int_a^b} f\,dx; \underline{\int_a^b} kf\,dx = k\underline{\int_a^b} f\,dx\,(k>0)$

(ii) $\overline{\int_a^b} kf\,dx = k\underline{\int_a^b} f\,dx; \underline{\int_a^b} kf\,dx = k\overline{\int_a^b} f\,dx.(k<0)$.

Deduce that if f is bounded and integrable over [a, b], then so is kf and

$\int_a^b kf\,dx = k\int_a^b f\,dx$ *(k being a constant).*

Solution:

Let M_i and m_i be the bounds of f in $[x_{i-1}, x_i]$.

Then (kM_i, km_i) are the bounds of kf in $[x_{i-1}, x_i]$ for k > 0.

Now, $U(P, kf) = \Sigma\, kM_i\,\Delta x_i = kU(P, f)$.

$\therefore \overline{\int_a^b} kf\,dx = \inf\{U(P, kf)\} = k\inf\{U(P, f)\} = k\overline{\int_a^b} f\,dx$

Similarly $\underline{\int_a^b} kf\,dx = k\underline{\int_a^b} f\,dx.$ (k > 0)

For K < 0, (km_i, kM_i) are the bounds of kf in $[x_{i-1}, x_i]$.

Now, $U(P, kf)\} = \Sigma km_i\,\Delta x_i = kL(P, f)$.

$\therefore\ U(P, kf)\} = \inf\{kL(P, f)\} = k\sup\{L(P, f)\}$ ($\because$ k < 0)

$$\overline{\int_a^b} kf\,dx = k\underline{\int_a^b} f\,dx.$$

Similarly, $\underline{\int_a^b} kf\,dx = k\overline{\int_a^b} f\,dx.$ (k < 0)

If f is integrable then

$$\underline{\int_a^b} f\,dx = \overline{\int_a^b} f\,dx. = \int_a^b f\,dx.$$

Thus, $\overline{\int_a^b} kf\,dx. = k\int_a^b f\,dx = \underline{\int_a^b} kf\,dx.$ (by part (i))

Hence, $\int_a^b kf\,dx = k\int_a^b f\,dx.$

Example 43:

Let f be the function defined by

$$f(x) = \begin{cases} 1+x, & \text{when } 0 \le x < \frac{1}{2}, \frac{1}{2} < x \le 1 \\ 0\ , & \text{when } x = \frac{1}{2}. \end{cases}$$

Show that f is integrable over [0, 1] and $\int_0^1 f\,dx = \frac{3}{2}$.

Solution:

The given function is clearly bounded and has only one point of discontinuity at $x = \frac{1}{2}$

Thus, f is integrable on [0, 1]

Consider the partition of [0, 1] as follows.

$$P = \left\{0, \frac{1}{n}, \frac{2}{n}, \ldots, \frac{n}{n} = 1\right\}, \Delta x_r = \frac{1}{n} \text{ for } r = 1, 2, \ldots, n.$$

We know,

$$\int_0^1 f\,dx = \lim_{\mu(P)\to 0} \sum_{r=1}^{n} f(\xi_r)\,\Delta x_r,\ \xi_r = \frac{r}{n} \in \left[\frac{r-1}{n}, \frac{r}{n}\right]$$

$$= \lim_{n\to\infty} \sum_{r=1}^{n} f\left(\frac{r}{n}\right).\frac{1}{n} = \lim_{n\to\infty} \sum_{r=1}^{n} \frac{1}{n}\left(1 + \frac{r}{n}\right)$$

$$= \lim_{n\to\infty} \left[\frac{1}{n}\left\{\left(1 + \frac{1}{n}\right) + \ldots + \left(1 + \frac{n}{n}\right)\right\}\right]$$

$$= \lim_{n\to\infty} \left[\frac{1}{n}\left\{(1 + 1 + \ldots + 1) + \frac{1}{n}(1 + 2 + \ldots + n)\right\}\right]$$

$$= \lim_{n\to\infty} \frac{1}{n}\left\{n + \frac{n(n+1)}{2n}\right\} = \lim_{n\to\infty} \frac{3n+1}{2n} = \frac{3}{2}.$$

Example 44:

If $G(x, \xi) = \begin{cases} x(\xi - 1), & \text{when } x \le \xi, \\ \xi(x-1), & \text{when } \xi < x, \end{cases}$

and if f is continuous function of x in [0, 1] and if

$$g(x) = \int_0^1 f(\xi)\,G(x, \xi)\,d\xi,$$

Show that $g''(x) = f(x)\ \forall \in [0, 1]$ *and find g(0) and g(1).*

Solution:

For all $x \in [0, 1]$, we have

$$g(x) = \int_0^1 f(\xi)\, G(x, \xi)\, d\xi$$

$$= \int_0^x f(\xi)\, \xi(x-1)\, d\xi + \int_x^1 f(\xi)\, x(\xi - 1)\, d\xi. \qquad ...(1)$$

$$= (x-1) \int_0^x \xi f(\xi)\, d\xi + x \int_x^1 (\xi - 1) f(\xi)\, d\xi$$

$$= x \int_0^x \xi f(\xi)\, d\xi - \int_0^x \xi f(\xi)\, d\xi + x \int_x^1 \xi f(\xi)\, d\xi - x \int_x^1 \xi f(\xi)\, d\xi$$

$$= x \left[\int_0^x \xi f(\xi)\, d\xi + \int_x^1 \xi f(\xi)\, d\xi \right] - \int_0^x \xi f(\xi)\, d\xi - x \int_x^1 \xi f(\xi)\, d\xi$$

$$= x \int_0^1 \xi f(\xi)\, d\xi - \int_0^x \xi f(\xi)\, d\xi - x \int_x^1 \xi f(\xi)\, d\xi$$

$$\therefore\ g'(x) = \int_0^1 \xi f(\xi)\, d\xi - xf(x) \left[\int_x^1 \xi f(\xi)\, d\xi - xf(x) \right]$$

$$= \int_0^1 \xi f(\xi)\, d\xi - \int_x^1 \xi f(\xi)\, d\xi.$$

Now, $g''(x) = 0 - (-f(x)) = f(x)\ \forall\ x \in [0, 1]$.

From (1), we see that

$g(0) = 0 = g(1)$. **Ans.**

Example 45(a):

If f is a non-negative continuous function on [a, b] such that $\int_a^b f dx = 0$, *then show that* $f(x) = 0\ \forall\ x \in [a, b]$.

Solution:

Let $f(c) > 0$ for some $c \in [a, b]$.

Let $\varepsilon = \frac{1}{2} f(c) > 0$. Since f is continuous at c, then there exists some $\delta > 0$ such that $|f(x) - f(c)| < \varepsilon\ \forall\ x \in\ ]c - \delta, c + \delta[$

$\Rightarrow f(c) - \varepsilon < f(x) < f(c) + \varepsilon \Rightarrow f(x) > f(c) - \varepsilon$

$\Rightarrow f(x) > \frac{1}{2} . f(c)\ \forall\ x \in [c - \delta, c + \delta]$. ...(1)

Since f is continuous on [a, b], so f is integrable on [a, b].

$$\text{Now, } \int_a^b f dx = \int_a^{c-\delta} f dx + \int_{c-\delta}^{c+\delta} f dx + \int_{c+\delta}^b f dx$$

$$\geq \int_{c-\delta}^{c+\delta} f dx\ [\because f(x) \geq 0\ \forall\ x \in [a, b]]$$

$$> \frac{1}{2} f(c) \int_{c-\delta}^{c+\delta} dx = \delta f(c) > 0, \text{ using (1)}$$

Thus, $\int_a^b f dx > 0$, which is a contradiction.

Similarly if $f(c) < 0$.

Hence, $f(x) = 0 \; " \; x \in [a, b]$.

Example 45(b):

Show that the function [x], where [x] denotes the greatest integer not greater than x, is integrable over [0, 3] and

$$\int_0^3 [x]dx = 3$$

Solution:

We have

$$\begin{aligned}[x] &= 0 \text{ if } 0 \le x < 1,\\ &= 1 \text{ if } 1 \le x < 2,\\ &= 2 \text{ if } 2 \le x < 3,\\ &= 3 \text{ if } x = 3.\end{aligned}$$

It follows that the function $f(x) = [x]$ is bounded and has only three points of discontinuity at 1, 2, 3.

Thus, [x] is integrable over [0, 3] and so

$$\int_0^3 [x]dx = \int_0^1 [x]dx + \int_1^2 [x]dx + \int_2^3 [x]dx$$

$$= \int_0^1 0\,dx + \int_1^2 1\,dx + \int_2^3 2\,dx = 3. \qquad \textbf{Ans.}$$

Example 45(c):

Show that the function defined as follows:

$$f(x) = \frac{1}{2^n} \text{ when } \frac{1}{2^{n+1}} < x \le \frac{1}{2^n}, \; (n = 0, 1, 2, 3, ...)$$

$$f(0) = 0$$

is integrable in [0, 1] and that $\int_0^1 f dx = \frac{2}{3}$.

Solution:

We have

$$f(x) = 1, \quad \text{when } \frac{1}{2} < x \le 1,$$

$$= \frac{1}{2}, \quad \text{when } \frac{1}{2^2} < x \le \frac{1}{2}$$

$$= \frac{1}{2^2}, \text{ when } \frac{1}{2^3} < x \leq \frac{1}{2^3}$$

....

Now, it is clear that the points of discontinuity of f are

$$0, \frac{1}{2}, \left(\frac{1}{2}\right)^2, \left(\frac{1}{2}\right)^3, \ldots\ldots$$

which have only one limit point viz. 0 and so f is integrable on [0, 1].

Thus, we have

$$\int_{1/2n}^{1} f\,dx = \int_{1/2}^{1} f\,dx + \int_{1/2^2}^{1/2} f\,dx + \int_{1/2^3}^{1/2^2} f\,dx + \ldots + \int_{1/2^n}^{1/2^{n-1}} f\,dx$$

$$\Rightarrow \int_{1/2}^{1} 1.dx + \frac{1}{2}\int_{1/2^2}^{1/2} dx + \frac{1}{2^2}\int_{1/2^3}^{1/2^2} dx + \ldots + \frac{1}{2^{n-1}}\int_{1/2^n}^{1/2^{n-1}} dx$$

$$\Rightarrow \frac{1}{2} + \frac{1}{2}\left(\frac{1}{2} - \frac{1}{2^2}\right) + \frac{1}{2^2}\left(\frac{1}{2^2} - \frac{1}{2^3}\right) + \ldots + \frac{1}{2^{n-1}}\left(\frac{1}{2^{n-1}} - \frac{1}{2^n}\right)$$

$$\Rightarrow \frac{1}{2} + \frac{1}{2^2} + \frac{1}{2^5} + \ldots + \frac{1}{2^{2n-1}}$$

$$\Rightarrow \frac{1}{2}\left[1 + \frac{1}{2^2} + \left(\frac{1}{2^2}\right)^2 + \ldots + \left(\frac{1}{2^2}\right)^{n-1}\right]$$

$$\Rightarrow \frac{1}{2}\,\frac{1 - \left(\frac{1}{4}\right)^n}{1 - \frac{1}{4}} = \frac{2}{3}\left(1 - \frac{1}{4^n}\right)$$

$$\therefore \int_0^1 f\,dx = \lim_{n\to\infty}\int_{1/2n}^{1} f\,dx = \lim_{n\to\infty}\frac{2}{3}\left(1 - \frac{1}{4^n}\right) = \frac{2}{3}.$$ **Ans.**

Example 46:

If f is bounded and integrable in the internal [a, b], how that

$$\lim_{n\to\infty}\int_a^b f(x)\cos nx\,dx = 0.$$

Solution:

Let $I_n = \int_a^b f(x)\cos nx\,dx.$

Let ε be a positive number. Since f is bounded and integrable in [a, b], there exists a partition

$P = \{a= x_0 < x_1 < x_2 < .... < x_{r-1} < x_r < .... < x_p = b\}$ such that

$$U(P, f) - L(P, f) = \sum_{r=1}^{p}(M_r - m_r)\,\Delta x_r < \frac{1}{2}\varepsilon, \qquad ...(1)$$

M_r, m_r being the bounds of f in $[x_{r-1}, x_r]$.

Now, $I_n = \sum_{r=1}^{p} \int_{x_{r-1}}^{x_r} f(x)\cos nx\, dx$

$$\Rightarrow \sum \int_{x_{r-1}}^{x_r} [f(x_{r-1}) + f(x) - f(x_{r-1})] \cos nx\, dx$$

$$\Rightarrow \sum f(x_{r-1}) \int_{x_{r-1}}^{x_r} \cos nx\, dx + \sum \int_{x_{r-1}}^{x_r} [f(x) - f(x_{r-1})] \cos nx\, dx$$

$$\Rightarrow |I_n| \le \sum |f(x_{r-1})| \left| \int_{x_{r-1}}^{x_r} \cos nx\, dx \right| + \sum \left| \int_{x_{r-1}}^{x_r} \{f(x) - f(x_{r-1})\} \cos nx\, dx \right| \qquad ...(2)$$

We have $\forall\, x \in [x_{r-1}, x_r]$,

$|f(x) - f(x_{r-1})| \le M_r - m_r$ $(r = 1, 2, ..., p)$

$$\Rightarrow | [f(x) - f(x_{r-1})] \cos nx | \le M_r - m_r. \qquad ...(3)$$

$(\because |\cos nx| \le I)$

Also

$$\left| \int_{x_{r-1}}^{x_r} \cos nx\, dx \right| = \frac{1}{n} |(\sin nx_r - \sin nx_{r-1})|.$$

$$\le \frac{1}{n} [|\sin nx,| + |\sin nx_{r-1}] \le \frac{2}{n}. \qquad ...(4)$$

Using (3) and (4) in (2), we get

$$|I_n| \le \frac{2}{n} \sum |f(x_{r-2})| + \sum (M_r - m_r)(x_r - x_{r-1})$$

$$\le \frac{2}{n} \sum |f(x_{r-2})| + \frac{\varepsilon}{2}, \text{ using (1).}$$

Keeping the partition P fixed, we see that $\Sigma\, |f(x_{r-1})|$ is fixed. We now, choose a positive integer m such that $\forall\, n \ge m$

$$\frac{2}{n} \sum |f(x_{r-1})| < \frac{\varepsilon}{2}.$$

Thus, $|I_n| < ½\,\varepsilon + ½\,\varepsilon = \varepsilon\ \forall\, n \ge m$.

Hence, $\lim_{n\to\infty} I_n = 0.$

Remark:

It may similarly be shown that

$$\lim_{n\to\infty} \int_a^b f(x) \sin nx \, dx = 0 .$$

Example 47:

Verify the second mean value theorem and the generalized first mean value theorem for

$$f(x) = g(x) = x^2 \ \forall \ x \in [-1, 1].$$

Solution:

It is clear that f, g are both integrable on [–1, 1], as they are continuous functions on [–1, 1]. However, these functions are not monotonic on [–1, 0]. Indeed each function is monotonically decreasing on [–1, 0] and monotonically increasing on [0, 1]. If the second mean value is true, then there exists some $\xi \in [-1, 1]$ such that

$$\int_{-1}^{1} fg \, dx = f(-1) \int_{-1}^{\xi} g \, dx + f(1) \int_{\xi}^{1} g \, dx. \qquad ...(1)$$

$$\text{L.H.S. of (1)} = \int_{-1}^{1} x^4 \, dx = \frac{2}{5}.$$

$$\text{R.H.S. of (1)} = 1 \int_{-1}^{\xi} x^2 \, dx + 1 \int_{\xi}^{1} x^2 dx = \frac{1}{3}(\xi^2 + 1) + \frac{1}{3}(1-\xi^2) = \frac{2}{3}$$

Since $\frac{2}{5} \neq \frac{2}{3}$, the second mean value theorem is not true.

(ii) Again f and g are integrable on [–1, 1] and $g(x) \geq 0$ for all x in [–1, 1] i.e., g keeps the same sign over [–1, 1]. By generalized first mean value theorem, $\exists$ some μ lying between the bounds of f such; that

$$\int_{-1}^{1} fg \, dx = \mu \int_{-1}^{1} g \, dx$$

$$\Rightarrow \int_{-1}^{1} x^4 \, dx = \mu \int_{-1}^{1} x^2 dx \Rightarrow \frac{2}{5} = \mu . \frac{2}{3} \Rightarrow \mu = \frac{3}{5}.$$

Hence, the generalized first mean value theorem is verified.

Example 48:

Verify the two mean value theorems in [–1, 1] for the functions :

$$f(x) = e^x, \ g(x) = x.$$

Solution:

Since f and g are continuous in [–1, 1], so these functions are integrable in [–1, 1].

Also $f(x) = e^x$ is monotonically increasing in [–1, 1].

By second mean value theorem, there exists a point $x \in [-1, 1]$ such that

$$\int_{-1}^{1} fg\,dx = f(-1)\int_{-1}^{\xi} g\,dx + f(1)\int_{\xi}^{1} g\,dx$$

$$\text{or } \int_{-1}^{1} xe^x\,dx = e^{-1}.\frac{1}{2}(\xi^3 - 1) + e.\frac{1}{2}(1-\xi^2)$$

$$\text{or } \left|xe^x - e^x\right|_{-1}^{1} = \frac{1}{2}(1-\xi^2)(e-e^{-1})$$

$$0 - (-e^{-1} - e^{-1}) = \frac{(1-\xi^2)(e^2-1)}{2e}$$

$$\frac{2}{e} = \frac{(1-\xi^2)(e^2-1)}{2e} \Rightarrow 1 - \xi^2 = \frac{4}{e^2-1}$$

$$\therefore\ \xi = \pm\sqrt{\frac{e^2-5}{e^2-1}} \in [-1, 1].$$

(ii) Again f and g are integrable in [–1, 1]. But $g(x) = x$ does keep the same sign in [–1, 1]. If generalized first mean value theorem is true, then

$$\int_{-1}^{1} fg\,dx = \mu\int_{-1}^{1} g\,dx$$

$$\frac{2}{e} = \mu\int_{-1}^{1} x\,dx = 0, \text{which is impossible.}$$

Hence, the generalized first mean value theorem is not true.

Example 49:

If a function f is continuous on [0, 1], show that

$$\lim_{n\to\infty}\int_0^1 \frac{nf(x)}{1+n^2x^2}\,dx = \frac{\pi}{2}f(0).$$

Solution:

We write

$$\int_0^1 \frac{nf(x)dx}{1+n^2x^2} = \int_0^{1/\sqrt{n}} \frac{nf(x)\,dx}{1+n^2x^2} + \int_{1/\sqrt{n}}^1 \frac{nf(x)\,dx}{1+n^2x^2}. \quad ...(1)$$

By generalized first mean value theorem,

$$\int_0^{1/\sqrt{n}} \frac{nf(x)\,dx}{1+n^2x^2} = f(\alpha_n) \int_0^{1/\sqrt{n}} \frac{n}{1+n^2x^2}dx,\ 0 \le \alpha_n \le \frac{1}{\sqrt{n}}$$

$$\Rightarrow f(\alpha_n) \int_0^{1/\sqrt{n}} \frac{dt}{1+t^2}, t = nx$$

$$\Rightarrow f(\alpha_n) \tan^{-1} \sqrt{n} \to \frac{\pi}{2} f(0) \text{ as } n \to \infty. \qquad ...(2)$$

($\because \alpha_n \to 0 \Rightarrow f(\alpha_n) \to f(0)$, by the continuity of f).

Since f is continuous on [0, 1], it is bounded and so there exists same $K > 0$ such that $| f(x) | \le K\ \forall\ x \in [0, 1]$

$$\therefore \left| \int_{1/\sqrt{n}}^{1} \frac{nf(x)}{1+n^2x^2} dx \right| \le K \left| \int_{1/\sqrt{n}}^{1} \frac{ndx}{1+n^2x^2} \right|.$$

$$= K (\tan^{-1} n - \tan^{-1} \sqrt{n}).$$

$$= K (\tfrac{1}{2}\pi - \tfrac{1}{2}\pi) = 0 \text{ as } n \to \infty. \qquad ...(3)$$

Using (2) and (3) in (1), the result is proved.

EXERCISES

1. $\int_0^4 x d[x] = 10.$

2. Let $f(x) = \begin{cases} 1, & 0 \le x \le \frac{1}{2} \\ 2, & \frac{1}{2} < x \le 1 \end{cases}$ and $a(x) = \begin{cases} 3, & 0 \le x \le \frac{1}{2} \\ 4, & \frac{1}{2} < x \le 1. \end{cases}$

   Let $P = \left\{0, \frac{1}{3}, \frac{2}{3}, 1\right\}$. *Show that* $f \in \Re(\alpha)$ *although lim* $S(P, f, \alpha)$ *does not exist as* $\mu(P) \to 0$.

   Show that

3. $\int_0^n 2xd(x[x]) = n^3 + \sum_{r=1} r^2.$

12. Show that $\int_0^2 f\,d\alpha = 0$, where $f(x) = \begin{cases} 0, & 0 \le x \le 1 \\ 1, & 1 < x \le 2 \end{cases}$ and $\alpha(x) = \begin{cases} 0, & 0 \le x \le 1 \\ 1, & 1 \le x \le 2 \end{cases}.$

4. $\int_0^5 (x^2 + 1)\, d[x] = 60$

5. Show that f (x) = x is integrable on [0, 3] and $\int_0^1 x\,dx = \frac{1}{2}$.

7. $\int_0^3 x^2 d([x] - x) = 5.$

8. Show that $\int_0^1 f\,d\alpha = \frac{29}{2}$, where f (x) = $x^3 + 7$

and $\alpha(x) = \begin{cases} 1 + x^3, 0 < x \le 1 \\ 0, x = 0 \end{cases}$

9. Show that x^2 is integrable on [0, 1] and $\int_1^2 x^2\,dx = \frac{1}{3}$.

10. Show that (2x + 3) is integrable on (1, 2) and $\int_1^2 (2x+3)\,dx = 6$.

11. Show that $\int_0^2 f\,d\alpha = 0$, where $f(x) = \begin{cases} 0, 0 \le x \le 1 \\ 1, 1 < x \le 2 \end{cases}$.

and $\alpha(x) = \begin{cases} 0, 0 \le x < 1 \\ 1, 1 \le x \le 2. \end{cases}$

12. Evaluate $\int_0^2 x\,d\alpha$, where $a(x) = \begin{cases} x, 0 \le x \le 1 \\ 2, x, 1 < x \le 2. \end{cases}$

13. Show that $\int_0^1 f d\alpha = \frac{29}{2}$, where $f(x) = x^3 + 7$ and

$a(x) = \begin{cases} 1 + x^3, 0 < x \le 1 \\ 0, x = 0 \end{cases}$

14. Define f on [–1, 1] by f (x) = 0, x < 0 and f (x) = 1, x ≥ 0.
Let the integrator α be defined by α(x) = 0, x ≤ 0; α(x) = 1, x > 1
Show that f ∈ R(α) although lim S(P, f, α) does not exist as μ (P) → 0.

15. Evaluate $\int_0^2 x\,d\alpha$, $a(x) = \begin{cases} x, 0 \le x \le 1 \\ 2 + x, 1 < x \le 2. \end{cases}$

16. Define f on [– 1, 1] by f (x) = 0, x < 0 and f (x) = 1, x ≥ 0.
Let the integrator α be defined by α(x) = 0, x ≥ 0; α(x) = 1, x > 0.

Show that $f \in \Re(\alpha)$ although him $S(P, f, \alpha)$ does not exist as $\mu(P) \to 0$.

17. $\int_0^4 x d[x] = 10$. 2. $\int_0^4 x d([x]-x) = 2$.

18. $\int_2^5 [5-x] d(\log[x]) = \log 3$.

19. Show that $f(x) = 3x + 1$ is integrable on [1, 2] and $\int_1^2 (3x + 1)du = 11/2$.

20. $\int_0^3 x^2 d([x]-x) = 5$. 4. $\int_0^x d[t] = [x] \; \forall \; x \in R$.

21. $\int_0^5 (x^2+1) d[x] = 60$.

22. Give an example to prove that a bounded function need to R-integrable.

23. Give an example of discontinuous function which is R-integrable on [0, 1].

24. A function f(x) is defined on [0,3] as follows :

$f(x) = \dfrac{n}{n+2}$ when $\dfrac{1}{n+1} < x \leq \dfrac{1}{n}$ (n = 1, 2, 3 . . .) and f(x) = 1,

when x = 0.

Show that f(x) is R-integrable on [0, 1] and $\int_0^1 f(x)du = \dfrac{1}{2}$.

25. Let f(x) be a function bounded on [a, b] and let P_1 and P_2 be two partition of [a, b] such that $P_1 < P_2$, then prove that:

$$U(P_1, f) - L(P_1, f) \geq U(P_1, f) - L(P_2, f)$$

2

Improper Integrals

INTRODUCTION

The definite integral $\int_a^b f(x)\,dx$ as defined by Riemann is meaningful only when the range of integration i.e., [a, b] is finite and the integration f(x) is bounded in the range. However Caushy has extended the definition of integral $\int_a^b f(x)\,dx$ to include the following two cases.

When the range of integration, [a, b] is not finite but the integrand f is bounded in that range e.g., $\int_1^\infty \frac{1}{x^{5/2}}, \int_{-\infty}^\infty \frac{dx}{1+x^2}$ etc. is called an improper integral of the first kind or an infinite integral. Thus $\int_0^\infty \frac{dx}{1+x^2}$ is improper integral of first kind since the upper limit of integration is infinite and the integrand $\frac{1}{1+x^2}$ is bounded. In case the interval (a, b) is infinite and the integral f(x) is bounded we define:

(i) $\int_0^\infty f(x)dx = \lim_{x\to\infty} \int_a^x f(x)dx$ provided that the limit exist finitely, i.e., the limits is equal to definite real number.

(ii) $\int_{-\infty}^b f(x)dx = \lim_{x\to\infty} \int_{-x}^b f(x)dx$ provided that the limit exist finitely.

(iii) $\int_{-\infty}^\infty f(x)dx = \lim_{x\to\infty} \int_{-x_1}^c f(x)du + \lim_{x\to\infty} \int_{-x_1}^c f(x)du$ provided that both these limit exist finitely.

Definition:

The integral $\int_a^\infty f(x)\,dx$ is said to converse to the value I, if for any arbitrarily chosen positive number $\in$, however small but not zero. There exist a corresponding positive number N, such that

$$\left|\int_a^b f(x)\,dx - I\right| < \in \text{ for all values of } b \geq N$$

Similarly, we can define the convergence of an integral when the lower limit is infinite or when the integrand become finite at the upper and lower limit.

CONVERGENCE OF IMPROPER INTEGRAL

When the limit of an improper integral as define above, is a definite finite number we say that the given integral is convergent and the value of the integral is equal to the value of that limit when the limit is as $-\infty$. *The integral is said to be divergent i.e. The value of integral does not exist.*

In case the limit is neither a definite number nor ∞ *or* $-\infty$. *The integral is said to be oscillatory and in this case also the value of the integral does not exist i.e. The integral is not convergent.*

Def 1:

Convergence at the Lseft End-Point : Let a be the only point of infinite discontinuity of f so that according to the assumption made in the last section the integral $\int_{a+\lambda}^b f\,dx$ exit for every $\lambda, < \lambda < b - a$ the improper integral $\int_a^b f\,dx$ is defined as the limit of $\int_{a+\lambda}^b f\,dx$ when $\lambda \to 0+$, so that $\int_a^b f\,dx = \lim_{\lambda \to 0+} \int_{a+\lambda}^b f\,dx$.

If this limit exists and is finite, the improper integral $\int_a^b f\,dx$ is said to exist or converge at a if otherwise, it is called divergent i.e. If the limit does not exist, then $\int_a^b f\,dx$ is divergent.

Remark :

For any value c between a and b, $a < c < b$,

$$\int_a^b f\,dx = \int_a^c f\,dx + \int_c^b f\,dx$$

$\int_c^b f\,dx$ is a proper integral so that the two integrals $\int_a^b f\,dx$ and $\int_a^c f\,dx$ converge and diverge together. Thus while testing the integral $\int_a^b f\,dx$ for convergence at a, it may be replaced by $\int_a^c f\,dx$ for any convenient c such that $a < c < b$.

Def 2:

Convergence at the Right-End Point : Let b be the only point of infinite discontinuity of f, the improper integral is then defined by the relation

$$\int_a^b f\,dx = \lim_{\mu \to 0+} \int_a^{b-\mu} f\,dx,\ 0 < \mu < b - a$$

If the limit exists, the improper integral is said to be convergent at botherwise it is called *divergent* or it the limit does not exist the improper integral is called divergent.

Remark:

For the same reason as above, when testing the integral $\int_a^b f\,dx$ *for convergence at b, it may be replaced by* $\int_c^b f\,dx$ *for any convenient c between a and b.*

Def 3:

Convergence at Both the End Points : Let the end-points a and b be the only points of infinite discontinuity of a function f, Let c be any point such that a < c < b.

If the improper integrals

$\int_a^c f(x)\,dx$ and $\int_c^b f(x)\,dx$

both converge at a and b respectively, we say that the improper integral $\int_a^b f(x)\,dx$ converges and it can be written as

$$\int_a^b f(x)\,dx \int_a^c f(x)\,dx + \int_c^b f(x)\,dx.$$

Def 4:

(Convergence at ∞, –∞): The improper integral $\int_a^\infty f(x)\,dx$ is defined as

$$\int_a^\infty f(x)\,dx = \lim_{t \to \infty} \int_a^b f(x)\,dx$$

provided the limit exists. If the limit does not exist, we say that the improper integral is divergent.

Similarly $\int_{-\infty}^b f(x)\,dx = \lim_{t \to \infty} \int_t^b f(x)\,dx$ provided the limit exists.

If c is any number and $\int_{-\infty}^c f(x)\,dx$, $\int_c^\infty f(x)\,dx$

both exist, then we say that $\int_{-\infty}^{\infty} f(x)\,dx$ exists (convergent) and we write

$$\int_{-\infty}^{\infty} f(x)\,dx = \int_{-\infty}^{c} f(x)dx + \int_{c}^{\infty} f(x)dx.$$

COMPARISON TEST

Let f(x) and g(x) be two functions which are bounded and integrable in the interval (a, ∞) Also let g(x) be positive and $|f(x)| \le g(x)$ when $x \ge a$ then, if $\int_a^{\infty} f(x)\,dx$ is convergent $\int_a^{\infty} f(x)\,dx$ is also convergent.

Similarly if $|f(x)| \ge g(x)$ for all values of x greater than some number $x_0 > a$ $\int_a^{\infty} f(x)\,dx$ is divergent, then $\int_a^{\infty} f(x)$ is also divergent.

Alternative Form of the Above Comparison Test

If $\lim_{x\to\infty} \frac{f(x)}{g(x)}$ is a definite number, other then zero the integrals $\int_a^{\infty} f(x)\,dx$ and $\int_a^{\infty} f(x)\,dx$ either both converge or both diverge.

Note : While applying comparison test we generally take $g(x) = \frac{1}{x^n}$ i.e. $\int_a^{\infty} f(x)\,dx$ is generally taken as the comparison integral.

Remarks:

For l = 0, ∞, we have the following results :

1. If l = 0 and $\int_a^b g(x)$ converges, then $\int_a^b f(x)\,dx$ converges.
2. If l = ∞ and $\int_a^b g(x)$ diverges, then $\int_a^b f(x)\,dx$ diverges.

Similarly, results are also true for $\int_a^{\infty} f(x)\,dx$.

PRINCIPAL AND GENERAL VALUES OF IMPROPER INTEGRALS

Suppose that f(x) has an infinite discontinuity only at a point $x = c \in (a, b)$ and at all outer points of [a, b] f(x) is bounded. Enclose the point x = c in a small internal $(c - \epsilon_1, c + \epsilon_2)$. Where ϵ_1 and ϵ_2 are arbitrary positive numbers and are independent to each other.

Now the function f is bounded in $[a, c - \epsilon_1]$ and $[c + \epsilon_2, b]$ thus both the integrals.

$\int_a^{c-\epsilon_1} f(x)\,dx$ and $\int_{c+\epsilon_2} f(x)\,dx$ exist

then $\int_a^b f(x)\,dx \lim_{\epsilon_1\to 0} 0 \int_a^{c-\epsilon_1} f(x)\,dx + \lim_{\epsilon\to 0}\int_{c+\epsilon_2}^b f(x)dx$

provided both the limits of right hand side exist. This value is called the general value of the integral

If $\epsilon_1 = \epsilon_2$ the value of above limit is called the principal value of the integral principal value of $\int_a^b f(x)\,dx$ is denoted by $p\int_a^b f(x)\,dx$.

Note : It may be noted that the existence of the principal value does not imply the existence of general value but on the other hand it the general value exists then it is equal to the principal values.

CAUCHY'S TEST FOR CONVERGENCE AT ∞

Show that integral $\int_a^\infty f(x)\,dx$ converges at ∞ iff to every $\varepsilon > 0$, there exists a positive number X_0 such that

$$\left|\int_{X_1}^{X_2} f(x)\,dx\right| < \varepsilon \;\forall\; X_1, X_2 > X_0.$$

Proof:

Let $F(x) = \int_a^X f(x)\,dx$, Then

$\int_a^\infty f(x)\,dx$ exists $\Leftrightarrow \lim_{x\to\infty} F(X)$ exists finitely.

$\Leftrightarrow$ To each $e > 0$, some $X_0 > 0$ such that
$|F(X_1) - F(X_2)| < \varepsilon \;\forall\; X_1, X_2 > X_0$

(By Cauchy's principle for finite integral)

$$\Leftrightarrow \left|\int_a^{X_1} f(x)\,dx - \int_a^{X_2} f(x)\,dx\right| < \varepsilon$$

$$\Leftrightarrow \left|\int_{X_1}^{X_2} f(x)\,dx\right| < \varepsilon \;\forall\; X_1, X_2 > X_0$$

Hence Proved.

ABEL'S TEST

If ϕ is bounded and monotonic in $[a, \infty[$ and $\int_a^\infty f\,dx$ is convergent at ∞, then $\int_a^\infty f\,\phi\,dx$ is convergent at ∞,

Proof:

Since ϕ is monotonic in $[a, \infty[$, so ϕ is integrable in $[a, X]$ $\forall X \geq a$. Since $\int_a^\infty f\,dx$ is convergent at ∞, so f is integrable in $[a, X]$.

By Second Mean Value Theorem, we have

$$\int_{X_1}^{X_2} f\,\phi\,dx = \phi(X_1)\int_{X_1}^{\xi} f\,dx + \phi(X_2)\int_{\xi}^{X_2} f\,dx, \quad ...(1)$$

for $a < X_1 \leq \xi \leq X_2$.

Let $\varepsilon > 0$ be arbitrary.

Since ϕ is bounded in $[a, \infty[$, so there exists some $K > 0$ such that $|\phi(x)| \leq K$ $\forall x \geq \alpha$. For $X_1 > a_2$ $X_2 > a_2$ we have

$$|\phi(X_1)| \leq K, \ |\phi(X_2)| \leq K. \quad ...(2)$$

Since $\int_a^\infty f\,dx$ is convergent at to by Cauchy Test, there exists a number X_0 such that

$$\int_{X_1}^{X_2} f\,dx < \frac{\varepsilon}{2K'}\ \forall X_1, X_2 \geq X_2 \quad ...(3)$$

Since $X_1 \leq \xi \leq X_2$ and $X_1 \geq X_0$, so $\xi \leq X_0$, From (3),

$$\left|\int_{X_1}^{\xi} f\,dx\right| < \frac{\varepsilon}{2K}, \left|\int_{\xi}^{X_2} f\,dx\right| < \frac{\varepsilon}{2K} \quad ...(4)$$

From (1), (2), (4); we obtain

$$\left|\int_{X_1}^{X_2} f\,\phi\,dx\right| \leq |\phi(X_1)|.\left|\int_{X_1}^{\xi} f\,dx\right| + |\phi(X_2)|\left|\int_{\xi}^{X_2} f\,dx\right|$$

$$\leq K.\frac{\varepsilon}{2K} + K.\frac{\varepsilon}{2K} = \varepsilon\ \forall\ X_1 X_2 \geq X_0$$

Hence, by Cauchy's Test, $\int_a^\infty f\,\phi\,dx$ is convergent at ∞.

Hence Proved.

THE IMPROPER INTEGRAL

$$\int_a^b \frac{dx}{(x-a)^n}$$

converges if and only if $n < 1$.

Proof:

If $n = 1$, then

$$\int_{a+\epsilon}^{b} \frac{dx}{(x-a)} = \{\log (x-a)\}_{a+\epsilon}^{b} = \log \frac{b-a}{\epsilon}.$$

Since R.H.S. tends to ∞ as $\varepsilon \to 0$ therefore given integral is divergent for $n = 1$.

Again if $n \neq 1$

$$\int_{a+\epsilon}^{b} \frac{dx}{(x-a)^n} = \left\{ \frac{1}{n-1} (x-a)^{n+1} \right\}_{a+\epsilon}^{b}$$

$$= \frac{1}{(1-n)} \left\{ \frac{1}{(b-a)^{n-1}} - \frac{1}{\epsilon^{n-1}} \right\}$$

Now if $n > 1$

$$\lim_{\epsilon \to 0} \int_{a+\epsilon}^{b} \frac{dx}{(x-a)^n} = \lim_{\epsilon \to 0} \frac{1}{(1-n)} \left\{ \frac{1}{(b-a)^{n-1}} - \frac{1}{\epsilon^{n-1}} \right\}.$$

Thus, the given integral is also divergent for $n > 1$.

But if $n < 1$ above limit tends to $\frac{1}{1-n} \cdot \frac{1}{(b-a)^{n-1}}$ i.e. the given improper integral is convergent for $n < 1$.

Hence the theorem.

ABSOLUTE CONVERGENCE

An improper integral $\int_a^\infty f(x)\ dx$ is said to be absolutely convergent if $\int_a^\infty |f(x)|\ dx$ is convergent. An improper integral which is convergent but not absolutely convergent is called conditionally convergent.

Theorem:

Show that every absolutely convergent improper integral is convergent.

Proof:

Let $\int_a^\infty f(x)\ dx$ be absolutely convergent so that

$\int_a^\infty |f(x)|\ dx$ is convergent.

By Cauchy's Test, for any $\varepsilon > 0$, there exists a positive number X_0 such that

$$\left|\int_{X_1}^{X_2} |f(x)|\, dx\right| < \varepsilon \quad \forall\ X_1, X_2 > X_0.$$

Now, $\left|\int_{X_1}^{X_2} f(x)\, dx\right| \le \int_{X_1}^{X_2} |f(x)|\, dx \le \left|\int_{X_1}^{X_2} |f(x)|\, dx\right| < \varepsilon$

$$\therefore \quad \left|\int_{X_1}^{X_2} f(x)\, dx\right| < \varepsilon \qquad \forall\ X_1, X_2 > X_0.$$

Hence, $\int_a^{\infty} f(x)\, dx$ is convergent, by Cauchy's Test.

DIRICHLET'S TEST

If ϕ is bounded and monotonic in $[a, \infty[$ and tends to 0 as $x \to \infty$, and $\int_a^X f\, dx$ is bounded for $X \ge a$, then $\int_a^{\infty} f\phi\ dx$ is convergent at ∞.

Proof:

Since ϕ is monotonic, ϕ is integrable in $[a, X]$ $\forall$ $X \le a$. Also f is integrable in $[a, X]$. By Second Mean Value Theorem.

$$\int_{X_1}^{X_2} f\,\phi\ dx = \phi(X_1)\int_{X_1}^{\xi} f\ dx + \phi(X_2)\int_{\xi}^{X_2} f\ dx, \qquad ...(1)$$

for $a < X_1 \le \xi \le X_2$.

Since $\int_a^X f\, dx$ is bounded, so there exists a positive number K such that

$$\left|\int_a^X f\, dx\right| \le K \qquad \forall\ X \le a.$$

Now, $\left|\int_{X_1}^{\xi} f\, dx\right| = \left|\int_a^{\xi} f\, dx - \int_a^{X_1} f\ dx\right|$

$$\le \left|\int_a^{\xi} f\, dx\right| + \left|\int_a^{X_1} f\ dx\right|$$

$$\le K + K = 2K \quad \forall\ X_1, \xi \ge a. \qquad ...(2)$$

Similarly $\left|\int_{\xi}^{X_2} f\, dx\right| \le 2K, \qquad \forall\ X_2, \xi \ge a. \qquad ...(3)$

Let $\varepsilon > 0$ be arbitrary. Since $\lim_{x\to\infty} \phi(x) = 0$ so there exists a positive number X_0 such that

$$|\phi(x)| < \frac{\varepsilon}{4K} \forall x \geq X_0$$

$$\therefore \quad |\phi(X_1)| < \varepsilon/4K, |\phi(X_2)| < \varepsilon/4K \qquad ...(4)$$

$$\forall X_1, X_2 \geq X_0.$$

Using (2) to (4) in (1), we get

$$\left|\int_{X_1}^{X_2} f\phi\, dx\right| \leq |\phi(X_1)| \left|\int_{X_1}^{\xi} f\, dx\right| + |\phi(X_2)| \left|\int_{\xi}^{X_1} dx\right|$$

$$< \frac{\varepsilon}{4K}.2K + \frac{\varepsilon}{4K}.2K = \varepsilon \quad \forall X_1, X_2 \geq X_0.$$

Hence, by Cauchy's Test, $\int_a^\infty f\phi\, dx$ is convergent at ∞.

TWO COMPARISON INTEGRALS

(i) $\int_0^\infty \frac{dx}{x^n} (n > 0)$ converges if $n > 1$.

Proof:

For $n \neq 1$. Then we have

$$\int_a^t \frac{dx}{x^n} = -\frac{1}{(1-n)} \left|x^{1-n}\right|^t = \frac{1}{(1-n)}\left[\frac{1}{t^{n-1}} - \frac{1}{a^{n-1}}\right]$$

$$= \infty \qquad \text{if } n < 1$$

$$\therefore \lim_{t\to\infty} \int_a^t \frac{dx}{x} = \log\frac{t}{a} \to \infty \text{ as } t \to \infty$$

Hence, the given improper integral converges if $n > 1$.

(ii) $\int_a^b \frac{dx}{(x-a)^n}$ converges iff $n < 1$.

Proof:

Here a is the only point of infinite discontinuity of the integrand.

For $n = 1$, we have

$$\int_{a+\varepsilon}^{b} \frac{dx}{(x-a)^n} = \frac{1}{(1-n)} \left|(x-a)^{1-n}\right|_{a+\varepsilon}^{b}$$

$$= \frac{1}{(1-n)}\left[\frac{1}{(b-a)^{n-1}} - \frac{1}{\varepsilon^{n-1}}\right]$$

$$\therefore \lim_{\varepsilon \to 0+} \int_{a+\varepsilon}^{b} \frac{dx}{(x-a)^n} = \frac{1}{(1-n)(b-a)^{n-1}}, \text{ if } n < 1.$$

$$= +\infty, \text{ if } n > 1.$$

For n = 1, $\int_{a+\varepsilon}^{b} \frac{dx}{x-a} = \log(b-a) - \log \varepsilon \to +\infty$ as $\varepsilon \to 0+$.

Hence the given improper integral converges iff n > 1.

SOLVED EXAMPLES

Example 1:

Show that $\int_0^\infty \left(\frac{1}{1+x} e^{-n}\right) \frac{dx}{x}$ is convergent.

Solution:

Let $f(x) = \left(\frac{1}{1+x} - e^{-n}\right)\frac{1}{x} = \left(\frac{1}{1+x} - \frac{1}{e^a}\right)\frac{1}{x}$

or $f(x) = \frac{e^x - (1+x)}{x(1+x)e^x} > 0$ for $x > 0$ $\quad [\because f(x) = 1 + x\frac{x^2}{2!} + ...1$

Since $\lim_{x \to 0} f(x) = 0.$ 0 is not a point of infinite discontinuity.

We have $\int_0^\infty f(x)\,dx = \int_0^1 f(x)\,dx + \int_0^\infty f(x)\,dx, \int_0^1 f(x)\,dx$ being proper.

Now, we shall investigate the convergence of $\int_1^\infty f(x)\,dx$.

Let $g(x) = \frac{1}{x^2}$ $\quad \lim_{x \to \infty} \frac{f(x)}{g(x)} = \frac{e^x - (1+x)}{e^x}\left(\frac{x}{1+x}\right)$

$$= \lim_{x \to \infty} \frac{e^x - (1+x)}{e^x} = 1 - \lim_{x \to \infty} \frac{1+x}{e^x} = 1 - \lim_{x \to \infty} \frac{1}{e^x} = 1$$

Thus, $\int_1^\infty f(x)\,dx$ and $\int_1^\infty g(x)\,dx$ converge or diverge together.

But $\int_1^\infty g(x)\,dx = \int_1^\infty \frac{dx}{x^2}$ converges. So. $\int_1^\infty f(x)\,dx$ converges

Hence $\int_1^\infty f(x)\,dx$ is convergent.

Example 2: (Gamma Function)

Show that $\int_0^\infty x^{n-1} e^{-x}\, dx$ *is convergent iff* $n > 0$.

Solution:

Let $f(x) = x^{n-1} e^{-m}$. Then 0 is a point of infinite discontinuity of f, if $n < 1$. Thus, we have to examine convergence at 0 and ∞. We take

$$\int_0^\infty f(x)\, dx = \int_0^1 f(x)\, dx + \int_1^\infty f(x)\, dx.$$

Convergence at 0:

Let $g(x) = \dfrac{1}{x^{1-n}}$. Then $\lim\limits_{x\to 0} \dfrac{f(x)}{g(x)} = \lim\limits_{x\to 0} e^{-n} = 1$.

So, $\int_0^1 f(x)\, dx$, $\int_0^1 g(x)\, dx$ have identical behaviour.

Now, $\int_0^1 g(x)\, dx$ converges iff $(1 - n) < 1 \Rightarrow 0 < n$.

Convergence at ∞ :

For given n and for large values of x_1 then we have

$$e^x > x^{n+1} \Leftrightarrow \frac{1}{e^{-x}} > x^{n-1}.x^2$$

$$\Leftrightarrow x^{x-1} e^{-n} < 1/x^2$$

Since $\int_1^\infty \dfrac{1}{x^2}\, dx$ converges, so $\int_1^\infty x^{n-1} e^{-n}\, dx$ converges $\forall$ n

Hence $\int_1^\infty x^{n-1} e^{-x} dx$ is convergent iff $n > 0$.

Remark:

The integral $\int_1^\infty x^{n-1} e^{-n} dx$ is called Gamma *Function. It is denoted by F(n).*

Example 3:

Discuss the convergence of $\int_{-\infty}^\infty e^{-(x-a/x)^2}\, dx$.

Solution:

Let $f(x) = e^{-(x-a/x)^2}$. Then

$$\int_{-\infty}^{\infty} f(x)\,dx = \int_{-\infty}^{0} f(x)\,dx + \int_{0}^{\infty} f(x)\,dx$$

Putting x = – y, we see that

$$\int_{-\infty}^{0} e^{-(x-a/x)^2}\,dx = -\int_{\infty}^{0} e^{-(-y+a/y)^2}\,dy$$

$$= \int_{\infty}^{0} e^{-(y-a/y)^2}\,dy$$

$$\therefore \int_{-\infty}^{\infty} f(x)\,dx = 2\int_{0}^{\infty} e^{-(x-a/x)^2}\,dx$$

$$= 2\left[\int_{0}^{1} e^{-(x-a/x)^2}\,dx + \int_{1}^{\infty} e^{-(x-a/x)^2}\,dx\right]$$

We have $\lim\limits_{x\to 0} e^{-(x-a/x)^2} = 0$

$\int_{0}^{1} e^{(x-a/x)^2}\,dx$ is a proper integral.

Now we discuss convergence at ∞ only.

We know $e^{(x-a/x)^2} > (x - a/y)^2 \;\forall\; x \in R$.

$$\therefore \int_{1}^{\infty} e^{-(x-a/x)^2}\,dx < \int_{1}^{\infty} \frac{dx}{(x-a/x)^2}$$

Now $\displaystyle\int_{1}^{\infty} \frac{dx}{(x-a/x)^2} = \int_{1}^{\infty} \frac{x^2\,dx}{(x^2-a)^2}$

Let $g(x) = x^2/(x^2 - a)^2$,

$h(x) = \dfrac{1}{x^2}$. Then

$$\lim_{x\to\infty} \frac{g(x)}{h(x)} = \lim_{x\to\infty} \frac{x^4}{(x^3-a)^2} = \lim_{x\to\infty} \frac{1}{(1-a/x^2)^2} = 1$$

Since $\int_{1}^{\infty} h(x)\,dx$ is convergent, so

$$\int_{1}^{\infty} \frac{1}{(x-a/x)^2}\,dx \text{ is convergent}$$

$$\Rightarrow \quad \int_{1}^{\infty} e^{-(x-a/x)^2}\,dx \text{ is convergent.}$$

Hence $\int_{-\infty}^{\infty} e^{-(x-a/x)^2} dx$ is convergent.

Hence by (1), $\int_1^{\infty} \frac{x^{a-1} \log x}{1+x} dx$ is convergent iff n < 0.

Example 4:

Find the general and principal values of the integral $\int_{-1}^{1} \frac{dx}{x^4}$.

Solution:

Since the integrand $\frac{1}{x^4}$ becomes infinite only at x = 0, therefore, if $(-\epsilon_1, \epsilon_2)$ is a small interval containing x = 0, we have

$$\int_{-1}^{1} \frac{dx}{x^4} = \lim_{\epsilon_1 \to 0} \int_{-1}^{-\epsilon_1} \frac{1}{x^4} dx + \lim_{\epsilon_2 \to 0} \int_{\epsilon_2}^{1} \frac{dx}{x^4}$$

$$= \lim_{\epsilon_1 \to 0} \left[\frac{x^{-3}}{-3}\right]_{-1}^{-\epsilon_1} = \lim_{\epsilon_2 \to 0} \left[\frac{x^{-3}}{-3}\right]_{\epsilon_2}^{1}$$

Obviously this limit does not exist and so the general value of $\int_{-1}^{1} \frac{dx}{x^4}$ does not exist.

But the principal value is given by

$$p\int_{-1}^{1} \frac{1}{x^4} dx = \lim_{\epsilon \to 0} \left[\int_{-1}^{-\epsilon} \frac{1}{x^4} dx + \int_{\epsilon}^{1} \frac{1}{x^4} dx\right]$$

$$= \lim_{\epsilon \to 0} \left[\left\{-\frac{1}{3x^4}\right\}_{-1}^{-\epsilon} + \left\{-\frac{1}{3x^4}\right\}_{\epsilon}^{1}\right]$$

$$= \lim_{\epsilon \to 0} \left[-\frac{1}{3\epsilon^4} + \frac{1}{3} - \frac{1}{3} + \frac{1}{3\epsilon^4}\right]$$

$$= 0.$$ **Ans.**

Example 5:

Find the general and principal values of the integral $\int_0^3 \frac{dx}{(x-1)^3}$.

Solution:

Since the integrand $\frac{1}{(x-1)^3}$ becomes infinite at x = 1, we have

$$\int_0^3 \frac{dx}{(x-1)^3} = \lim_{\epsilon_1 \to 0} \int_0^{1-\epsilon_1} \frac{dx}{(x-1)^3} + \lim_{\epsilon_2 \to 0} \int_{1+\epsilon_2}^{3} \frac{dx}{(x-1)^3}$$

$$= \lim_{\epsilon_1 \to 0}\left[-\frac{1}{2(x-1)^2}\right]_0^{1-\epsilon_1} + \lim_{\epsilon_2 \to 0}\left[-\frac{1}{2(x-1)^2}\right]_{1+\epsilon_2}^{3}$$

$$= -\lim_{\epsilon_1 \to 0}\left[\frac{1}{2\epsilon_2^2}-\frac{1}{2}\right] - \lim_{\epsilon_2 \to 0}\left[\frac{1}{2(2^2)}-\frac{1}{2\epsilon_2^2}\right]$$

$$= \left(\tfrac{1}{2}-\tfrac{1}{8}\right) - \lim_{\epsilon_1 \to 0}\left(\frac{1}{2_{\epsilon_1}^{\ 2}}\right) + \lim_{\epsilon_2 \to 0}\left(\frac{1}{2_{\epsilon_2}^{\ 2}}\right)$$

This limit does not exist for it is of the form ∞ – ∞. But the principal value

$$= \tfrac{1}{2} - \tfrac{1}{8} + \lim_{\epsilon \to 0}\left(\frac{1}{2\epsilon^2}-\frac{1}{2\epsilon^2}\right) = \tfrac{3}{8}\,.$$ **Ans.**

Example 6(a):

Examine the convergence of : (i) $\int_0^1 \frac{dx}{x^2}$ (ii) $\int_0^1 \frac{dx}{\sqrt{1-x}}$.

Solution:

(i) 0 is the only point of infinite discontinuity of the integrand in [0, 1]. Thus,

$$\int_0^1 \frac{dx}{x^2} = \lim_{\lambda \to 0+} \int_\lambda^1 \frac{dx}{x^2}, 0 < \lambda < 1$$

$$= \lim_{\lambda \to 0+}\left(\frac{1}{\lambda}-1\right) = \infty$$

Thus, the improper integral is divergent.

(ii) Since 1 is the only point of infinite discontinuity of the integrand in [0, 1], we put

$$\int_0^1 \frac{dx}{\sqrt{1-x}} = \lim_{\mu \to 0+} \int^{1-\mu} \frac{dx}{\sqrt{1-x}}, 0 < \mu < 1$$

$$= \lim_{\mu \to 0+}\left[-2\sqrt{1-x}\right]_0^{1-\mu}$$

$$= -\lim_{\mu \to 0+} 2(\sqrt{\mu}-1) = 2$$

Thus, the improper integral exists and is equal to 2.

Example 6(b):

Examine the convergence of $\int_0^2 \frac{dx)}{(2x-x^2)}$.

Solution:

Both the end-points 0, 2 are the points of infinite discontinuity of the integrand and are in fact the only such points in [0, 2].

Thus for any point, say 1, within [0, 2], we put

$$\int_0^2 \frac{dx}{2x-x^2} = \lim_{\lambda \to 0+} \int_\lambda^1 \frac{dx}{x(2-x)} + \lim_{\mu \to 0+} \int_1^{2-\mu} \frac{dx}{x(2-x)}$$

$$= \tfrac{1}{2} \lim_{\lambda \to 0+} \left[\log \frac{x}{2-x} \right]_\lambda^1 + \tfrac{1}{2} \lim_{\mu \to 0+} \left[\log \frac{x}{2-x} \right]_1^{2-\mu}$$

$$= -\tfrac{1}{2} \lim_{\lambda \to 0+} \log \frac{\lambda}{2-\lambda} + \tfrac{1}{2} \lim_{\mu \to 0+} \log \frac{2-\mu}{\mu}$$

$$= \infty.$$

Thus, the given integral diverges.

Example 7:

Examine the convergence of :

(i) $\int_0^1 \frac{x^n}{1+x} dx$ (ii) $\int_0^\pi \frac{\sqrt{x}}{\sin x} dx$

Solution:

(i) Consider $\int_0^1 \frac{x^n}{1+x} dx$.

This is a proper integral if $n \geq 0$ and an improper integral if $n < 0$ (0 being the only point of infinite discontinuity of the integrand). For $n < 0$, we may take $n = -m (m > 0)$.

$$\therefore \int_0^1 \frac{x^n}{1+x} dx = \int_0^1 \frac{dx}{x^m(1+x)}$$

Let $f(x) = \frac{1}{x^m(1+x)}$ and $g(x) = \frac{1}{x^m}$. Then

$$\lim_{x \to 0+} \frac{f(x)}{g(x)} = \text{and} \lim_{x \to 0+} \frac{1}{1+x} = 1$$

Now $\int_0^1 g(x)\,dx = \int_0^1 \frac{dx}{x^n}$ convergence m < 1 or –n < 1 or n > – 1. Hence the given integral is convergent if n > –1.

(ii) Here 0 and π are the only points of infinite discontinuity of the integrand. Now

$$\int_0^\pi \frac{\sqrt{x}}{\sin x}\,dx = \int_0^1 \frac{\sqrt{x}}{\sin x}\,dx + \int_1^\pi \frac{\sqrt{x}}{\sin x}\,dx$$ convergence at 0

Let $f(x) = \frac{\sqrt{x}}{\sin x}$ and $g(x) = \frac{1}{\sqrt{x}}$. Then, we have

$$\lim_{x\to 0+} \frac{f(x)}{g(x)} = \lim_{x\to 0+} \frac{x}{\sin x} = 1$$

Since $\int_0^1 g(x)\,dx = \int_0^1 \frac{dx}{\sqrt{x}}$ converges,

Hence, $\int_0^1 \frac{\sqrt{x}}{\sin x}\,dx$ also converges,

Convergence at π

Let $f(x) = \frac{\sqrt{x}}{\sin x}$ and $g(x) = \frac{1}{x-\pi}$. Then

$$\lim_{x\to\pi-} \frac{f(x)}{g(x)} = \lim_{x\to\pi-} \frac{\sqrt{x}(x-\pi)}{\sin x}.\left(\frac{0}{0}\right)$$

$$= \lim_{x\to\pi-} \frac{\frac{3}{2}x^{1/2} - \frac{1}{2}\pi x^{-1/1}}{\cos x} = -\sqrt{\pi}$$ (By L' Hospital's rule)

Since $\int_1^\pi g(x)\,dx = \int_1^\pi \frac{dx}{(x-\pi)^1}$ is divergent,

so $\int_1^\pi \frac{\sqrt{x}}{\sin x}\,dx$ is divergent.

Hence, the given integral $\int_0^\pi \frac{\sqrt{x}}{\sin x}\,dx$ is divergent.

Example 8:

Examine the convergence of:

(i) $\int_0^\infty \frac{\sin^2 x}{x^2}\,dx$ (ii) $\int_0^\infty \frac{x^{n-1}}{1+x}\,dx$ (iii) $\int_1^\infty \frac{x^{n-1}\log x}{1+x}\,dx$

Solution:

(i) Let $f(x) = \sin^2 x/x^2$. Then

$$\int_0^\infty f(x)\,dx = \int_0^1 f(x)\,dx + \int_1^\infty f(x)\,dx$$

Now, $\int_0^1 \frac{\sin^2 x}{x^2}\,dx$ is a proper integral,

since $\lim_{x\to 0} \frac{\sin^2 x}{x^2} = 1$ and so 0 is not a point of infinite discontinuity.

We have $\frac{\sin^2 x}{x^3} < \frac{1}{x^4}$ $\forall\, x \in [1, \infty[$.

Since $\int_1^\infty \frac{1}{x^2}\,dx$ converges, so $\int_1^\infty \frac{\sin^2 x}{x^2}\,dx$ also convergence

Hence, the given Integral $\int_1^\infty \frac{\sin^2 x}{x^2}\,dx$ is convergent.

(ii) $\int_1^\infty \frac{x^{m-1}}{1+x}\,dx = \int_0^1 \frac{x^{m-1}}{1+x}\,dx + \int_1^\infty \frac{x^{m-1}}{1+x}\,dx$ convergence at 0

Let $f(x) = \frac{x^{m-1}}{1+x}$, $g(x) = \frac{1}{x^{1-m}}$

$\therefore \lim_{x\to 0} \frac{f(x)}{g(x)} = \lim_{x\to 0}\left(\frac{1}{1+x}\right) = 1$

Now $\int_0^1 g(x)\,dx$ converges if $1-m < 1$ i.e., $0 < m$.

Hence the given integral $\int_0^1 f(x)\,dx$ converges if $0\ \Delta m$

Convergence at ∞

Let $g(x) = \frac{1}{x^{2-m}}$

$\therefore \lim_{x\to\infty} \frac{f(x)}{g(x)} = \lim_{x\to\infty}\left(\frac{x}{1+x}\right) = 1$

Since $\int_1^\infty \frac{dx}{2^{2-m}}$ converges if $2 - m > 1$ i.e., $1 > m$.

So $\int_0^\infty f(x)\,dx$ converges if $m < 1$.

Hence the given integral $\int_1^{\infty} \frac{x^{m-1}}{1+x}\,dx$ converges if $0 < m < 1$.

For large values of x, $\log x < x$.

$$\therefore \int_1^{\infty} \frac{x^{a-1}\log x}{1+x}\,dx < \int_1^{\infty} \frac{x^a\,dx}{1+x} \qquad ...(1)$$

Let $f(x) = \frac{x^a}{1+x}$ and $f(x) = \frac{1}{x^{1-a}}$

$$\therefore \lim_{x\to\infty} \frac{f(x)}{g(x)} \lim_{x\to\infty} \frac{x}{1+x} = 1$$

Now $\int_1^{\infty} g(x)\,dx$ is convergent iff $1 - a > 1$

$\Rightarrow \int_1^{\infty} f(x)\,dx$ is convergent iff $0 > a$

Hence by (i) $\int_1^{\infty} \frac{x^{n-1}\log x}{1+x}\,dx$ is convergent iff $a < 0$.

Example 9:

Examine the convergence of:

(i) $\int_0^{\pi 2} \frac{\sin x}{x^n}\,dx$ (ii) $\int_0^1 x^{n-1}e^{-x}\,dx$

Solution:

(i) Here 0 is the only point of infinite discontinuity of the integrand.

Let $f(x) = \frac{\sin x}{x^n}$ and $g(x) = \frac{1}{x^{n-2}}$

$$\therefore \lim_{x\to 0+} \frac{f(x)}{g(x)} = \lim_{x\to 0+} \frac{\sin x}{x} = 1$$

Now $\int_0^{\frac{1}{2}} g(x)\,dx = \int_0^{\frac{1}{2}} \frac{dx}{x^{n-1}}$ converges if $(n - 1 < 1)$ a – 1 i.e., $n < 2$

By second comparison test, the given improper integral is convergent if $n < 2$.

(ii) Consider $\int_0^1 \frac{e^{-x}\,dx}{x^{1-a}}$

Here 0 is the only point of infinite discontinuity of the integrand if $a < 1$.

Let $f(x) \dfrac{e^{-x}}{x^{1-a}}$ and $g(x) = \dfrac{1}{x^{1-a}}$

$$\therefore \qquad \lim_{x\to 0+} \frac{f(x)}{g(x)} = \lim_{x\to 0+} e^{-x} = 1$$

Now $\int_0^1 \dfrac{dx}{x^{1-a}}$ converges if $1 - a <$ i.e., $0 < a$.

Hence, the given integral converges if $a > 0$.

Example 10:

Examine the convergence of (i) $\int_0^1 \dfrac{dx}{\sqrt{x-x^2}}$ (ii) $\int_0^1 \dfrac{dx}{\sqrt{1-x^2}}$ (iii) $\int_0^\pi \dfrac{dx}{\sin e}$

Solution:

(i) Notice that 0 and 1 are the only points of infinite discontinuity of the integrand in [0, 1]. We take 0 < ½ < 1.

Putting $x = \sin^2\theta$, $\displaystyle\int \frac{dx}{\sqrt{x-x^2}} = 2\,\theta = 2\sin^{-1}\sqrt{x}$

$$= \lim_{\varepsilon\to 0} \int_\varepsilon^{1/2} \frac{dx}{\sqrt{x-x^2}} = \lim_{\varepsilon\to 0+} \left[2\sin^{-1}(\sqrt{\tfrac{1}{2}}) - 2\sin^{-1}\sqrt{\varepsilon}\right]$$

$$= 2\left(\frac{\pi}{4}\right) = \frac{\pi}{2}.$$

$$\lim_{\varepsilon\to 0} \int_{\frac{1}{2}}^{1-\varepsilon} \frac{dx}{\sqrt{x-x^2}} = \lim_{\varepsilon\to 0+}[2\sin^{-1}\sqrt{(1-\varepsilon)} - 2\sin^{-1}(\sqrt{\tfrac{1}{2}})]$$

$$= 2\left(\frac{\pi}{2} - \frac{\pi}{4}\right) = \frac{\pi}{2}$$

Hence, the given integral is convergent and

$$\int_0^1 \frac{dx}{\sqrt{x-x^2}} = \frac{\pi}{2} + \frac{\pi}{2} = \pi$$ **Ans.**

(ii) It is easy to see that 1 is the only point of limits discontinuity of the integrand in [0, 1] we have

$$\int_0^{1-\varepsilon} \frac{dx}{\sqrt{1-x^2}} = \left|\sin^{-1} x\right|_0^{1-\varepsilon} = \sin^{-1} = \frac{\pi}{2}$$

$$\therefore \lim_{\varepsilon\to 0} \int_0^{1-\varepsilon} \frac{dx}{\sqrt{1-x^2}} = \lim_{\varepsilon\to 0+} \sin^{-1}(1-\varepsilon) = \sin^{-1} 1 = \frac{\pi}{2}$$

Thus, the given improper integral is convergent with value ½π.

(iii) Notice that 0 and π are the only points of infinite discontinuity of the integrand in [0, π]. We take 0 < 1 < π. Now

$$\int_0^{1-\varepsilon} \frac{dx}{\sin x} = \left|\log \tan \tfrac{1}{2}x\right|_\varepsilon^1 = (\log \tan \tfrac{1}{2} - \log \tan \tfrac{1}{2}\varepsilon)$$

$$\to \infty \text{ as } \varepsilon \to 0+.$$

Thus $\int_0^1 \frac{dx}{\sin x}$ is divergent and consequently the given improper integral is divergent.

Example 11:

Discuss the convergence of

$$\int_0^\infty \left(\frac{1}{(e^x - 1)} - \frac{1}{x} + \frac{1}{2}\right)\frac{e^{-kx}}{x}\, dx.$$

Solution:

Given integral = $\int_0^1 f(x)\, dx + \int_1^\infty f(x)\, dx$.

Convergence at 0.:

We have $\lim_{x\to 0+}\left(\frac{1}{e^x - 1} - \frac{1}{x} + \frac{1}{2}\right)\frac{e^{-kx}}{x}$

$$\Rightarrow \lim_{x\to 0+}\left[\frac{2x - 2(e^x - 1) + x(e^x - 1)}{2x(e^x - 1)}\right]\frac{e^{-kx}}{x}$$

$$\Rightarrow \lim_{x\to 0+}\frac{x - 2e^x + xe^x + 2}{2x^2(e^x - 1)} \cdot \lim_{x\to 0+} e^{-kx}$$

$$\Rightarrow \lim_{x\to 0+}\left[\frac{1 - 2e^x + e^x + xe^x}{4x(e^x - 1) + 2x^2 e^x}\right].1$$

$$\Rightarrow \lim_{x\to 0+}\left[\frac{-e^x + (e^x + xe^x)}{4(e^x - 1) + 8xe^x + 2x^2 e^x}\right]$$

$$\Rightarrow \lim_{x\to 0+}\left[\frac{e^x(x+1)}{12e^x + 12xe^x + 2x^2 e^x}\right] = \frac{1}{12}$$

So 0 is not a point of infinite discontinuity.

Convergence at ∞

We have $f(x) = \dfrac{(x-2e^a+xe^a+2)}{2x(e^a-1)}.\dfrac{e^{-ka}}{x}$

Let $\qquad g(x) = \dfrac{e^{-ka}}{x}$

$\therefore \lim\limits_{x\to\infty} \dfrac{f(x)}{g(x)} = \dfrac{1}{2}$

Thus $\int_1^\infty f(x)\,dx$ and $\int_1^\infty g(x)\,dx$ have identical behaviour.

Since $e^{ka} > kx \ \forall\ x \in R$, so

$$\int_1^\infty \frac{e^{-kx}}{kx^2} \int_1^\infty \frac{1}{kx^2}\,dx.$$

Since, $\int_1^\infty \dfrac{1}{kx^2}\,dx$ converges, so $\int_1^\infty \dfrac{e^{-ka}}{x}\,dx$ converges.

Hence, the given integral $\int_1^\infty f(x)\,dx$ is convergent and so the given integral is convergent.

Example 12:

Show that $\int_0^1 x^{m-1}(1-x)^{n-1}dx$ exists iff $m > 0$ and $n > 0$.

Solution:

The given integral is a proper integral if $m \geq 1$. $n \geq 1$. If $m < 1$, then 0 is the only point of infinite discontinuity and if $n < 1$, then 1 is the only point of infinite discontinuity.

For $m < 1$ and $n < 1$, we take $0 < ½ < 1$.

We have $\int_0^1 f\,dx = \int_0^{1/2} f\,dx + \int_{1/2}^1 f dx, f(x) = m^{m-1}(1-x)^{n-1}$.

Convergence at 0 (m < 1):

Let $f(x) = \dfrac{(1-x)^{n-1}}{x^{1-m}}$ and $g(x) = \dfrac{1}{x^{1-m}}$

$\therefore \lim\limits_{x\to 0+} \dfrac{f(x)}{g(x)} = \lim\limits_{x\to 0+} (1-x)^{n-1} = 1$

Since $\int_0^{1/2} g(x)dx = \int_0^{1/2} \frac{dx}{x^{e-m}}$ converges at iff 1 – m < 1 i.e. m > 0

so $\int_0^{1/2} x^{m-1}(1-x)^{x-1}\, dx$ converges at zero iff m > 0

Convergence at 1 (n < 1):

Let $f(x) = \dfrac{x^{m-1}}{(1-x)^{1-n}}$ and $g(x) = \dfrac{1}{(1-x)^{1-n}}$

$\therefore \lim\limits_{x\to 1} \dfrac{f(x)}{g(x)} = \lim\limits_{x\to 1} x^{m-1} = 1.$

Since $\int_{1/2}^{1} g(x)\, dx = \int_{1/2}^{1} \dfrac{dx}{(1-x)^{1-n}}$ converges iff 1 – n < 1 i.e., n > 0.

so $\int_{1/2}^{1} x^{m-1}(1-x)^{n-1}\, dx$ converges at 1 iff n > 0.

Hence, $\int_0^1 x^{m-1}(1-x)^{m-1}$ exists iff m > 0 and n > 0.

Remark:

The integral $\int_0^1 x^{m}(1-x)^{n-1}$ *dx is called Beta.*

Function:

It is denoted by β(m, n).

Example 13(a):

Show that $\int_0^{\pi/2} \sin^{m-1} x \cos^{n-1} dx$ *is convergent if m >, n > 0.*

Solution:

We see that 0 is a point of infinite discontinuity if m – 1 < 0 and π/2 is a point of infinite discontinuity if n – 1 < 0. Then we write

$$\int_0^{\pi/2} f(x)\, dx = \int_0^{\pi/4} f(x)\, dx + \int_{\pi/4}^{\pi/2} f(x)\, dx,\ f(x) = \sin^{m-1} x \cos^{n-1} x.$$

Convergence at 0:

Let $g(x) = \dfrac{1}{x^{2-m}}$ so that $\dfrac{f(x)}{g(x)} = \dfrac{\sin^{m-1}}{x^{m-1}} \cos^{n-1} x \to 1$ as $x \to 0$

Since $\int_0^{\pi/4} g(x)\,dx$ converges if $1 - m < 1$ i.e., $m > 0$.

so $\int_0^{\pi/4} g(x)$ converges if $m > 0$.

Convergence at π/2:

Let $$g(x) = \frac{1}{\left(x - \frac{1}{2}\pi\right)^{1-n}} \text{ so that}$$

$$\frac{f(x)}{g(x)} = \sin^{m-1} x \left(\frac{\cos x}{x - \frac{1}{2}\pi}\right)^{n-1} \to (-1)^{n-1} \text{ as } x \to \tfrac{1}{2}\pi$$

As $\int_{\pi/4}^{\pi/2} g(x)\,dx$ converges if $1 - n < 1$ i.e., $n > 0$.

so $\int_{\pi/4}^{\pi/2} g(x)\,dx$ converges if $n > 0$.

Hence, the given integral is convergent if $m > 0$, $n > 0$.

Example 13(b):

Examine the convergence of

(i) $\int_0^1 \frac{\log x}{\sqrt{x}}$, (ii) $\int_0^{\pi/2} \frac{\log x}{x^4}\,dx\,(a < 1)$.

Solution:

(i) Since $\log x \mid \sqrt{x}$ is negative in]0, 1], we take

$$f(x) = -\frac{\log x}{\sqrt{x}} = \frac{\log 1/x}{\sqrt{x}}$$

Here 0 is the only point of infinite discontinuity.

Let $g(x) = \frac{1}{x^{3/4}}$ so that we have

$$\lim_{x \to 0+} \frac{f(x)}{g(x)} = \lim_{x \to 0+} x^{1/4} \log\frac{1}{x} = \lim_{x \to 0+} \frac{\log 1/x}{x^{-1/4}}$$

$$= \lim_{x \to 0+} \frac{x(-1/x^2)}{(-1/4)x^{-5/4}} \text{ is } \lim_{x \to 0+} x^{1/4} = 0$$

Since $\int_0^1 g(x)\,dx = \int_0^1 \frac{dx}{x^{3/4}}$ is convergent, so the given integral is convergent.

(ii) Since log x/x^a is negative when $0 < x <^1$, so we have

$$f(x) = \frac{\log x}{x^a} = \frac{\log(1/x)}{x^a}$$

Let k be a fixed number such that a < k < 1.

Let $g(x) = \frac{1}{x^2}$ so that $\frac{f(x)}{g(x)} = x^{k-a} \log\left(\frac{1}{x}\right) \to 0$ as $x \to 0+$

Since, $\int_0^1 g(x)\,dx = \int_0^1 \frac{dx}{g(x)} = (k < 1)$ is donvergent, so $\int_0^1 f(x)\,dx$ and hence, $\int_0^1 \frac{\log x}{x^a}\,dx$ is convergent if a < 1.

Again, $\int_0^{\pi/2} \frac{\log x}{x^a} dx = \int_0^1 \frac{\log x}{x^a} dx + \int_1^{\pi/2} \frac{\log x}{x^a} dx$

where the second integral on the R.H.S. is proper.

Hence, the given integral is convergent for a < 1.

Example 13(c):

Examine for convergence the integrals

(i) $\int_1^\infty \frac{\tan^{-1} x}{x^2} dx$, (ii) $\int_1^\infty \frac{x \tan^{-1} x}{(1+x^4)^{1/3}} dx$,

Solution:

(i) Let $f(x) = \tan^{-1} x/x^2$ and $g(x) = 1/x^2$ so that

$$\lim_{x\to\infty} \frac{f(x)}{g(x)} = \lim_{x\to\infty} \tan^{-1} x = \frac{\pi}{2}$$

Since $\int_1^\infty g(x)\,dx$ converges, so $\int_1^\infty f(x)\,dx$ converges

(ii) Let $f(x) = \frac{x \tan^{-1} x}{(1+x^4)^{1/3}} = \frac{x \tan^{-1} x}{x^{1/2}(1+x^{-5})^{1/3}}, g(x) = \frac{x}{x^{4/3}} = \frac{1}{x^{1/2}}$

$$\therefore \lim_{x\to\infty} \frac{f(x)}{g(x)} = \lim_{x\to\infty} \frac{\tan^{-1} x}{(1+x^{-4})} = \frac{\pi}{2}$$

Since $\int_1^\infty \frac{dx}{x^{1/3}}$ diverges, so $\int_1^\infty \frac{x \tan^{-1} x\,dx}{(1+x^4)^{1/3}}$ also diverges.

Example 14:

Show that the improper integral

$$\int_0^\infty \log(1+2\,\text{sec hx})\,dx \text{ is convergent.}$$

Solution:

$$\int_0^\infty f(x)\,dx = \int_0^1 f(x)dx + \int_1^\infty f(x)dx,$$

where $\quad f(x) = \log(1 + 2 \text{ sec hx})$

We know see hx $= \dfrac{2}{e^2+e^{-2}} = \dfrac{2e^n}{e^{2n}+1}$

Convergence at 0:

$$\lim_{x\to 0+} \log(1+2\,\text{sec hx}) = \lim_{x\to 0+} \log 1+\frac{4e^x}{e^{2x}+1} = \log 3.$$

Thus 0 is not a point of infinite discontinuity and so $\int_0^1 f(x)\,dx$ is a proper integral.

Convergence at ∞:

We know $\log(1 + x) < x$ for all $x \geq 1$

$\Rightarrow \log(1 + 2 \text{ sec hx}) < 2 \text{ sec hx}$

$$\Rightarrow \int_0^\infty \log(1+2\,\text{sec hx})\,dx < \int_1^\infty 2\,\text{sec hx}\,dx$$

$$= \int_1^\infty \frac{4e^x}{e^{2x}+1}dx = 4\int_e^\infty \frac{dy}{1+y^2},\ y = e^x$$

$$< 4\int_e^\infty \frac{dy}{y^2}, \text{ which is convergent.}$$

Hence $\int_1^\infty f(x)\,dx$ is convergent and so the given improper integral is convergent.

Example 15:

Discuss the convergence of

$$\int_0^\infty \frac{x^{2m}}{1+x^{2n}}(m > 0, n > 0).$$

Solution:

Clearly 0 is not a point of infinite discontinuity of the integrand. We shall examine convergence at ∞ only.

Let $\quad f(x) = \dfrac{x^{2m}}{1+x^{2n}}$ and $g(x) = \dfrac{x^{2m}}{x^{2n}} = \dfrac{1}{x^{2(n-m)}}$.

$$\therefore \lim_{x\to\infty} \frac{f(x)}{g(x)} = \lim_{x\to\infty} \frac{x^{2n}}{1+x^{2n}} = \lim_{x\to\infty} \frac{1}{1+(1/x^{an})} = 1$$

Since $\int_0^\infty g(x)\,dx$ converges iff 2 (n–m)>1, so

$$\int_0^\infty \frac{x^{2m}}{1+x^{2m}} \text{ converges iff } (n - m) > ½.$$

Example 16:

Test for convergence the integral

$$\int_0^1 x^p \left(\log\frac{1}{x}\right)^q dx.$$

Solution:

Taking $\log\left(\dfrac{1}{x}\right) = t$ or $x = e^{-1}$, we get

$$\int_0^1 x^p \left(\log\frac{1}{x}\right)^q dx = \int_0^\infty t^q e^{-(p+1)t}\, dt.$$

Here 0 is a point of infinite discontinuity, so we take

$$\int_0^\infty t^q e^{(p+1)}\, dt = \int_0^1 t^q e^{-(p+1)t}\, dt + \int_1^\infty t^q e^{-(p+1)t}\, dt.$$

Convergence at 0:

Let $\quad f(t) = t^q e^{-(p+1)t} = \dfrac{e^{-(p+1)t}}{t^{-q}}$

Comparing with $\int_0^1 \dfrac{dt}{t^{-q}}$, we see that $\int_0^1 e^{-(p+1)t}\, t^q\, dt$ converges at 0 if $-q < 1$ or $q > -1$, for all values of p.

Convergence at ∞ :

Let $f(f) = t^q\, e^{-(p+1)t}$

$g(t) = \dfrac{1}{t^2}$

Now $\dfrac{f(t)}{g(t)} = \dfrac{t^{q+a}}{e^{(p+1)t}} \to 0$ as $t \to \infty$ for $p + 1 > 0$ and for all q.

Since $\int_1^\infty \dfrac{dt}{t^2}$ converges, so $\int_1^\infty t^q e^{-(p+1)t}\, dt$

converges for q and $p > -1$.

Hence, $\int_0^1 x^p \left(\log\dfrac{1}{x}\right)^q dx$ is convergent for $p > -1, q > -1$ and divergent for all other values of q.

Example 17:

Examine for convergence the integral

(i) $\displaystyle\int_0^\infty \frac{\sqrt{x}}{(1+x)^2}dx$, (ii) $\displaystyle\int_{-\infty}^\infty \frac{dx}{(1+x^3)^2}$.

Solution:

(i) Let $f(x) = \dfrac{\sqrt{x}}{(1+x)^2}$ and $g(x) = \dfrac{1}{x\sqrt{x}} = \dfrac{1}{x^{3/2}}$ so that

$$\lim_{x\to\infty}\frac{f(x)}{g(x)} = \lim_{x\to\infty}\frac{x^2}{(1+x)^2} = 1$$

Thus, $\int_1^\infty f(x)\, dx$ and $\int_1^\infty g(x)\, dx$ have identical behaviour.

But $\int_1^\infty g(x)\, dx = \int_1^\infty \dfrac{dx}{x^{3/2}}$ converges, so $\int_1^\infty f(x)\, dx$ also converges.

(ii) $$\int_{-\infty}^\infty \frac{dx}{(1-x^2)^2} = \int_{-\infty}^0 \frac{dx}{(1+x^2)} + \int_0^\infty \frac{dx}{(1+x^2)^2}$$

$$= 2\int_0^\infty \frac{dx}{(1+x^2)^2}$$

(Take $x = -y$ in the first integral)

Now, $2\displaystyle\int_0^t \frac{dx}{(1+x^2)^2} = 2\int \cos^2\theta\, d\theta = \int (1+\cos 2\theta)\, d\theta,\ x = \tan\theta$

$$= 0 + \frac{1}{2}(\sin 2\theta) = \theta + \frac{\tan\theta}{1+\tan^2\theta}$$

$$= \tan^{-1} x + \frac{x}{1+x^2} = \tan^{-1} t + t(1+t^2)$$

$\therefore \lim_{t\to\infty} \frac{dx}{(1+x^3)^2} = \frac{\pi}{2}$. Hence $\int_{-\infty}^{\infty} \frac{dx}{(1+x^3)^2}$ converges.

Example 18:

Show that $\int_0^{\pi/2} \log$ *sin x dx is convergent.*

Solution:

The function f(x) = log sin x is non-positive in [0, π/2], we consider the function –f(x) for examining the convergence at 0.

Let $g(x) = \frac{1}{x^m}$, where m < 1. Then we have

$$\lim_{x\to 0} \frac{-f(x)}{g(x)} = \lim_{x\to 0}(-x^m \log \sin x) = 0$$

So $\int_0^{\pi/2} -f(x)\,dx$ and $\int_0^{\pi/2} g(x)\,dx$ converge or diverge together.

Since $\int_0^{\pi/2} g(x)\,dx = \int_0^{\pi/2} \frac{dx}{x^m}$ is convergent for m < 1, so

$- \int_0^{\pi/2} \log \sin x\,dx$ is convergent.

Hence $\int_0^{\pi/2} \log\ \sin x\,dx$ is convergent.

Example 19:

Show that $\int_{-\infty}^{\infty} e^{-x^2}\,dx$ converges.

Solution:

$$\int_{-\infty}^{\infty} e^{-x^2}\,dx = \int_{-\infty}^{0} e^{-x^2}\,dx + \int_0^{\infty} e^{-x^2}\,dx$$

Putting x = – y, we obtain

$$\int_{-\infty}^{\infty} e^{-x^2}\,dx = -\int_{\infty}^{0} e^{-y^4}\,dy + \int_0^{\infty} e^{-x^2}\,dx$$

$$= 2\int_0^{\infty} e^{-x^2}\,dx$$

Hence $\int_{-\infty}^{\infty} e^{-x^2}\,dx$ is convergent.

Example 20(a):

Show that $\int_0^1 x^{n-1} \log x\, dx$ *converges iff n > 0.*

Solution:

Since log x is negative in [0, 1], so we consider

$$\int_0^1 -x^n \log x\, dx = \int_0^1 x^{n-1} \log\left(\frac{1}{x}\right) dx$$

Now, $\lim_{x\to 0} x^{n-1} \log\left(\frac{1}{x}\right) = 0$ if $n-1>0$

Thus, the given integral is proper if n – 1 > 0

Let n – 1 ≤ 0. In this case 0 is a point of infinite discontinuity.

Again, Let m be a positive number such that (m + n – 1) > 0.

Then, $\lim_{x\to 0} [-x^{m+n-1} \log x] = 0$

It follows that for any ε > 0. ∃ some δ > 0 such that

$-x^{m+n-1} \log x < \varepsilon$ for $|x| < \delta$

$\Rightarrow -x^{m+n-1} \log x < \varepsilon/x^m$ for $|x| < \delta$

We know $\int_0^1 \frac{\varepsilon}{x^m} dx$ converges iff m < 1.

It is possible to choose a number m, 0 < m < 1, such that

(m + n – 1) = (m – 1) + n > 0 iff n > 0.

Thus, $\int_0^1 -x^{n-1} \log x\, dx$ converges if n > 0.

For n = 0, given integral = $\int_0^1 \frac{\log x}{x} dx$. Now

$$\int_0^1 \frac{\log x}{x} dx = -\tfrac{1}{2}(\log \varepsilon)^2 \to -\infty \text{ as } \varepsilon \to 0+$$

Thus, the given integral does not converge if n = 0,

For n < 0, we have

$x^{-1} < x^{n-1}\ \forall\, x \in [0, 1]$

$\Rightarrow x^{-1} \log x < x^{n-1} \log x\ \forall\, x \in [0, 1]$

Since, $\int_0^1 x^{-1} \log x\, dx$ diverges, so

$\int_0^1 x^{n-1} \log x \, dx$ also diverges for $n < 0$.

Hence, $\int_0^1 x^{n-1} \log x \, dx$ converges iff $n > 0$.

Example 20(b):

Discuss the convergence of

$\int_0^1 x^{m-1} (1-x)^{n-1} \log x \, dx$.

Solution:

Since the integrand is negative in [0, 1], so we shall test the convergence of

$$\int_0^1 -x^{m-1} (1-x)^{n-1} \log x \, dx$$

$$= \int_0^1 x^{m-1} (1-x)^{n-1} \log \frac{1}{x} \, dx$$

Here 0 and 1 are the possible points of infinite discontinuity of $f(x) = -x^{m-1} (1 - x)^{n-1} \log x$. We take

$$\int_0^1 f(x) \, dx = \int_0^{1/2} f(x) \, dx + \int_{1/2}^1 f(x) \, dx.$$

Convergence at 0:

The integral $\int_0^{1/2} f(x) \, dx$ is proper if m – > 0, since

$$\lim_{x \to 0} -x^{m-1} \log x = 0$$

For $m \le 1$, let $g(x) = 1/x^p$. Then

$$\int_0^{1/2} g(x) \, dx \text{ is convergent iff } p < 1.$$

Also, $\dfrac{f(x)}{g(x)} = x^{p+m-1}(1-x)^{n-1} \log \dfrac{1}{x} \to 0$ as $x \to 0+$,

If $\quad P + m - 1 > 0 \quad$ or $\quad m > 1 - p > 0$

Thus, $\int_0^{1/2} f(x) \, dx$ converges if $m > 0$.

Convergence at 1:

Clearly $\lim_{x \to 1} -(1-x)^{n-1} \log x$ exists finitely when $n \ge 0$.

Thus, $\int_{1/2}^{1} f(x)\ dx$ is a proper integral if $n \geq 0$.

For $n < 0$, let $g(x) = \dfrac{1}{(1-x)^q}$. Then

$\int_{1/2}^{1} g(x)\ dx$ is convergent iff $q < 1$.

Also $\dfrac{f(x)}{g(x)} = \dfrac{x^{m-1}\log(1/x)}{(1-x)^{1-n-q}}$,

which tends to a finite limit as $x \to 0$, if

$1 - n - q \leq 1$ or $n \geq -q > -1$

Thus $\int_{1/2}^{1} f(x)\ dx$ converges if $n > -1$.

Hence, the given integral is convergent if $m > 0$, $n > -1$.

Example 21:

Examine the convergence of $\int_0^2 \dfrac{\sqrt{x}}{\log x}\, dx$.

Solution:

Here 1 is the only point of infinite discontinuity.

Let $f(x) = \dfrac{\sqrt{x}}{\log x}$ and $g(x) = \dfrac{1}{x-1}$. Then

$$\lim_{x\to 1+} \frac{f(x)}{g(x)} = \lim_{x\to 1+} \frac{(x-1)\sqrt{x}}{\log x}\left(\frac{0}{0}\right)$$

$$= \lim_{x\to 1+} \frac{\frac{3}{2}x^{1/2} - \frac{1}{2}x^{-1/2}}{g(x)} = \lim_{x\to 1+}\left(\tfrac{3}{2}x^{3/2} = \tfrac{1}{2}x^{1/2}\right) = 1$$

Since $\int_1^2 g(x)\ dx = \int_1^2 \dfrac{dx}{(x-1)}$ diverges $(\because n = 1)$,

so $\int_1^2 f(x)\ dx = \int_1^2 \dfrac{\sqrt{x}}{\log x}$ diverges.

Example 22:

Show that $\int_0^{1/2} x^m$ *cosec*^{n} *x dx exists iff* $n < (m + 1)$.

Solution:

We have

$$f(x) = x^m \operatorname{cosec}^n x = \frac{x^m}{\sin^n x} = \left(\frac{x}{\sin x}\right)^n x^{m-n}$$

or $$f(x) = \left(\frac{x}{\sin x}\right)^n \frac{1}{x^{n-m}}$$

$$\therefore \lim_{x\to 0+} f(x) = \begin{cases} 1 \,.\, 0 = 0 & \text{if } n-m<0, \\ 1 \,.\, 1 = 1 & \text{if } n-m=0, \\ \infty & \text{if } n-m>0, \end{cases}$$

Thus the given integral is proper if $n - m \le 0$ and improper if $n - m > 0$ i.e., $n > m$.

Let $g(x) = \dfrac{1}{x^{n-m}} (n > m)$. Then

$$\lim_{x\to 0+} \frac{f(x)}{g(x)} = \lim_{x\to 0+} \left(\frac{x}{\sin x}\right)^n = 1$$

So $\int_0^{\pi/2} f(x)\, dx$ and $\int_0^{\pi/2} g(x)\, dx$ have identical behaviour.

Now $\int_0^{\pi/2} g(x)\, dx = \int_0^{\pi/2} \dfrac{dx}{x^{n-m}}$ converges iff $n - m < 1$ or $n < m + 1$.

Hence the given integral exists iff $n < m + 1$.

Example 23:

Show that $\int_0^{\infty} \left(\dfrac{1}{x} - \dfrac{1}{\sin hx}\right) \dfrac{dx}{x}$ *is convergent.*

Solution:

Let

$$f(x) = \left(\frac{1}{x} - \frac{1}{\sin hx}\right)\frac{1}{x} = \frac{\sin hx - x}{x^2 \sin hx} \qquad ...(1)$$

$$\lim_{x\to 0} f(x) = \lim_{x\to 0} \frac{\sin hx - x}{x^2 \sin hx} = \lim_{x\to 0} \frac{\cos hx - 1}{2x \sin hx + x^2 \cos hx}$$

$$= \lim_{x\to 0} \frac{\sin hx}{2 \sin hx + 4x \cos hx + x^2 \sin hx}$$

$$= \lim_{x \to 0} \frac{\cos hx}{\{(2+x^2)\cos hx + 2x \sin hx\} + 4(x \sin hx + \cos hx)}$$

$$= \frac{1}{(2+0).1+0+4(0+1)} = \frac{1}{6}.$$

Thus, 0 is not a point of infinite discontinuity and so $\int_0^1 f(x)\, dx$ exists.

Now, we discuss convergence at ∞ only.

Let $g(x) = \dfrac{1}{x^2}$. $\quad \therefore \dfrac{f(x)}{g(x)} = \dfrac{\sin hx - x}{\sin hx}$

$$\lim_{x \to \infty} \frac{f(x)}{g(x)} = \lim_{x \to \infty} \frac{\sin hx - x}{\sin hx} = \lim_{x \to \infty} \frac{\cos hx - 1}{\cos hx}$$

$$= \lim_{x \to \infty} \frac{\sin hx}{\sin hx} = 1$$

So, $\int_1^\infty f(x)\, dx$ and $\int_1^\infty g(x)\, dx$ converge or diverge together.

Since, $\int_1^\infty f(x)\, dx = \int_1^\infty \dfrac{g(x)}{x^2}$ is convergent, so $\int_1^\infty f(x)\, dx$ is convergent.

Hence, $\int_0^\infty f(x)\, dx = \int_0^1 f(x)\, dx + \int_1^\infty f(x)\, dx$ is convergent.

Example 24:

Discuss the convergence of $\displaystyle\int_0^\infty \frac{x \log x}{(1+x^2)^2}\, dx$.

Solution:

We have

$$\int_0^\infty f(x)\, dx = \int_0^1 f(x)\, dx + \int_1^\infty f(x)\, dx,\ f(x) = \frac{x \log x}{(1+x^2)^2}$$

Putting $x = 1/t \Rightarrow dx = -dt/t^2$, we obtain

$$\int_1^\infty \frac{x \log x}{(1+x^2)^2} dx = -\int_1^0 \frac{t^4 . \log(1/t) dt}{t^3(1+t^2)^2} = \int_0^1 \frac{t \log t}{(1+t^2)^2} dt$$

$$\therefore \int_0^\infty \frac{x \log x\, dx}{(1+x^2)^2} 2 \int_0^1 \frac{x \log x\, dx}{(1+x^2)^2}$$

$$= 2 \lim_{\varepsilon \to 0+} \int_{\varepsilon}^{1} \frac{x \log x \, dx}{(1+x^2)^2} = 0$$

Since $\lim_{\varepsilon \to 0+} (x \log x) = 0.$

Example 25:

Show that the integral $\int_0^{\pi/2} \frac{\sin^m x}{x^n} dx$ *exists iff* $n < m + 1$.

Solution:

Let $f(x) = \frac{\sin^m x}{x^n} = \frac{1}{x^{n-m}} \left(\frac{\sin x}{x} \right)^m$

Now, $x \to 0+ \Rightarrow \begin{cases} f(x) \to 1 \text{ if } n-m=0, \\ f(x) \to 0 \text{ if } n-m<0, \\ f(x) \to \infty \text{ if } n-m>0. \end{cases}$

Thus, the given integral is proper if $n \le m$ and improper if $n > m$.

Let $g(x) = \frac{1}{x^{n=m}} (n > m)$. Then

Let $f(x) = \frac{\sin^m x}{x^n} = \frac{1}{x^{n-m}} \left(\frac{\sin x}{x} \right)^m$

Also, $\int_0^{\pi/2} g(x)\, dx = \int_0^{\pi/2} \frac{d(x)}{x^{n-m}}$ is convergent iff $n - m < 1$.

Hence, $\int_0^{\pi/2} \frac{\sin^m x}{x^n} dx$ converges iff $n < m + 1$.

Example 26:

Show that $\int_0^{\pi/2} \sin x \log \sin x \, dx$ *is convergent with value* $\log (2/e)$.

Solution:

Integrating by parts,

$$\int_{\varepsilon}^{\pi/2} \sin x \log \sin x \, dx = \left| -\cos x \log \sin x \right|_{\varepsilon}^{\pi/2} + \int_{\varepsilon}^{\pi/2} \frac{\cos^2 x}{\sin x} dx$$

$$= \cos \varepsilon \log \sin \varepsilon + \int_{\varepsilon}^{\pi/2} (\operatorname{cosec} x - \sin x)\, dx$$

$$= \cos \varepsilon \log \sin \varepsilon + \left| \log \tan \frac{1}{2} x + \cos x \right|_{\varepsilon}^{\pi/2}$$

$= \cos \varepsilon \log \sin \varepsilon + (0 - \log \tan \frac{1}{2}\varepsilon) - \cos \varepsilon$

$= \cos \varepsilon \log 2 \sin \frac{1}{2} \varepsilon \cos \frac{1}{2} \varepsilon - (\log \sin \frac{1}{2} \varepsilon - \log \cos \frac{1}{2} \varepsilon) - \cos \varepsilon$

$= \cos \varepsilon (\log 2 \cos \frac{1}{2} \varepsilon + \log \sin \frac{1}{2}\varepsilon) - \log \sin \frac{1}{2} \varepsilon + \log \cos \frac{1}{2} \varepsilon - \cos \varepsilon$

$= -(1 - \cos \varepsilon) \log \sin \frac{1}{2} \varepsilon + \cos \varepsilon \log 2 \cos \frac{1}{2} \varepsilon - \cos \varepsilon - \cos \varepsilon + \log \cos \frac{1}{2} \varepsilon$ . . . (1)

As $\varepsilon \to 0$, $\lim (1 - \cos \varepsilon) \log \sin \frac{1}{2} \varepsilon$

$= \lim (2 \sin^2 \frac{1}{2} \varepsilon \log \sin \frac{1}{2} \varepsilon) = \tan (2t^2 \log)$,

$= 0$, when $t \to 0$ $\quad (t = \sin \frac{1}{2} \varepsilon)$

Again $\log \cos \frac{1}{2} \varepsilon \to 0$ as $\varepsilon \to 0$, From (1), we get

$$\int_0^{\pi/2} \sin x \log \sin x \, dx = 0 + \log 2 - 1 + 0 = \log 2 - \log e.$$

Example 27(a):

Test the convergence of $\int_0^2 \frac{\log x \, dx}{\sqrt{2-x}}$,

Solution:

It is clear that 0 and 2 are the only points of infinite discontinuity.

We have $\int_0^1 f dx = \int_0^1 f dx + \int_1^2 f dx$.

Convergence at 0:

Let $0 < p < 1$. Then

$$\lim_{x \to 0} \left[\frac{-x^p \log x}{\sqrt{2-x}} \right] = 0.$$

So for any $\varepsilon > 0$, there exists some $\delta < 0$ such that

$$\left| \frac{-x^p \log x}{\sqrt{2-x}} - 0 \right| < \varepsilon. \text{ when } 0 < | x - 0 | < \delta.$$

$$\Rightarrow \frac{x^p \log x}{\sqrt{2-x}} < \varepsilon \qquad \Rightarrow \frac{\log x}{\sqrt{2-x}} < \frac{3}{x^p}$$

$$\therefore \int_0^1 \frac{\log x \, dx}{\sqrt{2-x}} < \int_0^1 \frac{\varepsilon}{x^p} dx (p<1)$$

By comparison test,

$$\int_0^1 \frac{\log x}{\sqrt{2-x}} dx \text{ is convergent.}$$

Let $f(x) = \dfrac{\log x}{\sqrt{2-x}}, g(x) = \dfrac{1}{\sqrt{2-x}}$

$$\therefore \lim_{\varepsilon-2} \frac{f(x)}{g(x)} = \lim_{x\to 2} \log x = \log 2.$$

Now, $\int_1^2 g(x)\, dx = \int_1^2 \dfrac{dx}{(2-x)^{1/2}}$ converges, so

$$\int_1^2 f(x)\, dx \text{ converges}$$

Hence, $\int_1^2 f(x)\, dx$ converges.

Ex.:

Discuss the convergence of the integral

$$\int_0^1 x^{m-1}(1-x)^{n-1} \log\left(\frac{1}{x}\right) dx.$$

Example 27(b):

Show that $\int_0^\infty e^{-x^2} dx$ *converges.*

Solution:

We see that 0 is not a point of infinite discontinuity and so we shall examine convergence at ∞ only.

We have $\int_0^\infty e^{-x^2} dx = \int_0^1 e^{-x^2} dx + \int_1^\infty e^{-x^2} dx$,

where the first integral on the R.H.S. is propers.

We know the for all real x, $e^2 > x^2 \Rightarrow e^{-x^2} < 1/x^2$.

Since $\int_0^\infty \frac{1}{x^2} dx$ converges at ∞. so $\int_1^\infty e^{-x^2} dx\ dx$ converges.

Hence, $\int_0^\infty e^{-x^2} dx$ is convergent.

$= \log (2/e)$.

Example 28:

Discuss the convergence of

$$\int_2^\infty \frac{\cos x}{\log x}\,dx$$

Solution:

Let $\phi(x) = 1/\log x$ so that $\phi(x)$ is a monotonically decreasing function for $x \geq 2$ and further $\phi(x) \to 0$ as $x \to \infty$. Now

$$\left|\int_2^t \cos x\,dx\right| = |\sin t - \sin 2| \leq 1 + 1 = 2$$

$\therefore \int_2^t \cos x\,dx$ is bounded $\forall\, t \geq 2$.

Hence, by Dirichlet's test,

$\int_2^\infty \frac{\cos x}{\log x}\,dx$ is convergent.

Ex.: *Examine the convergence of* $\int_2^\infty \frac{\sin x}{\log x}\,dx$.

Ex.: *Show that* $\int_0^\infty e^{-ax}$ cos dx *is convergent.*

Example 29:

Show that $\int_1^\infty \frac{\sin x}{x^p}\,dx$ *converges absolutely if p > 1.*

Solution:

We have $\left|\frac{\sin x}{x^p}\right| = \frac{|\sin x|}{x^p} \leq \frac{1}{x^p} \qquad \forall\, x \leq 1.$

Since $\int_1^\infty \frac{dx}{x^p}$ converges if $p > 1$. So by comparison test,

$\int_1^\infty \left|\frac{\sin x}{x^p}\right|$ converges if $p > 1$.

Hence $\int_1^\infty \frac{\sin x}{x^p}$ converges absolutely if $p > 1$.

Note : For $p = 1$, the integral is not absolutely convergent.

Example 30:

Show that $\int_1^\infty \frac{\sin x}{x^p}\,dx$ *is convergent for p > 0.*

Solution:

For $p > 0$, $1/x^p$ is monotonically decreasing for $x > 1$ and $\to 0$ as $x \to \infty$.

Also $\left|\int_1^t \sin x\, dx\right| = |\cos 1 - \cos t| \le |\cos 1| + |\cos t| \le 2 \ \forall\ t \ge 1.$

$\therefore \int_1^t \sin x\, dx$ is bounded " $t \ge 1$.

Hence, by Dirichlet's test,

$$\int_1^\infty \frac{\sin x}{x^p}\, dx \text{ is convergent for } p > 0.$$

Example 31:

Test the convergence of

$$\int_0^\infty e^{-a^2x^2} \cos bx\, dx.$$

Solution:

Let $\phi(x) = e^{-a^2x^2}$ so that ϕ is monotonically decreasing for $x \ge 0$. Further $\phi(x) \to 0$ as $x \to \infty$.

Now, $\left|\int_0^t \cos bx\, dx\right| = \left|\frac{1}{b}\sin bt\right| \le \left|\frac{1}{b}\right| \ \forall\ t \ge 0.$

Thus, $\int_0^t$ cos cos bx dx is bounded $\forall\ t \ge 0$.

Hence by Dirichelt's test $\int_0^\infty e^{-a^2x^2} \cos bx\, dx$ is convergent.

Example 32:

Show that $\int_a^\infty \frac{\sin x}{x}$ *is convergent.*

Solution:

Since $\lim\limits_{x\to 0} \frac{\sin x}{x} = 1$, so 0 is not a point of infinite discontinuity. We write.

$$\int_a^\infty \frac{\sin x}{x}\, dx = \int_0^1 \frac{\sin x}{x}\, dx + \int_1^\infty \frac{\sin x}{x}\, dx.$$

It is clear that $\int_0^1 \frac{\sin x}{x}\, dx$ is a proper integral.

Convergence at ∞:

Let $\varepsilon > 0$ be given. Let $X_0 = 2/\varepsilon$.

Let $X_1, X_2 > X_0$.

Consider $\int_{X_1}^{X_2} \frac{\sin x}{x} dx = \left| -\cos x. \frac{1}{x} \right|_{X_1}^{X_2} - \int_{X_1}^{X_2} \frac{\cos x}{x^2} dx$

(Integrate by parts)

$$\therefore \left| \int_{X_1}^{X_2} \frac{\sin x}{x} dx \right| \leq \left| \frac{\cos X_1}{X_1} = \frac{\cos X_2}{X_2} \right| + \left| \int_{X_1}^{X_2} \frac{\cos x}{x^2} dx \right|$$

$$\leq \frac{|\cos X_1|}{X_1} + \frac{|\cos X_2|}{X_2} + \int_{X_1}^{X_2} \frac{|\cos x|}{x^2} dx$$

$$\leq \frac{1}{X_1} + \frac{1}{X_2} + \int_{X_1}^{X_2} \frac{1}{x^2} dx \qquad [\because |\cos \theta| \leq 1 \ \forall \ \theta \in R]$$

$$= \frac{1}{X_1} + \frac{1}{X_2} = \left| \frac{1}{x} \right|_{X_1}^{X_2}$$

$$= \frac{1}{X_1} + \frac{1}{X_2} - \frac{1}{X_2} + \frac{1}{X1} = \frac{2}{X_1}$$

$$< \frac{2}{X_0} = \varepsilon \left(\because X_1 > X_2 \text{ and } X_0 = \frac{2}{\varepsilon} \right)$$

Hence by Cauchy's Test, $\int_1^{\infty} \frac{\sin x}{x} dx$ is convergent and so $\int_0^{\infty} \frac{\sin x}{x} dx$ is convergent.

Example 33(a):

Test for convergence $\int_1^{\infty} \sin x^p \ dx$.

Solution:

We can write

$$\int_1^{\infty} \sin x^p \ dx = \int_1^{\infty} \frac{1}{px^{p-1}} \sin x^p \ dx.$$

Now for $p > 1$, $1/px^{p-1}$ is monotonically decreasing for $x \leq 1$ and $\rightarrow 0$ as $x \rightarrow \infty$. Also

$$\left| \int_1^t px^{p-1} \sin x^p \ dx \right| = |-\cos t^p + \cos 1| < 1 + 1 = 2.$$

Thus, $\int_1^t px^{p-1} \sin x^p \ dx$ is bounded $\forall \ t \geq 1$.

Hence, by Dirichlet's test,

$$\int_1^{\infty} \frac{1}{px^{p-1}} px^{p-1} \sin x^p \, dx = \int_1^{\infty} \sin x^p \, dx \text{ is convergent if } p > 1.$$

Example 33(b):

Show that $\int_0^{\infty} \frac{\sin x}{x} dx$ *is convergent.*

Solution:

$$\int_0^{\infty} \frac{\sin x}{x} dx = \int_0^{1} \frac{\sin x}{x} dx + \int_1^{\infty} \frac{\sin x}{x} dx$$

Since, $\lim_{x \to 0} \frac{\sin x}{x} = 1$, so 0 is not a point of infinite discontinuity.

Thus, $\int_0^{1} \frac{\sin x}{x} dx$ is a proper integral.

Now, we consider $\int_1^{\infty} \frac{\sin x}{x} dx$. Let $\phi(x) = \frac{1}{x}$.

Then $\phi(x)$ is bounded and monotonically decreasing for $x \geq 1$ and $\phi(x) \to 0$ as $x \to \infty$.

Now, $\left| \int_1^{t} \sin x \, dx \right| = | - \cos t + \cos 1 | < | \cos t | + | \cos 1 |$

$\leq 1 + 1 = 2 \; \forall \; t \geq 1.$

Thus, $\int_1^{t} \sin x \, dx$ is bounded $\forall \; t \geq 1$.

By Dirichlet's test, $\int_1^{\infty} f \phi \, dx = \int_1^{\infty} \frac{\sin x}{x} dx$ is convergent.

Hence, $\int_0^{\infty} \frac{\sin x}{x} dx$ is convergent.

Ex.: *Test the following for convergence :*

(i) $\int_0^{\infty} \frac{\sin x}{\sqrt{x}} dx$, (ii) $\int_0^{\infty} \frac{\sin kx}{x} dx$, (iii) $\int_1^{\infty} \frac{\sin x}{x^2} dx$.

Solution:

Do your self.

Example 34:

Show that $\int_1^{\infty} e^{-ax} \frac{\sin x}{x}$ *dx (a ≥ 0) is convergent.*

Solution:

For $a \geq 0$, e^{-ax} is monotonically decreasing and bounded $\forall\ x \geq 0$.

Also $\int_1^{\infty} (\sin x/x)\ dx$ is convergent .

Hence, by Abel's Test, $\int_1^{\infty} e^{-ax}(\sin x/x)\ dx$ is convergent for $a \geq 0$.

Example 35:

Test for convergence the integral

$$\int_0^{\infty} \sin x^2\ dx.$$

Solution:

$$\int_0^{\infty} \sin x^2\ dx = \int_0^{1} \sin x^2\ dx + \int_0^{\infty} \sin x^2\ dx$$

The first integral on the R.H.S. is a proper integral.

Now $\int_1^{\infty} \sin x^2\ dx = \int_1^{\infty} \frac{1}{2x}.\, 2x \sin x^2\ dx.$

Clearly, $\phi(x) = \frac{1}{2x}$ is bounded and monotonically decreasing for $x \geq 1$ and $\phi(x) \to 0$ as $x \to \infty$.

Now, $\left|\int_1^{t} 2x \sin x^2\ dx\right| = \left|-\cos x^2\right|_1^t = |-\cos t^2 + \cos 1| \leq |\cot t^2|$

$+ |\cos 1| \leq 1 + 1 = 2.$

Thus, $\int_1^{t} 2x \sin x^2\ dx$ is bounded $\forall\ t \geq 1$.

By Dirichlet's Test, $\int_1^{\infty} f\phi\ dx = \int_1^{\infty} \frac{1}{2x}.\, 2x \sin x^2\ dx$

$= \int_1^{\infty} \sin x^2\ dx$ is convergent.

Hence, $\int_0^{\infty} \sin x^2\ dx$ is convergent.

Example 36:

Test for convergence $\int_0^{\infty} \frac{\sin (x + x^2)}{x^n}\ dx.$

Solution:

$$\int_0^{\infty} f(x)\ dx = \int_0^{1} f(x)\ dx + \int_0^{\infty} f(x)\ dx, \text{ where}$$

$$f(x) = \frac{\sin(x + x^2)}{x^n}.$$

Now $\int_0^1 \frac{\sin(x+x^2)}{x^n}\,dx \le \int_0^1 \frac{x + x^2}{x^n}\,dx$

$$= \int_0^1 \frac{dx}{x^{n-1}} + \int_0^1 \frac{dx}{x^{x-2}}$$

We know $= \int_0^1 \frac{dx}{x^{n-1}}$ and $\int_0^1 \frac{dx}{x^{n-2}}$ converge if $n - 1 < 1$ and $n - 2 < 1$

$\Rightarrow n < 2$ and $n < 3$ respectively.

Thus, both these integrals converge if $n < 2$.

Hence, $\int_0^1 f(x)\,dx$ is convergent if $n < 2$.

Now, we consider $\int_1^\infty \frac{\sin(x+x^2)}{x^n}\,dx.$

Let $g(x) = (1 + 2x)\sin(x + x^2)$, $\phi(x) = \frac{1}{(1 + 2x)x^n}$

For $n > -1$, the function $\phi(x)$ is monotonically decreasing $\forall\ x \ge 1$ and tends to 0 as $x \to \infty$.

Also $\left|\int_1^t g(x)\,dx\right| = |\cos 2 - \cos(t+t^2)|$

$$\le 1 + 1 = 2.$$

Thus, $\int_1^t g(x)\,dx$ is bounded $\forall\ t \ge 1$. Hence, by Dirichlet's Test,

$\int_1^\infty g(x)\,\phi(x)\,dx = \int_1^\infty g(x)\,dx$ is convergent for $n > -1$.

Consequently the given integral is convergent if $-1 < n < 2$.

Example 37:

Examine the convergence of

$$\int_0^\infty \frac{\cos x}{\sqrt{x+x^2}}\,dx.$$

Solution:

$$\int_0^\infty \frac{\cos x}{\sqrt{x+x^2}}\,dx = \int_0^1 \frac{\cos x}{\sqrt{x+x^2}}\,dx + \int_1^\infty \frac{\cos x}{\sqrt{x+x^2}\,dx}$$

Convergence at 0:

Let $f(x) = \dfrac{\cos x}{\sqrt{x}\sqrt{1+x}}$ and $g(x) = \dfrac{1}{\sqrt{x}}$.

$\therefore \lim\limits_{x\to 0} \dfrac{f(x)}{g(x)} = \lim\limits_{x\to 0} \dfrac{\cos x}{\sqrt{1+x}} = 1$

Since $\int_0^1 g\, dx = \int_0^1 \dfrac{1}{x^{1/2}}\, dx$ is convergent,

so $\int_0^1 f\, dx$ is convergent.

Convergence at ∞:

Let $\phi(x) = \dfrac{1}{\sqrt{x+x^2}}$. Then $\phi(x)$ is bounded and monotonically decreasing for $x \geq 1$. Also $\phi(x) \to 0$ as $x \to \infty$.

Now, $\left|\int_1^X \cos x\, dx\right| = |\sin X - \sin 1| \leq |\sin X| + |\sin 1| \leq 1 + 1 = 2\ \forall\ X \geq 1$.

Thus $\int_1^X \cos x\, dx$ is bounded $\forall\ X \geq 1$.

By Dirichlet's Test,

$$\int_1^\infty \cos \theta(x)\, dx = \int_1^\infty \frac{\cos x}{\sqrt{x+x^2}} \text{ is convergent.}$$

Hence, $\displaystyle\int_0^\infty \frac{\cos x}{\sqrt{x+x^2}}$ is convergent.

Example 38(a):

Show that $\displaystyle\int_0^\infty \frac{x\, dx}{1+x^6 \sin^2 x}$ is *convergent.*

Solution:

We write

$$\int_0^{n\pi} \frac{x\, dx}{1+x^6 \sin^2 x} = \sum_{r=0}^{n} \int_{(r-1)\pi}^{r\pi} \frac{x\, dx}{1+x^6 \sin^2 x}$$

Now $\forall\ x \in [(r-1)\pi, r\pi]$, we have

$$\frac{x}{1+x^6 \sin^2 x} \leq \frac{r\pi}{1+(r-1^6\pi^6 \sin^2 x}$$

$$\therefore \int_{(r-1)\pi}^{r\pi} \frac{x\,dx}{1+x^6 \sin^2 x} \int_{(r-1)\pi}^{r\pi} \frac{r\pi\,dx}{1+(r-1)^6\pi^6 \sin^2 x}$$

Putting $x = (r - 1)\pi + y$, we see that

$$\int_0^{\pi} \frac{r\pi\,dy}{1+(r-1)^6\pi^6 \sin^2 y} = 2\int_0^{r\pi} \frac{r\pi\,dx}{1+(r-1)^6\pi^6 \sin^2 y}$$

$$= 2r\pi \int_0^{\pi/2} \frac{dy}{1+A \sin^2 y} = 2r\pi \int_0^{r\pi} \frac{\operatorname{cosec}^2 y\,dy}{A+1+\cot^2 y}$$

where $A = (r - 1)^6\pi^6$, $\operatorname{cosec}^2 y = 1 + \cos^2 y$

$$\frac{2r\pi}{\sqrt{(A+1)}} \left| \tan^{-1} \frac{\cot y}{\sqrt{(A+1)}} \right|_0^{\pi/2} = \frac{\pi}{2} . \frac{2r\pi}{\sqrt{(A+1)}}$$

$$= \frac{r\pi^2}{\sqrt{[(r-1)^6\pi^6+1]}}$$

$$\therefore \int_{(r-1)\pi}^{r\pi} \frac{x}{1+x^6 \sin^2 x}\,dx \le \frac{r\pi^2}{\sqrt{[1+(r-1)^6\pi^6]}}$$

$$\le \frac{r}{(r-1)^2}\frac{1}{\pi}\ (r \ne 1)$$

Since $\frac{r}{(r-1)^2} = \frac{1}{(r-1)^3} + \frac{1}{(r-1)^3}$, and

$$\sum_{r=2}^{\infty} \frac{1}{(r-1)^2} \text{ and } \sum_{r=2}^{\infty} \frac{1}{(r-1)^3}$$

are both convergent, it follows that

$\sum_{r=2}^{\infty} \frac{r}{(r-1)^3} . \frac{1}{\pi}$ is a convergence series.

Since $\lim_{n\to\infty} \int_0^{n\pi} \frac{x\,dx}{1+x^6 \sin^2 x} \le \lim_{n\to\infty} \sum_{r=2}^{\infty} \frac{r}{(r-1)^3} . \frac{1}{\pi}$,

so $\int_0^{\infty} \frac{x\,dx}{1+x^6 \sin^2 x}$ is convergent.

Example 38(b):

Show that

$$\int_0^\infty \frac{dx}{1+x^4\cos^2 x} \text{ is } \textit{convergent.}$$

Solution:

We have

$$\int_0^{n\pi} \frac{dx}{1+x^4\cos^2 x} = \sum_{r-1}^{n} \int_{(r-1)\pi}^{r\pi} \frac{dx}{1+x^4\cos^2 x}$$

Now $\int_{(r-1)\pi}^{r\pi} \frac{dx}{1+x^4\cos^2 x} \le \int_{(r-1)\pi}^{r\pi} \frac{dx}{1+(r-1)^4\pi^4\cos^2 x}$

$$\forall\ x \in [r-1)\pi,\ r\pi]$$

$$\le \int_0^{\pi} \frac{dy}{1+(r-1)^4\pi^4\cos^2 y}, x=(r-1)\pi+y$$

$$=2\int_0^{\pi/2} \frac{\sec^2 y\, dy}{1+\tan^2 y+(r-1)^4\pi^4}$$

$$=\frac{2}{\sqrt{[1+(r-1)^4\pi^4]}}\left|\tan^{-1}\frac{\tan y}{\sqrt{[1+(r-1)^4\pi^4]}}\right|_0^{\pi/2}$$

$$=\frac{\pi}{\sqrt{[1+(r-1)^4\pi^4]}}$$

$$\lim_{x\to\infty}\int_0^{n\pi}\frac{dx}{1+x^4\cos^2 x} \le \lim_{x\to\infty}\sum_{x}^{n}\frac{\pi}{\sqrt{[1+(r-1)^4\pi^4]}}$$

But $\sum_{r=1}^{n}\frac{\pi}{\sqrt{[a+(r-1)^4\pi^4]}}\left(\sim\sum_{r=1}^{\infty}\frac{1}{r^2}\right)$ is a convergent series, so

$$\int_0^\infty \frac{dx}{1+x^4\cos^2 x} \text{ is convergent.}$$

Example 39:

The function f is defined on [0, ∞[by

$f(x) = (-1)^{n-1}$, $n-1 < x < n$, $n \in N$.

Show that $\int_0^\infty f(x)\,dx$ *does not converge.*

Solution:

We have

$$\int_0^{2n} f(x)\,dx = \int_0^1 (-1)^0\,dx + \int_1^2 (-1)\,dx + \int_2^3 (-1)^2\,dx + ...$$

$$+ \int_{2n-1}^{2n} (-1)^{2n-1}\,dx$$

$$= 1 - 1 + 1 - 1 + ... + 1 - 1 = 0,$$

$$\int_0^{2n+1} f(x)\,dx = \int_0^1 + \int_1^2 + ... \int_{2n-1}^{2n} .....+ \int_{2n}^{2n+1} (-1)^{2n}\,dx(-1)^{2n}\,dx$$

$$= 0 + 1 = 1.$$

Thus, $\lim_{n\to\infty} \int_0^{2n} f(x)\,dx = 0$ and $\lim_{n\to\infty} \int_0^{2n+1} f(x)\,dx = 1.$

Hence, $\int_0^\infty f(x)\,dx$ does not exist and so it is not convergent.

Example 40:

Show that $\int_0^\infty \frac{x\,dx}{1+x^4\cos^2 x}$ *is divergent.*

Solution:

$$\int_{(r-1)\pi}^{r\pi} \frac{(r-1)\pi}{1+(r\pi)^4\cos^2 x}\,dx$$

$$= 2(r-1)\pi \int_0^{\pi/2} \frac{dy}{1+(r\pi)^4\cos^2 y}$$

$$= 2(r-1)\pi \int_0^{\pi/2} \frac{\sec^2 y}{1+\tan^4 y+(r\pi)^4} (t = \tan y)$$

$$= \frac{2(r-1)\pi}{\sqrt{[1+(r\pi)^4]}} \tan^{-1} \left| \frac{t}{\sqrt{[1+(r\pi)^4]}} \right|_0^\infty = \frac{2(r-1)\pi}{\sqrt{[1+(r\pi)^4]}} \cdot \frac{\pi}{2}$$

$$= \frac{(r-1)\pi^2}{\sqrt{[1+(r\pi)^4]}}$$

Since $\sum_{r=1}^{\infty} \frac{(r-1)\pi^2}{\sqrt{[1+(r\pi)^4]}}$ diverges $\left(\sim \sum \frac{1}{r}\right)$

so the given integral is divergent.

Example 41(a):

Show that the improper integral

$$\int_0^\infty \frac{\sin x}{x}\, dx \text{ is not absolutely convergent.}$$

Solution:

We have to show that $\int_0^\infty \frac{|\sin x|}{x} dx$ is not convergent. Consider the proper integral

$$\int_0^{n\pi} \frac{|\sin x|\, dx}{x}, \text{ (n being a positive integer)}$$

We have $\int_0^{n\pi} \frac{|\sin x|}{x} dx = \sum_{r=1}^{n} \int_{(r-1)\pi}^{r\pi} \frac{+\sin x|}{x} dx$ ...(1)

Let $x = (r - 1)\pi + y$ so that y varies in $[0, \pi]$. We have

$|\sin [(r - 1)\pi + y]| = |(-1)^{r-1} \sin y| = \sin y.$

$$\therefore \int_{(r-1)\pi}^{r\pi} \frac{|\sin x|}{x} dx = \int_0^{\pi} \frac{\sin y}{(r-1)\pi + y} dx \quad ...(2)$$

Since $r\pi$ is the maximum value of $[(r - 1)\pi + y]$ when $y \in [0, \pi]$,

$$\therefore \int_0^{\pi} \frac{\sin y}{(r-1)\pi + y} dy \geq \frac{1}{r\pi} \int_0^{\pi} \sin y\, dy = \frac{2}{r\pi} \quad ...(3)$$

From (1), (2), (3); we obtain

$$\int_0^{n\pi} \frac{|\sin x|}{x} dx \geq \sum_1^n \frac{2}{r\pi} = \frac{2}{\pi} \sum_1^n \frac{1}{r}$$

Since $\sum_1^n \frac{1}{r} \to \infty$ as $n \to \infty$, therefore

$$\int_0^{n\pi} \frac{|\sin x|}{x} dx \to \infty \text{ as } n \to \infty. \quad ...(4)$$

Let t be any real number. There exists a positive integer n such that $n\pi \geq t < (n + 1)\pi$.

We have

$\int_0^{t} \frac{|\sin x|}{x} dx \geq \int_0^{n\pi} \frac{|\sin x|}{x} dx$ Using (4), we get

$$\lim_{t\to\infty} \int_0^t \frac{|\sin x|}{x} dx \to \infty \; [\because t \to \infty \Rightarrow n \to \infty]$$

Hence, $\int_0^\infty \frac{|\sin x|}{x} dx$ is not convergent.

Example 41(b):

Show that $\int_0^{\pi/2} \sin x \log \sin x \, dx$ *is convergent and its value is log (2/e).*

Solution:

Here the integrand has an infinite discontinuity at x = 0 in (0, π/2.)

Let $\phi(x) = \sin x. \log (\sin x)$

$$\lim_{x\to 0} f(x) = \lim_{x\to 0}\left[\frac{\log \sin x}{\operatorname{cosec} x}\right] \quad \left(\text{Form } \frac{\infty}{\infty}\right)$$

$$= \lim_{x\to 0}\left[\frac{\cot x}{-\operatorname{cosec} x \cot x}\right]$$

$$= \lim_{x\to 0} [-\sin x] = 0$$

Hence, $\int_0^{\pi/2} \sin x \log \sin x \, dx$ is convergent.

$$\int_0^{\pi/2} \sin \; \sin x \log (\sin x) \, dx$$

$$= \lim_{\epsilon\to 0} \int_\epsilon^{\pi/2} \sin \; x \log \sin x \, dx$$

$$= \lim_{\epsilon\to 0}\left[\{\log \sin x \, (-\cos x)\}_\epsilon^{\pi/2} + \int_\epsilon^{\pi/2} \cot x \cos x \, dx\right]$$

$$= \lim_{\epsilon\to 0}\left[-\cos x \log \sin x + \log \tan \frac{x}{2} + \cos x\right]_\epsilon^{\pi/2}$$

$$\text{for } \int_\epsilon^{\pi/2} \cot x \cos x \, dx = \int_\epsilon^{\pi/2} (\operatorname{cosec} x - \sin x) \, dx$$

$$= \lim_{\epsilon\to 0}\left[\cos \epsilon \log \sin \epsilon - \log \tan \frac{\epsilon}{2} - \cos \epsilon\right]$$

$$= \lim_{\epsilon\to 0}\left[-(1-\cos \epsilon) \log \sin \frac{\epsilon}{2} + \cos \epsilon \log \left(2 \cos \frac{\epsilon}{2}\right) - \cos \epsilon + \log \cos \frac{\epsilon}{2}\right]$$

But $\lim_{\epsilon\to 0}\left[(1-\cos\epsilon)\log\sin\frac{\epsilon}{2}\right]$

$$= \lim_{\epsilon\to 0}\left[2\sin^2\frac{\epsilon}{2}\log\sin\frac{\epsilon}{2}\right]$$

$$= \lim_{l\to 0}[2\,l^2\log l] \qquad \text{where } l = \sin\varepsilon/2$$

$= 0.$

Hence, $\int_0^{\pi/2}\sin x\log\sin x\,dx = \log_e(2) - 1\log(2) - \log_e(e)$

$$= \log_e\left(\frac{2}{e}\right).$$

Example 41(c):

Show that $\int_0^{\infty}\frac{x\,dx}{1+x^4\sin^2 x}$ *is divergent.*

Solution:

We write

$$\int_0^{n\pi}\frac{x\,dx}{1+x^4\sin^2 x} = \sum_{r=1}^{n}\int_{(r-1)\pi}^{r\pi}\frac{x\,dx}{1+x^4\sin^2 x}$$

Now $\forall\ x \in [(r-1)\pi, r\pi]$, we have

$$\frac{(r-1)\pi}{1+(r\pi)^4\sin^2 x} \le \frac{x}{1+x^4\sin^2 x}$$

$$\therefore\ \int_{(r-1)\pi}^{r\pi}\frac{(r-1)\pi}{1+(r\pi)^4\sin^2 x}\,dx \quad \int_{(r-1)\pi}^{r\pi}\frac{x}{1+x^4\sin^2 x}\,dx$$

Let $x = (r-1)\pi + y$ so that y varies in $[0, \pi]$. Then

$$\int_{(r-1)\pi}^{r\pi}\frac{(r-1)\pi}{1+(r\pi)^4\sin^2 x}\,dx = (r-1)\pi\int_0^{\pi}\frac{dy}{1+(r\pi)^4\sin^2 y}$$

$$= 2(r-1)\pi\int_0^{\pi/2}\frac{dy}{1+(r\pi)^4\sin^2 y} \qquad \text{...(1)}$$

$$= 2(r-1)\pi\int_0^{\pi/2}\frac{\sec^2 y\,dy}{\sec^2 y+(r\pi)^4\tan^2 y}$$

$$= 2(r-1)\pi \int_0^{\infty} \frac{dt}{1+(1+r\pi^4)t^2}, \; t = \tan y$$

$$= \frac{2(r-1)\pi}{\sqrt{[1+(r\pi)^4]}} \left| \tan^{-1} \sqrt{[1+(r\pi^4)]} \right|_0^{\infty}$$

$$= 2(r-1)\pi \frac{\pi}{2} \frac{1}{\sqrt{[1+(r\pi)^4]}} = \frac{(r-1)\pi^2}{\sqrt{[1+(r\pi)^4]}}$$

$$\therefore \lim_{n\to\infty} \int_0^{n\pi} \frac{x\,dx}{1+x^4 \sin^2 x} \geq \lim_{n\to\infty} \sum_{r=1}^{n} \frac{(r-1)\pi^2}{\sqrt{[1+(r\pi)^4]}}$$

The infinite series $\sum \frac{(r-1)\pi^2}{\sqrt{[1+(r\pi)^4]}}$ diverges $\left(\sim \sum \frac{1}{r}\right)$.

Hence the given integral is divergent.

Example 42:

Test the convergence of the improper integral

$$\int_0^{\pi} \frac{dx}{\sin x}.$$

Solution:

Here the integrand becomes infinite at $x = 0$ and at $x = \pi$. We have

$$\int_0^{\pi} \frac{dx}{\sin x} = \int_0^{\pi/2} \frac{dx}{\sin x} + \int_{\pi/2}^{\pi} \frac{dx}{\sin x}.$$

The integral $\int_0^{\pi} \frac{dx}{\sin x}$ will exist if $\int_0^{\pi/2} \frac{dx}{\sin x}$ converges at zero and $\int_{\pi/2}^{\pi} \frac{dx}{\sin x}$ converges at π.

Thus, $\int_0^{\pi} \frac{dx}{\sin x} = \lim_{\epsilon\to 0} \int_{\epsilon}^{\pi/2} \frac{dx}{\sin x} \lim_{\epsilon\to 0} \int_{\pi/2}^{\pi} \frac{dx}{\sin x}$

$$= \lim_{\epsilon\to 0} \left[\log \tan \frac{x}{2} \right]_{\epsilon}^{\pi/2} + \lim_{\epsilon\to 0} \left[\log \tan \frac{x}{2} \right]_{\pi/2}^{\pi-\epsilon}$$

$$= \lim_{\epsilon\to 0} \left[\log \tan \frac{\epsilon}{2} \right] + \lim_{\epsilon\to 0} \left[\log \tan \left(\frac{\pi}{2} - \frac{\epsilon}{2} \right) \right]$$

$$= \lim_{\epsilon\to 0} \left[\log \cot \frac{\epsilon}{2} \right] + \lim_{\epsilon\to 0} \log \cot \left(\frac{\epsilon}{2} \right)$$

Since R.H.S. tends to infinity therefore general value of $\int_0^\pi \frac{dx}{\sin x}$ is infinite.

But if we put $\varepsilon = \varepsilon'$

$$P \int_0^\pi \frac{dx}{\sin x} = \lim_{\epsilon \to 0} \log\left[\cot\frac{\epsilon}{2}\cot\frac{\epsilon}{2}\right] = \infty.$$

Hence, the given integral is not convergent.

Example 43:

Test for convergence

$$\int_0^\infty \frac{\sin x^m}{x^n}\,dx.$$

Solution:

For m = 0, given integral = $\int_0^\infty \frac{dx}{x^n}$, which converges at 0 if n < 1, and converges at ∞ if n > 1. Thus the integral cannot converge at 0 and ∞ together. So it diverges for m = 0. Now we take $m \neq 0$.

$$\int_0^\infty \frac{\sin x^m}{x^n}\,dx = \int_0^1 \frac{\sin x^m}{x^n}\,dx = \int_1^\infty \frac{\sin x^m}{x^n}\,dx$$

Consider $\int_0^1 \frac{\sin x^m}{x^n}\,dx = \int_0^1 \left(\frac{\sin x^m}{x^m}\right)\frac{1}{x^{n-m}}\,dx$

Here 0 is the point of infinite discontinuity if n – m > 0.

Let $\quad f(x) = \frac{\sin x^m}{x^n};\ f(x) = \frac{1}{x^{n-m}}$

Then $\lim\limits_{x\to 0}\frac{f(x)}{g(x)} = 1$. Since $\int_0^1 \frac{dx}{x^{n-m}}$ converges if n – m < 1.

so $\int_0^1 \frac{\sin x^m}{x^n}\,dx$ converges if n < 1 + m.

Now, we consider $\int_1^\infty \frac{\sin x^m}{x^n}\,dx = \int_1^\infty \frac{mx^{m-1}\sin x^m}{mx^{m-1}.x^n}\,dx$

Let $\phi(x) = \frac{1}{mx^{m+n-1}}$ so that $\phi(x)$ is monotonically decreasing $\forall\ x \geq 1$ and $\phi(x) \to 0$ as $x \to \infty$ if m + n – 1 > 0.

Now, $\left|\int_1^t mx^{m-1}\sin x^m\,dx\right| = |\cos 1 - \cos t^m| \leq 2\ \forall\ t \geq 1$.

Thus, $\int_1^t mx^{m-1} \sin x^m \, dx$ is bounded $\forall\ t \geq 1$.

By Dirichlet's Test,

$$\int_1^\infty \frac{mx^{m-1} \sin x^m}{mx^{m+n-1}} dx \int_1^\infty \frac{\sin x^m}{x^n} dx \text{ is convergent, if } m + n - 1 > 0 \text{ i.e.,}$$

$n > 1 - m$.

Hence, the given integral is convergent if $1 - m < n < 1 + m$.

Example 44:

Show that $\int_0^\infty \frac{\sin x}{x}$ *dx is convergent but it is not absolutely convergent.*

Solution:

Here $\int_0^\infty \frac{\sin x}{x} dx = \int_0^a \frac{\sin x}{x} dx + \int_a^\infty \frac{\sin x}{x} dx$ ...(1)

were $a > 0$.

Since $\lim_{x \to 0} \frac{\sin x}{x} = 1$, $\int_0^a \frac{\sin x}{x} dx$ is a proper integral and so convergent.

Now, we consider the integral $\int_a^\infty \frac{\sin x}{x} dx$

Let $\phi(x) = 1/x$ which is bounded and monotonic in (a, ∞)

and $\lim_{x \to \infty} \frac{1}{x} = 0$

Let $f(x) = \sin x$.

We have

$$\left| \int_a^\infty f(x)\, dx \right| = \left| \int_a^\infty \sin x\, dx \right| = | \cos a - \cos x |$$

$$\leq | \cos a | + | \cos x |$$

$$< 2. \qquad \forall\ x \text{ (finite)}$$

Thus, $\int_a^\infty \sin x \, dx$ is bounded for all finite values of $n \geq 1$.

Hence, by Dirichlet's test $\int_a^\infty \frac{\sin x}{x} dx$ is convergent.

Hence, $\int_0^\infty \frac{\sin x}{x}$ dx is convergent.

In order to prove that $\int_0^\infty \frac{\sin x}{x} dx$ is not absolutely convergent we have to show that $\int_0^\infty \frac{|\sin x|}{x} dx$ is divergent; i.e. to show that

$$\lim_{x\to\infty} \int_0^{n\pi} \frac{|\sin x|}{x} dx \text{ is divergent.}$$

$$\int_0^{n\pi} \frac{|\sin x|}{x} dx = \sum_{r=1}^{n} \int_{(r-1)\pi}^{r\pi} \frac{|\sin x|}{x} dx$$

Let $x = (r - 1)\pi + y$

$$= \sum_{r=1}^{n} \int_0^{\pi} \frac{\sin y\, dy}{(r-1)\pi + y}$$

$$> \sum_{r=1}^{n} \frac{1}{r\pi} \int_0^{\pi} \sin y\, dy$$

$$> \sum_{r=4}^{n} \frac{1}{r\pi} [-\cos y]_0^{\pi}$$

$$> \sum_{r=1}^{n} \frac{2}{r\pi}$$

$$> \frac{2}{\pi} \sum_{r=1}^{n} \frac{1}{r}$$

Therefore, $\int_0^\infty \frac{|\sin x|}{x} dx = \lim_{n\to 0} \int_0^{n\pi} \frac{|\sin x|}{x} dx > \frac{2}{4} \frac{1}{r} = \infty$

Thus, $\int_0^\infty \frac{\sin x}{x} dx$ is divergent and so the given integral is not absolutes divergent.

Example 45:

Show that $\int_2^\infty \frac{\cos x}{\log x} dx$ *is not absolutely convergent.*

Solution:

We have to show that

$$\int_2^\infty \frac{|\cos x|}{\log x} dx \quad \text{is not convergent}$$

We have

$$\int_2^\infty \frac{|\cos x|}{\log x}\,dx = \sum_{r=1}^{\infty} \int_{(2r-1)\pi/2}^{(2r+1)\pi/2} \frac{|\cos x|}{\log x}\,.\,dx \text{ Also}$$

$$\int_2^\infty \frac{|\cos x|}{\log x}\,dx = \int_2^{\pi/2} + \int_{\pi/2}^{3\pi/2} + \int_{3\pi/2}^{5\pi/2} + \ldots\ldots$$

$$= \underbrace{\sum_{r=1}^{\infty} \int_{(2r-1)\pi/2}^{(2r+1)\pi/2} \frac{|\cos x|}{\log x}\,dx}_{(I_1)} + \underbrace{\int_2^{\pi/2} \frac{|\cos x|}{\log x}\,dx}_{(I_2)}$$

Now $I_1 \geq \dfrac{1}{\log(2r+1)\pi/2}\left|\int_{(2r-1)\pi/2}^{(2r+1)\pi/2} \cos x\,dx\right|$

$$= \frac{1}{\log(2r+1)\pi/2}\left|\sin(2r+1)\frac{\pi}{2} - \sin(2r-1)\frac{\pi}{2}\right|$$

$$= \frac{|2(-1)|}{\log(2r+1)\pi/2} = \frac{2}{\log\,(2r+1)\pi/2}$$

$$\therefore \int_2^\infty \frac{|\cos x|}{\log\ x}\,dx \geq \sum_{r=1}^{\infty} \frac{2}{\log(2r+1)\pi/2} + \int_2^{\pi/2} \frac{|\cos x|}{\log x}\,dx$$

Since $\sum_{n=2}^{\infty} \dfrac{1}{\log x}$ is divergent and I_2 is a proper integral, so $\int_2^\infty \dfrac{|\cos x|}{\log x}\,dx$ is divergent.

Example 46:

Let the function f be defined on the interval $0 < x \leq 1$ as follows :

$f(x) = 2$ for $\frac{1}{2} < x \leq 1$

$f(x) = -3$ for $\frac{1}{3} < x \leq \frac{1}{2}$

$f(x) = 4$ for $\frac{1}{4} < x \leq \frac{1}{3}$

$f(x) = -5$ for $\frac{1}{5} < x \leq \frac{1}{4}$

and so on, the values being alternatively positive and negative.

Show that the infinite integral $\int_0^1 f(x)\,dx$ converges but not absolutely.

Solution:

$$\int_0^1 f(x)\,dx = \int_{1/2}^1 f(x)\,dx + \int_{1/3}^{1/2} f(x)\,dx + \int_{1/4}^{1/3} f(x)\,dx + \ldots$$

$$= \int_{1/2}^1 f(x)\,dx + \int_{1/3}^{1/2} (-3)\,dx + \int_{1/4}^{1/3} 4\,dx + \ldots$$

$$= 2(1-\tfrac{1}{2}) + (-3)(\tfrac{1}{2}-\tfrac{1}{3}) + \ldots$$

$$= 2(1-\tfrac{1}{2}) - 3(\tfrac{1}{2}-\tfrac{1}{3}) + 4(\tfrac{1}{3}-\tfrac{1}{4}) + \ldots$$

$$= 1 - \tfrac{1}{2} + \tfrac{1}{3} - \tfrac{1}{4} + \ldots$$

$$= \log.(1+1) = \log 2.$$

Hence, the given integral is convergent.

For absolute convergence, we have

$$\int_0^1 |f(x)|\,dx = \int_{1/2}^1 |2|\,dx + \int_{1/3}^{1/2} |-3|\,dx + \int_{1/4}^{1/3} |4|\,dx + \int_{1/5}^{1/4} |-5|\,dx + \ldots$$

$$= 1 + \tfrac{1}{2} + \tfrac{1}{3} + \tfrac{1}{4} + \tfrac{1}{5} + \ldots \text{ which is divergent.}$$

Hence the given integral is not absolutely convergent.

Example 47:

Discuss the convergence of $\int_0^\infty \frac{x^p \sin^2 x}{1+x^2}\,dx$.

Solution:

$$\int_0^\infty f(x)\,dx = \int_0^1 f(x)\,dx + \int_1^\infty f(x)\,dx, \text{ where}$$

$$f(x) = \frac{x^p \sin^2 x}{1+x^2}$$

For $p \geq 0$, $\int_0^1 f(x)\,dx$ is a proper integral.

For $p < 0$, we take $g(x) = \frac{1}{x^{-(p+2)}}$ so that

$$\lim_{x\to 0} \frac{f(x)}{g(x)} = \lim_{x\to 0} \frac{\sin^2 x}{x^2} \cdot \frac{1}{1+x^2} = 1$$

Now, $\int_0^1 g(x)\, dx$ is convergent iff $-(p + 2) < 1 \Leftrightarrow -3 < p$.

$\therefore \int_0^1 f(x)\, dx$ is convergent iff $-3 < p$.

Consider the integral $\int_1^\infty \frac{x^p \sin^2 x}{1+x^2} dx$. We have

$\int_1^\infty \frac{x^p \sin^2 x}{1+x^2} dx \le \int_1^\infty x^p \frac{\sin^2 x}{x^2} dx \int_1^\infty \frac{\sin^2 x \, dx}{x^{2-p}}$. We see that

$$\left|\frac{\sin^2 x}{x^{2-p}}\right| = \frac{|\sin^2 x|}{x^{2-p}} \le \frac{1}{x^{2-p}} \; \forall \; x \ge 1.$$

Now, $\int_1^\infty \frac{dx}{x^{2-p}}$ converges if $2 - p > 1$ i.e., $p < 1$.

By comparison test it follows that

$\int_1^\infty x^p \frac{\sin^2 x}{x^2} dx$ is convergent if $p < 1$.

$\Rightarrow \quad \int_1^\infty x^p \frac{\sin^2 x}{1+x^2} dx$ is convergent if $p < 1$.

Hence, $\int_1^\infty x^p \frac{\sin^2 x}{1+x^2} dx$ is convergent for $-3 < p < 1$.

Ex.: *Test* $\int_0^\infty \frac{x^m \cos ax}{1+x^n} dx$ *for convergence.*

[Ans. Convergent for $(n > 0, -1 < m < n)$ or $(n < 0, 0 > m > n - 1)$.**]**

3

Uniform Convergence

INTRODUCTION

A sequence of function $S_n(z)$ is said to converge uniformly in two domain D to the limiting function S(z) if given a positive small number $\in$ we can find a positive integer M($\in$) depending an $\in$ alone, such that for n $\geq$ m the in equality

$|S_n(z) - S(z)| < \in$ holds at each point Z of the domain D.

The sequence $S_1(z), S_2(z), S_3(z), ...$ of one valued functions each defined in a bounded closed domain D_1 is said to be convergent at a point z of D if the sequence of the complex numbers $S_1(z), S_2(z), S_3(z), ...$ tends to a definite limit S(z) and we write $S(z) = \lim_{n\to\infty} S_n(z)$.

Formally it means that given any point Z of D and any small positive number $\in$ we can find a positive integer m such that

$$|S_n(z) - S(z)| < \in \text{ for } n \geq m$$

The integer m depends, in general upon $\in$ and also on the particular point Z of D. Thus m = m ($\in$, z).

Now for a fixed $\in$ and Z taking all the values in the domain D, e get an aggregate of values of m($\in$, a) which in general will not have an upper bound but if this aggregate has an upper bound say m, such that m($\in$, z) $\leq$ M for every Z in D then the sequence $S_n(z)$ is said to be uniformly convergent in the domain D.

UNIFORM CONVERGENCE OF AN INFINITE SERIES

Definition: A series of functions Σf_n is said to converge uniformly on [a, b] if the sequence (S_n) of its partial sums defined by

$$S_n(x) = f_1(x) + f_2(x) + ... + f_n(x) = \sum_{i=1}^{n} f_i(x)$$

converge uniformly on [a, b].

In other words, a series of functions $\sum f_n$ converges uniformly to a function f on [a, b] if for any $\varepsilon > o$, and for all $x \in [a, b]$, there exists a positive integer m (dependent on ε only and independent of x) such that

$$| f_1(x) + f_2(x) + ... + f_n(x) - f(x) | < \varepsilon, \quad \forall\ n \geq m, \quad \forall\ x \in [a, b].$$

Necessary and Sufficient Condition

Theorem 1:

The necessary and sufficient condition for the uniform convergence of the series $\sum f_n(x)$ defined on E is that $\forall\ \varepsilon > 0$, however small, there exists a natural number N independent of x such that for all $n \geq N$ we have

$$| f_{n+1}(x) + f_{n+2}(x) + f_{n+3}(x) + ... + (f_{n+p}(x) | \leq \varepsilon$$

where p may have any positive integral value.

Proof:

The condition is necessary

Let $\sum f_n(x)$ be uniformly convergent on E. Then for a given $\varepsilon > 0$ we can find a natural number N such that $\forall\ n \geq N$ and $\forall\ x \in E$ we have

$$| s_n(x) - s(x) | \leq \tfrac{1}{2}\varepsilon \qquad ...(1)$$

where $s_n(x) = f_1(x) + f_2(x) + ... + f_n(x)$

and also $| s_{n+p}(x) - s(x) | \leq \tfrac{1}{2}\varepsilon \qquad ...(2)$

Thus,

$$\begin{aligned} | f_{n+1}(x) + f_{n+2}(x) + ... + f_{n+p}(x) | &= | s_{n+p}(x) - s_n(x) | \\ &\leq | s_{n+p}(x) - s(x) | \\ &\quad + |s_n(x) - s(x) | \\ &\leq \varepsilon/2 + \varepsilon/2 = \varepsilon. \end{aligned}$$

This proves the necessity part of the theorem.

The condition is also sufficient.

If the given condition is satisfied for all $x \in E$, we have

$$| s_{n+p}(x) - s_n(x) | \leq \tfrac{1}{2}\varepsilon$$

or $s_n(x) - \tfrac{1}{2}\varepsilon \leq s_{n+p}(x) \leq s_n(x) + \tfrac{1}{2}\varepsilon$

$$\forall\ n \geq N \text{ and } P \geq 0.$$

Now keeping n fixed and letting $p \to \infty$, so that $s_{n+p}(x) \to s(x)$, then

$$s_n(x) - \tfrac{1}{2}\varepsilon \leq s(x) \leq s_n(x) + \tfrac{1}{2}\varepsilon$$

i.e. $\quad | s_n(x) - s(x) | \leq \frac{1}{2} \varepsilon < \varepsilon \ \forall \ n \geq N$ and $\forall \ x \in E$.

This shows that the series $\Sigma f_n(x)$ is uniformly convergent.

Theorem 2:

A series of functions Σf_n defined on [a, b] converges uniformly on [a, b] iff for each $\varepsilon > 0$ and for all $x \in [a, b]$, there exists a positive integer m such that

$$| f_{n+1}(x) + f_{n+2}(x) + ... + f_{n+p}(x) | < \varepsilon \ \forall \ n \geq m, \ p \geq 1.$$

Proof:

Let $\{s_n\}$ be the sequence of partial sums of the series Σf_n. Then $s_n(x)$

$$= \sum_{i=1}^{n} f_i(x). \qquad ...(1)$$

Let $\varepsilon > 0$ be given.

Hence Σf_n is uniformly convergent on [a, b].

$\Leftrightarrow$ the sequence $\{s_n\}$ is uniformly convergent on [a, b]

$\Leftrightarrow |s_{n+p}(x) - s_n(x)| < \varepsilon \ \forall \ n \geq m, \ p \geq 1, \ \forall \ x \in [a, b]$.

(Cauchy's principle of uniform convergence)

$\Leftrightarrow |f_{n+1}(x) + f_{n+2}(x) + ... + f_{n+p}(x)| < \varepsilon$

$\forall \ n \geq m, \ \forall \ p \geq 1$ and $\forall \ x \in [a, b]$. [using (1)]

UNIFORM CONVERGENCE OF SEQUENCES OF FUNCTIONS

It is natural to expect that the value of N depends on ε as well as on x i.e. for a given ε the values of N will be different for different elements x of E.

If for each $\varepsilon > 0$ it is possible to determine an N such that (2) holds for all $x \in E$ then we say that $\{f_n\}$ is uniformly convergent on E.

Definition 3:

Let $\{f_n\}$ be a sequence of real valued functions on a set E. We say that $\{f_n\}$ converges uniformly to f on E if for a given $\varepsilon > 0$, there exists a positive integer N such that

$$| f_n(x) - f(x) \} \leq \varepsilon \qquad (n \geq N, \ x \in E).$$

obviously here N depends on ε but not x. It may also be noted that if f_n converge uniformly to f on E then $\{f_n\}$ is also pointwise convergent on E.

Example 1:

Show that the sequence $\{f_n\}$ where

$$f_n(x) = \frac{n}{n+x}, \quad x \geq 0$$

is uniformly convergent is any finite interval.

Solution:

Here, we have

$$f(x) = \lim_{n\to\infty} \frac{n}{n+x} = \lim_{n\to\infty} \frac{1}{x/n+1} = 1 \;\; \forall\, x \geq 0.$$

Then for an arbitrarily chosen positive number ε we have

$$| f_n(x) - f(x) | \leq e$$

if $$\left| \frac{n}{x+n} - 1 \right| \leq \varepsilon$$

i.e., if $$\left| \frac{-x}{x+n} \right| \leq \varepsilon$$

or if $$\frac{x}{x+n} \leq \varepsilon$$

or if $$n \geq x\left(\frac{1}{\varepsilon} - 1\right).$$

Obviously N increases with x and tends to ∞ as n tends ∞. Hence convergence is not uniform in $(0, \infty)$. But if $(0, k)$ is any finite interval where $k > 0$ however large then, N = any integer greater than or equal to $k\left(\frac{1}{\varepsilon} - 1\right)$ and $| f_n(x) - f(x) | \leq \varepsilon$ for every $n \geq N$ and every $x \in [0, k)$. This shows that the given sequence is uniformly convergent in $[0, k]$.

Example 2(a):

Find the point-wise limit of $\{f_n\}$, where

$$f_n(x) = \frac{n}{x+n} \qquad \forall\, x \geq 0.$$

Solution:

$$\lim_{n\to\infty} f_n(x) = \lim_{n\to\infty} \frac{n}{x+n} = \lim_{n\to\infty} \frac{1}{(x/n)+1} = 1.$$

Here the given sequence (f_n) is point-wise convergent for all $x \geq 0$ and the point-wise f is given by

$$f(x) = 1 \qquad \forall\, x \geq 0.$$

Example 2(b):

Show that the sequence $\{f_n\}$, $f_n(x) = x^n$, is point-wise convergent in [0, 1].

Solution:

$$\lim_{n\to\infty} f_n(x) = \lim_{n\to\infty} x^n$$

$$= \begin{cases} 0 & \text{for } 0 \le x < 1, \\ 1 & \text{for } x = 1. \end{cases}$$

Hence the given sequence is point-wise convergent in [0. 1] and the point-wise limit f is given by

$$f(x) = \begin{cases} 0 & \text{when } 0 \le x < 1, \\ 1 & \text{when } x = 1. \end{cases}$$

Example 2(c):

Show that the sequence $\{f_n\}$, where

$f_n(x) = n/(x + n)$

is uniformly convergent in [0, k] whatever k may be, but not uniformly convergent in [0, ∞].

Solution:

$$\lim_{n\to\infty} f_n(x) = \lim_{n\to\infty} \frac{x}{x+n} = \lim_{n\to\infty} \frac{1}{1+(x/n)}$$

$$= 1 \text{ for all } x \le 0.$$

Thus, the sequence $\{f_n\}$ is point-wise convergent for all $x \le 0$ and the point-wise limit is

$$f(x) = 1 \quad \forall\, x \ge 0.$$

Let $\varepsilon > 0$ be given. We have

$$|f_n(x) - f(x)| = \left|\frac{n}{x+n} - 1\right| = \frac{x}{x+n} < \varepsilon,$$

if $\dfrac{x+n}{x} > \dfrac{1}{\varepsilon}$ or $n > x\left(\dfrac{1}{\varepsilon} - 1\right)$.

Consider the interval [0, k]. Let m be a positive integer;

$m > k\left(\dfrac{1}{\varepsilon} - 1\right)$. Then

$$| f_n(x) - f(x) | < \varepsilon \quad \forall\, n \ge m \text{ and } \forall\, x \in [0, k].$$

Hence the given sequence is uniformly convergent in [0, k]. Now we show that $\{f_n\}$ is not uniformly convergent in [0, ∞]. Let m be a positive integer > $x\left(\frac{1}{\varepsilon}-1\right)$. Then m → ∞ as x → ∞ so that it is not possible to choose any positive integer m such that

$$\forall n \geq m \text{ and } \forall\, x \geq 0,\ |f_n(x) - f(x)\ \} < \varepsilon.$$

Hence the convergence is not uniform in [0, ∞[.

Example 3:

Show that the sequence $\{f_n\}$, where

$$f_n(x) = \frac{nx}{1+n^2x^2}, x \in R$$

is point-wise convergent but is not uniformly convergent in any interval containing zero.

Solution:

We know that 1, the given sequence is point-wise convergent with point-wise limit f, where

$$f(x) = 0 \qquad \forall\, x \in R.$$

Let, if possible, $\{f_n\}$ be uniformly convergent in [a, b] containing 0. By he above Remark, the point-wise limit is also the uniform limit. By the definition of uniform convergence, for ε = ½, there exists a positive integer m such that

$$\left|\frac{nx}{1+n^2x^2} - 0\right| < \frac{1}{4} \quad \forall\ n \geq m \text{ and } \forall x \in [a, b]. \qquad \text{...(1)}$$

We can choose a positive integer k such that

k ≤ m and 1/k ∈ [a, b].

Taking n = k and x = 1/k, we have

$$\left|\frac{nx}{1+n^2x^2} - 0\right| = \frac{1}{2} \qquad \text{...(2)}$$

As (1) and (2) are contradictory statements, it follows that the given sequence is not uniformly convergent in any interval [a, b] containing 0, although it is point-wise convergent therein.

Hence the problem.

Example 4:

Consider the series

$$\sum_{n=0}^{\infty} f_n, \text{ where } f_n(x) = \frac{x^2}{(1+x^2)^n} \text{ (x real)}.$$

Solution:

At x = 0, each $f_n(x) = 0$, so that the sum of the series f(0) = 0.

For x ≠ 0, it forms a geometric series with common ratio $1/(1 + x^2)$, so that its sum function $f(x) = 1 + x^2$.

Hence

$$f(x) = \begin{cases} 1+x^2, & x \neq 0 \\ 0 & , x = 0 \end{cases}$$

Each term of the series is continuous but the sum f is not.

Example 5(a):

Show that the sequence $\{f_n\}$, where

$$f_n(x) = \frac{nx}{1+n^2x^2}$$

converges point-wise to zero for all real x.

Solution:

$$\lim_{n\to\infty} f_n(x) = \lim_{n\to\infty} \frac{nx}{1+n^2x^2}$$

$$= \lim_{n\to\infty} \frac{x}{\frac{1}{n}+nx^2} = 0 \quad \forall\, x \in R.$$

Hence, the given sequence converges point-wise to zero for all real x.

Example 5(b):

Show that the sequence $\{f_n\}$, $f_n(x) = x^n$, is uniformly convergent in [0, k] k < 1 and only point-wise convergent in [0, 1].

Solution:

We know $\lim_{x\to\infty} x^n = 0, \quad \text{if } 0 \le x < 1,$

$= 1, \quad \text{if } x = 1.$

Thus the given sequence is point-wise convergent in [0, 1] and the point-wise limit f is given by

$$f(x) = \begin{cases} 0, & \text{if } 0 \le x < 1, \\ 1, & \text{if } x = 1. \end{cases}$$

Let us choose ε, $0 < \varepsilon < 1$. For $0 < x < 1$, we have

$$| f_n(x) - f(x) | = | x^n - 0 | = x^n < \varepsilon.$$

Now $x^n < \varepsilon \Leftrightarrow (1/x)^n > 1/\varepsilon \Leftrightarrow n \log(1/x) > \log(1/\varepsilon)$

$$\Leftrightarrow n > (\log 1/\varepsilon) \div \log(1/x)$$

For $x = 0$, $| f_n(x) - f(x) | = 0 < \varepsilon$, when $n \geq 1$.

Let m be a positive integer such that

$$m = \max\{1, (\log 1/\varepsilon) \div \log(1/x)\}. \quad ...(1)$$

Then $m \to \infty$ as $x \to 1$, so that there does not exists any positive integer such that $\forall\, n \geq m$, and $\forall\, x \in [0, 1]$,

$$| f_n(x) - f(x) | < \varepsilon.$$

Hence the convergence is not uniform in [0. 1].

Suppose k is any number such that $0 \leq k < 1$.

Then, by (1), $m = \log(1/\varepsilon) \div \log(1/k)$.

$\therefore | f_n(x) - f(x) | < \varepsilon \quad \forall\, n \geq m$ and $\forall\, x \in [o, k]$.

Hence, $\{f_n\}$ is uniformly convergent in [0, k], k < 1.

Example 6:

The sequence $\{f_n\}$, *where* $f_n(x) = \dfrac{\sin nx}{\sqrt{n}}$, *(x real), has the limit.*

Solution :

$$f(x) = \lim_{n\to\infty} f_n(x) = 0$$

$\therefore f'(x) = 0$, and so $f'(0) = 0$

But

$$f'_n(x) = \sqrt{n} \cos nx$$

so that

$$f'_n(0) = \sqrt{n} \to \infty \text{ as } n \to \infty$$

Thus, at $x = 0$, the sequence $\{f'_n(x)\}$ diverges whereas the limit function $f'(x) = 0$, i.e., the limit of differentials is not equal to the differential of the limit.

Let us consider few examples

Note 1: If a sequence converges point-wise to f then for a given $\varepsilon > 0$, each point x_i of [a, b] determines an integer N_i such that

$$| f_n(x_i) - f(x_i) | < \varepsilon, \text{ for } n \geq N_i$$

Consideration of all points of [a, b] gives rise to a sequence of integers $N_1, N_2, N_3, \ldots$

In case the sequence $\{N_i\}$ is bounded above, with supremum N, say, them (4) holds for all points of [a, b] when $n \geq N$ and so the given sequence $\{f_n\}$ converges uniformly on [a, b].

If no such N exists, the sequence $\{f_n\}$ is not uniformly convergent.

Note 2: *Uniform convergence ⇒ point wise convergence*

but not vice versa. However

Non-pointwise convergence ⇒ non-uniform convergence

i.e., a sequence which is not point-wise convergent cannot be uniformly convergent.

A series of functions Σf_n is said to converge uniformly on [a, b] if the sequence $\{S_n\}$ of its partial sums, defined by

$$S_n(x) = \sum_{i=1}^{n} f_i(x)$$

converges uniformly on [a, b].

Thus, a series of functions Σf_n converges uniformly to f on [a, b] if for $\varepsilon > 0$ and all $x \in [a, b]$ there exists an integer N (independent of x and dependent on ε) such that for all x in [a, b]

$$| f_1(x) + f_2(x) + \ldots + f_n(x) - f(x) | < \varepsilon, \text{ for } n \geq N$$

ABEL'S TEST FOR THE CONVERGENCE OF INTEGRAL OF A PRODUCT

If $\int_a^\infty f(x)\,dx$ converges and $\phi(x)$ is bounded and monotonic for $x > a$ then $\int_a^\infty f(x) \,.\, \phi(x)\,dx$ is convergent.

Example 1:

Test for the convergence of $\int_a^\infty e^{-x} \frac{\sin x}{x^2}\,dx$ *when* $a > 0$

Solution:

Let $f(x) = \frac{\sin x}{x^2}$ and $\phi(x) = e^{-x}$

Since $\left|\frac{\sin x}{x^2}\right| \leq \frac{1}{x^2}$ and $\int_a^\infty \frac{1}{x^2}\,dx$ is convergent.

Therefore by comparison test $\int_a^\infty \frac{\sin x}{x^2}\,dx$ is also convergent.

Again e^{-x} is monotonic decreasing and bounded function for x > a

Hence by Abel's Test $\int_a^\infty e^{-x} \frac{\sin x}{x^2} dx$ is convergent

Example 2:

Test the convergence of $\int_a^\infty (1 - e^{-x}) \frac{\cos x}{x^2}$ *dx where as 0*

Solution:

$f(x) = \frac{\cos x}{x^2}$ and $\phi(x) = 1 - e^{-x}$

Here we have $\left|\frac{\cos x}{x^2}\right| \le \frac{1}{x^2}$ as $|\cos x| \le 1$

Since $\int_a^\infty \frac{1}{x^2}$ is convergent, therefore by comparison test $\int_a^\infty \frac{\cos x}{x^2} dx$ is also convergent Again $\phi(x) = 1 - e^{-x}$ is monotonic increasing and bounded function for x > a.

Hence by Abel's Test $\int_a^\infty (1 - e^{-x}) \frac{\cos x}{x^2} dx$ is convergent

DIRICHLET'S TEST FOR THE CONVERGENCE OF INTEGRATE OF A PRODUCT

If f(x) is bounded and monotonic $a \le x < \infty$ f $\lim_{x\to\infty} f(x) = 0$ Then thee integral $\int_a^\infty f(x)\,\phi(x)\, dx$ converges provided $\left|\int_a^\infty \phi(x)\, dx\right|$ is bounded as x takes all finite values.

Example 1:

Show that the integral

$$\int_0^\infty e^{-ax} \frac{\sin x}{x} dx, \ a \le 0 \text{ convergent}$$

Solution:

$$\int_0^\infty e^{-ax} \frac{\sin x}{x} = \int_0^\infty e^{-ax} \frac{\sin x}{x} dx + \int_a^{\infty/e^{-ax}} \frac{\sin x}{x} dx$$

Where a > 0

Since $\lim_{x\to 0} e^{-ax} \frac{\sin x}{x} = 1$ the integral $\int_0^a e^{-ax} \frac{\sin x}{x} dx$ is a proper integral and hence convergent.

Now it remains to test the convergence of $\int_a^\infty e^{-ax} \frac{\sin x}{x} dx$

Let $f(x) = \frac{e^{-ax}}{x}$ and $\phi(x) = \sin x$

Obviously the function $f(x) = \frac{1}{x^{ax}}$ is bounded and monotonic decreasing for all $x \geq 0$ and $\lim_{x\to\infty} f(x) \lim_{x\to\infty} \frac{1}{xe^{ax}} = 0$

Moreover $\left|\int_a^x \phi(x)\, dx\right| = \left|\int_a^x \sin x\, dx\right|$

$= |\cos a = \cos x| \leq 2$ for all sin x value of x

Example 2:

Test the convergence of the integral

$\int_a^\infty \frac{\sin x}{\sqrt{x}} dx$, *where a > 0*

Solution:

Let $f(x) = \frac{1}{\sqrt{x}}$ and $\phi(x) = \sin$

Now $\frac{1}{\sqrt{x}}$ is bounded and monotonic decreasing for all $x \geq a$ and $\lim_{x\to\infty} \frac{1}{\sqrt{x}} = 0$

Also $\left|\int_a^x \phi(x)\, dx\right| = \left|\int_a^x \sin x\, dx\right| = |\cos a - \cos x| \leq 2$ for all finite value of x

[Note that the value of cone lies between –1 and 1]

$\left|\int_a^x f(x)\, dx\right|$ is bounded for all finite values of x

Hence by Drichlet's test the integral $\int_a^\infty \frac{\sin}{\sqrt{x}} dx$ is convergent

THE μ-TEST

Let f(x) be bounded and integrable in the interval (a, ∞) where $a > 0$

If there is number $\mu\ \phi > 1$ such that $\lim_{x\to\infty} x^\mu f(x)$ exists then $\int_a^\infty f(x)\, dx$ is convergent

If there is a number $\mu \leq 1$ such that $\lim_{x\to\infty} x^\mu f(x)$ exists and is non-zero then $\int_a^\infty f(x)\, dx$ is divergent and the same is true if $\lim_{x\to\infty} x^\mu f(x)$ is $+\infty$ or $-\infty$

While applying the μ test the value of μ is usually taken to be equal to the highest power of x in the denominator of the integrand minus the highest power of x in the numerator of the integrand.

ABSOLUTE CONVERGENCE

The infinite integral $\int_a^\infty f(x)$ is said to be absolute convergent if the integral $\int_a^\infty |f(x)|\ dx$ is convergent.

If the integral $\int_a^\infty f(x)\,dx$ is absolutely convergent it is necessarily convergent, but if the integral $\int_a^\infty f(x)\,dx$ convergent. If is not necessarily absolutely convergent. Thus absolute convergent gives a sufficient but not a necessary condition for the convergence of an infinite integral.

Example:

Show that $\int_1^\infty \frac{\sin x}{x^4}\,dx$ is absolutely convergent.

Solution:

The integral $\int_1^\infty \frac{\sin x}{x^4}\,dx$ will absolutely convergent if $\int_1^\infty \left|\frac{\sin x}{x^4}\right| dx$ is convergent.

Let $f(x) = \left|\frac{\sin x}{x^4}\right|$, then f(x) is bounded in the interval $(1, \infty)$ we have

$$f(x) = \left|\frac{\sin x}{x^4}\right| \leq 1/x^4 \sin x\ |\sin x| = 1$$

∴ By Comparison Test $\int_1^\infty f(x)\,dx$ is convergent if $\int_1^\infty \frac{1}{x^4}\,dx$ is convergent but the comparison integral $\int_1^\infty \frac{1}{x^4}\,dx$ is convergent because $\int_1^\infty$ Here $n = 4$ which is > 1

Hence $\int_1^\infty f(x)\,dx$ is convergent and so the given integral is absolutely convergent.

CAUCHY'S CRITERIA FOR UNIFORM CONVERGENCE OF SEQUENCE OF FUNCTIONS

The sequence of functions $\{f_n\}$, defined on a set E, is uniformly convergent on E if and only if for every $\varepsilon > 0$, $\exists$ a positive integer N s.t.

$$m \geq N,\ n \geq N \Rightarrow$$

$$|f_n(x) - f_m(x)| \le \varepsilon \quad \forall x \in E,$$

i.e. $\{f_n\}$ is uniformly convergent of E iff we are able to find a term fN after which if f_m and f_n are any two terms ($m \le N, n \le N$) then the difference of $f_n(x)$ and $f_m(x)$ is arbitrarily small.

Proof:

Let $\{f_n\}$ be uniformly convergent on E and let f be the limit function. Then by the definition there exists a positive integer N such that

$$|f_n(x) - f(x)| \le \varepsilon/2, \text{ for } n \le N \text{ and } \forall x \in E$$

In a similar way if $m \ge N$ then

$$|f_n(x) - f(x)| \ge \varepsilon/2 \quad \forall x \in E$$

Thus, $|f_n(x) - f_m(x)|$

$$= |[f_n(x) - f(x)] - \{f_m(x) - f(x)\}|$$

$$\le |f_n(x) - f(x)| + |f_m(x) - f(x)|$$

$$\le \varepsilon/2 + \varepsilon/2$$

$$\therefore |f_n(x) - f_m(x)| \le \varepsilon.$$

Conversely, if Cauchy's condition hold then the sequence of $f_n(x)$ converges for every x to a limit say f(x). Thus the sequence $\{f_n\}$ converges to f on E. We have to prove that the convergence is uniform.

Let $\varepsilon > 0$ be a given arbitrary number, we choose N such that

$$|f_n(x) - f_m(x)| \le \varepsilon \text{ for } n \ge N, m \ge N \text{ and } x \in E$$

Now keeping x fixed letting $m \to \infty$ over $f_m(x) \to f(x)$, and $|f_n(x) - f(x)| \le \epsilon$ for every $n \ge N$ and every $n \in E$.

Hence theorem.

Example 1:

Let $f_n(x) \quad \dfrac{x^n}{1+x^n} \quad (0 \le x \le 1)$

Show that $\{f_n\}$ converges point wise on [0, 1], if

$$f(x) = \lim_{n\to\infty} f_n(x) \quad (0 \le x \le 1)$$

is there on N such that

$$|f_n(x) - f(x)| \le ¼ \ (n \ge N).$$

Solution:

$$f(x) = \lim_{n\to\infty} f_n(x) = \lim_{n\to\infty} \frac{x^n}{1+x^n} = 0 \qquad \text{if } (0 \le x \le 1)$$

$$= \tfrac{1}{2} \quad \text{if } (x = 1)$$

since $\lim_{n\to\infty} x^n = 0$ when $0 \le x < 1$

and $\lim_{n\to\infty} x^n = 1$ when $x = 1$

Let ε be any arbitrary small positive real number.

We have $| f_n(x) - f(x) | \le \varepsilon \quad (0 \le x < 1)$

$$\text{If } \left| \frac{x^n}{1+x^n} - 0 \right| \le \varepsilon$$

or if $x^n \le (1 + x^n)\, \varepsilon$

or if $n^n (1 - \varepsilon) \le \varepsilon$

$$\text{or if } n^n \le \frac{1-\varepsilon}{\varepsilon}$$

$$\text{or if } \left(\frac{1}{x}\right) \ge \frac{1-\varepsilon}{\varepsilon}$$

$$\text{or if } n \log \left(\frac{1}{x}\right)^n \ge \log \frac{1-\varepsilon}{\varepsilon}$$

$$\text{or if } n \ge \frac{\log \dfrac{1-\varepsilon}{\varepsilon}}{\log\left(\dfrac{1}{x}\right)} \qquad (0 \le x < 1)$$

Thus, there exist a positive integer N for all x ∈ [0, 1] i.e. given sequence converges at every point of [0, 1].

Again when x = 1

$$\left| \frac{x^n}{x^n + 1} - \frac{1}{2} \right| \le \varepsilon \text{ for all values of n.}$$

Hence the given sequence is point wise convergent on [0, 1].

Again, when ε = ¼

$$\text{Then} \quad N = \frac{\log \dfrac{1 - \frac{1}{4}}{\frac{1}{4}}}{\log\left(\dfrac{1}{x}\right)}$$

$$= \frac{\log 3}{\log\left(\frac{1}{x}\right)}$$

Example 2:

Apply M_n test to show that $\{f_n\}$, the sequence of functions is non-uniformly convergent if

$f_n(x) - nx\,(1 - x)^n \qquad (0 \le x \le 1)$

Solution:

We have

$$f(x) = \lim_{n\to\infty} f_n(x) = \lim_{n\to\infty} nx(1-x)^n$$

$$= \lim_{n\to\infty} \frac{nx}{(1-x)^{-n}} \quad \left(\text{Form } \frac{\infty}{\infty}\right)$$

$$= \lim_{n\to\infty} \frac{nx}{-(1-x)^{-n}\log(1-x)}$$

$$= \lim_{n\to\infty} \frac{x(1-x)^n}{\log(1-x)} = 0$$

since $1 - x < 1$

Thus, $f(x) = 0 \quad \forall\, x \in [0, 1]$

Now let

$$y = |\,f_n(x) - f(x)\,| = nx\,(1 - x)^n$$

For maximum value of y, $\frac{dy}{dx} = 0$

$\therefore n(1 - x)^n - n^2x\,(1 - x)^{n-1} = 0$

or $\qquad x = \frac{1}{n+1}$

Also $\quad \frac{d^2y}{dx^2} = $ –ve quantity when $x = \frac{1}{x+1}$

Thus, $\qquad M_n$ = maximum value of y

$$= n\,\frac{1}{n+1}\left(1-\frac{1}{n+1}\right)^n$$

$$= \left(\frac{n}{n+1}\right)\left(\frac{n}{n+1}\right)^n$$

$$\text{Now} \quad \lim_{n\to\infty} M_n = \lim_{n\to\infty}\left(\frac{n}{n+1}\right)\left(\frac{1}{1+\frac{1}{n}}\right)^n = \frac{1}{e} \neq 0$$

Therefore, the given sequence is non-uniformly convergent.

Example 3:

Show that the sequence $\{f_n\}$, where $f_n(x) = x^n$ is uniformly convergent on [0, k], k < 1 and pointwise convergent on [0, 1].

Now

$$f(x) = \lim_{n\to\infty} f_n(x) = \begin{cases} 0, & 0 \le x < 1 \\ 1, & x = 1 \end{cases}$$

Thus the sequence converges pointwise to a discontinuous function on [0, 1].

Solution:

Let $\varepsilon > 0$ be given.

For $0 < x \le k < 1$, we have

$| f_n(x) - f(x) | = x^n < \varepsilon$

if $\left(\frac{1}{x}\right)^n > \frac{1}{\varepsilon}$

or if $n > \log(1/\varepsilon)/\log(1/x)$

This number, $\log(1/\varepsilon)/\log(1/x)$ increases with x, its maximum value being $\log(1/\varepsilon)/\log(1/k)$ in $]0, k]$, $k > 0$.

Now let N be an integer $\ge \log(1/\varepsilon)/\log(1/k)$.

$\therefore | f_n(x) - f(x) | < \varepsilon, \quad \forall\, n \ge N,\ 0 < x < 1$

Again at $x = 0$,

$| f_n(x) - f(x) | = 0 < \varepsilon, \quad \forall\, n \ge 1$

Thus for any $\varepsilon > 0$, $\exists$ N such that for all $x \in [0, k]$, $k < 1$

$| f_n(x) - f(x) | < \varepsilon, \quad n \ge N$

Therefore, the sequence $\{f_n\}$ is uniformly convergent in [0, k], k < 1.

However, the number $\log(1/\varepsilon)/\log(1/x) \to \infty$ as $x \to 1$ so that it is not possible to find an integer N such that $| f_n(x) - f(x) < \varepsilon$, all $n \ge N$ and all x in [0, 1].

Hence the sequence is not uniformly convergent on any interval containing 1 and in particular on [0, 1].

Note: A point, like x = 1 which is such that the sequence is not uniformly convergent in any interval containing x = 1, is called a *point of non-uniform convergence.*

A TEST FOR UNIFORM CONVERGENCE OF SEQUENCE

Theorem 3:

Let $\{f_n\}$ *be a sequence of functions such that*

$$\lim_{n\to\infty} f_n(x) = f(x) \quad \forall\ x \in [a, b].$$

Let $M_n = \sup \{ \mid f_n(x) - f(x) \mid : x \in [a, b]\}$ ...(1)

Then $f_n \to f$ *uniformly on* $[a, b]$ *iff* $m_n \to o$ *as* $n \to \infty$.

Proof:

(Necessary Condition) : Let $f_n \to f$ uniformly on [a, b].

Then for any $\varepsilon > 0$, there exists a positive integer m such that

$\mid f_n(x) - f(x) \mid < \varepsilon\ \forall\ n \geq m$ and $\forall\ x \in [a, b]$

$\Rightarrow \sup \{ \mid f_n(x) - f(x) \mid : x \in [a, b]\} < \varepsilon\ \forall\ n \geq m$

$\Rightarrow M_n < \varepsilon\ \forall\ n \geq m \Rightarrow \mid M_n - 0 \mid < \varepsilon\ \forall\ n \geq m.$

Hence $M_n \to 0$ as $n \to \infty$.

(Sufficient Condition) : Let $M_n \to 0$ as $n \to \infty$.

Then for any $\varepsilon > 0$, there exists some positive integer m such that

$\mid M_n - 0 \mid < \varepsilon\ \forall\ n \geq m \Rightarrow M_n < \varepsilon\ \ \forall\ n \geq m$

$\Rightarrow \sup \{ \mid f_n(x) - f(x) \mid : x \in [a, b]\} < \varepsilon\ \ \forall\ n \geq m$

$\Rightarrow \mid f_n(x) - f(x) \mid < \varepsilon\ \forall\ n \geq m, \quad \forall\ x \in [a, b].$

Hence, $f_n \to f$ uniformly on [a, b].

Theorem 4:

(Weierstrass's M-test). A series of functions Σf_n *will converge uniformly (and absolutely) on [a, b] if there exists a convergent series* ΣM_n *of positive numbers such that for all* $x \in [a, b]$

$\mid f_n(x) \mid \leq M_n$, *for all n*

Proof:

Let $\varepsilon > 0$ be a positive number,

Since ΣM_n is convergent, therefore there exists a positive integer N such that

$$| M_{n+1} + M_{n+2} + ... + M_{n+p} | < \varepsilon \quad \forall \quad n \geq N, \quad p \geq 1 \qquad ...(1)$$

Hence for all x ∈ [a, b] and for all n ≥ N, p ≥ 1, we have

$$| f_{n+1} + f_{n+2} + ... + f_{n+p} (x) \leq | f_{n+1} (x) | + | f_{n+2} (x) | + ...$$
$$+ | f_{n+p} (x) | \qquad ...(2)$$
$$\leq M_{n+1} + M_{n+2} + ... + M_{n+p} < \varepsilon \qquad ...(3)$$

(2) and (3) imply that Σf_n is uniformly and absolutely convergent on [a, b].

Remark 1:

The converse is not asserted and is, in fact, not true, i.e., non-convergence of ΣM_n does not imply anything as far as Σf_n is concerned.

Remark 2:

Series which satisfy the M-test have been called normally convergent by Baire, to emphasise the fact that such series are uniformly as well as absolutely convergent. The terminology has the additional advantage of emphasising the fact that the test can be applied to nearly all series in ordinary everyday use.

UNIFORM CONVERGENCE AND CONTINUITY

Theorem 5:

If $\{f_n\}$ is sequence of continuous functions on [a, b] and if $f_n \to f$ uniformly on [a, b], then f is continuous on [a, b].

Prove that the uniform limit of a uniformly convergent sequence of continuous functions is continuous.

Proof:

Let c be any point of [a, b]. Let ε > 0 be given.

Since $f_n \to f$ uniformly on [a, b], so there exists a positive integer m in such that ∀ x ∈ [a, b],

$$| f_n (x) - f(x) | < \tfrac{1}{3} \varepsilon \quad \forall \; n \geq m.$$

In particular, $| f_m(x) - f(x) | < \tfrac{1}{3} \varepsilon \quad \forall \; x \in [a, b]$ ...(1)

$\Rightarrow | f_m(c) - f(c) | < \tfrac{1}{3} \varepsilon.$ ...(2)

Since f_m is continuous at c, there exists some $\delta > 0$ such that

$$| f_m(x) - f_m(c) | < \tfrac{1}{3}\varepsilon \ \forall\ x \in]c - d, c + d[. \qquad ...(3)$$

Now

$$| f(x) - f(c) | = | f(x) - f_m(x) + f_m(x) - f_m(c) + f_m(c) - f(c) |$$
$$\leq | f(x) - f_m(x) | + | f_m(x) - f_m(c) | + |f_m(c) - f(c) |$$
$$< \tfrac{1}{3}\ \varepsilon + \tfrac{1}{3}\ \varepsilon + \tfrac{1}{3}\ \varepsilon = \varepsilon\ \forall\ x \in]\ c - \delta, c + \delta\ [$$

[using (1), (2), (3)]

$\therefore | f(x) - f(c) | < \varepsilon\ \forall\ x \in]\ c - \delta, c + \delta[.$

Thus, f is continuous at c. As $c \in [a, b]$ is arbitrary, f is continuous on [a, b].

Theorem 6:

If a series Σf_n converges uniformly to f in [a, b] where each f_n is continuous in [a, b], then f is continuous in [a, b].

Proof:

Since Σf_n converges uniformly to f on [a, b], so for $\varepsilon >$, there exists a positive integer m such that for all $x \in [a, b]$.

$$| S_n(x) - f(x) | < \tfrac{1}{3}\ \varepsilon\ \ \forall\ \ n \geq m, S_n(x) = \sum_{r=1}^{n} f_r (x).$$

Let c be any point of [a, b]. It follows that

$$| S_m(x) - f(x) | < \tfrac{1}{3}\ \varepsilon\ \ \forall\ x \in [a, b], \qquad ...(1)$$

and
$$| S_m(c) - f(c) | < \tfrac{1}{3}\ \varepsilon. \qquad ...(2)$$

Since each $f_n(x)$ is continuous at c, so

$$S_m(x) = \sum_{r=1}^{n} f_r (x) \text{ is continuous at c.}$$

For $\varepsilon > 0$, there exists some $\delta > 0$ such that

$$| S_m(x) - S_m(c) | < \tfrac{1}{3}\ \in, \forall\ x \in [c - \delta, c + \delta]. \qquad ....(3)$$

$$| S_m(x) + S_m(x) | = | f(x) - S_m(x) + S_m(x) - S_m(c) + S_m(c) - f(c) |$$
$$\leq | f(x) - S_m(x) | + | S_m(x) - S_m(c) | + |S_m(c) - f(c) |$$
$$< \tfrac{1}{3}\varepsilon + \tfrac{1}{3}\varepsilon + \tfrac{1}{3}\varepsilon = \varepsilon\ \forall\ x \in\ ]\ c - \delta, c + \delta[$$

[using (1), (2), (3)]

$\therefore |f(x) - f(c)| < \varepsilon \quad \forall x \in]c - \delta, c + \delta[$.

Hence, f is continuous at $c \in [a, b]$ and so in [a, b].

Notes:

1. The converse of Theorems 6 and 7 may not be true. For example, $\{f_n\}$ defined by $f_n(x) = n\ xe^{-nx^2}$ is a sequence of continuous functions is not uniformly convergent in [0, 1] but the point-wise limit f, given by $f(x) = 0\ x \in [0, 1]$, is a continuous function in [0, 1].
2. If the limit function of a sequence of continuous functions is not continuous on [a, b], then the convergence cannot be uniform.
3. If the sum function of a series of continuous functions is not continuous in [a, b], then the convergence cannot be uniform.

Theorem 7:

(Abel's test). If $b_n(x)$ is a positive, monotonic decreasing function of n for each fixed value of x in the interval [a, b], and $b_n(x)$ is bounded for all values of n and x concerned, and if the series $\sum u_n(x)$ is uniformly convergent on [a, b], then so also is the series $\sum b_n(x)u_n(x)$.

Proof:

Since $b_n(x)$ is bounded for all values of n and for x in [a, b], therefore there exists a number $K > 0$, independent of x and n, such that for all $x \in$ [a, b],

$$0 \le b_n(x) \le K, \qquad (\text{for } n = 1, 2, 3, \ldots) \qquad \ldots(1)$$

Again since $\Sigma u_n(x)$ converges uniformly on [a, b], therefore for any $\varepsilon > 0$, we can find an integer N such that

$$\left|\sum_{r=n+1}^{n+p} u_r(x)\right| < \frac{\varepsilon}{K}, \qquad \forall\, n \ge N,\ p \ge 1 \qquad \ldots(2)$$

Hence using Abel's Lemma, we get

$$\left|\sum_{r=n+1}^{n+p} b_r(x)u_r(x)\right| \le b_{n+1}(x) \max_{q=1,2,\ldots p} \left|\sum_{r=n+1}^{n+q} u_r(x)\right|$$

$$< K\,\frac{\varepsilon}{K} = \varepsilon, \text{ for } n \ge N,\ p \ge 1,\ a \le x \le b.$$

$\Rightarrow \Sigma\, b_n(x)U_n(x)$ is uniformly convergent on [a, b].

Corollary 1:

A uniformly convergent series $\Sigma u_n(x)$ remains uniformly convergent on [a, b], if its terms are each multiplied by a function $a_n(x)$, $a \le x \le b$, provided that the sequence $\{a_n(x)\}$ is uniformly bounded on [a, b] (ie., $\exists K > 0$, such that $| a_n(x) | \le K$, for all x in [a, b] and for all n), and monotonic in n, for each $x \in [a, b]$.

Proof:

Under the given conditions, $\{a_n(x)\}$ converges point-wise. Let us write for each $x \in [a, b]$,

$$b_n(x) = \left\{\lim_{n\to\infty} a_n(x)\right\} - a_n(x), \text{ or } a_n(x) - \lim_{n\to\infty} a_n(x),$$

according as $\{a_n(x)\}$ is monotonic increasing or decreasing. With this function $b_n(x)$, we deduce as above that the series $\Sigma\, b_n(x)\, u_n(x)$ converges uniformly on [a, b]. Also, since $\Sigma\, u_n(x)$ and hence $\Sigma\, [\lim a_n(x)]\, u_n(x)$ ($\because |\lim a_n(x)| \le K$, $a \le x \le b$) converges uniformly on [a, b]. The uniform convergence of $\Sigma\, a_n(x)\, u_n(x)$, then follows easily.

Corollary 2:

If $\sum_{n=1}^{\infty} a_n x^n$ is a (power) series which converges for all values of x, where $| x | < R$, then $\Sigma a_n x^n$ is uniformly convergent in [0, R] if and only if $\Sigma a_n R^n$ is convergent.

Proof:

Let $\Sigma\, a_n R^n$ be convergent, so that being a series of real numbers, it is uniformly convergent in [0, R].

Now, since $\Sigma\, a_n R^n$ is uniformly convergent, and $(x/R)^n$ is a positive monotonic decreasing bounded function of n, for each value of x in [0, R], therefore by Abel's test, the series $\Sigma\, a_n R^n (x/R)^n \quad \Sigma\, a_n x^n R^n$ is uniformly convergent on [a, b].

If the series $\Sigma\, a_n x^n$ is uniformly convergent in [0, R], it is obviously convergent at x = R.

Theorem 8:

If a series Σf_n converges uniformly to f on [a, b] and each f_n is continuous on [a, b], then f is integrable on [a, b] and

$$\int_a^b f\, dt = \sum_{n=1}^{\infty} \int_a^b f_n\, dt.$$

Proof:

By theorem 6, f is continuous on [a, b] and so f is integrable on [a, b]. Since each $f_n(x)$ is continuous on [a, b], so

$$S_n(x) = \sum_{r=1}^{n} f_r(x) \text{ is also continuous and integrable on } [a, b].$$

Further,

$$\sum_{r=1}^{n} \int_a^b f_r \, dt = \int_a^b \sum_{r=1}^{n} f_r dt. \quad ...(1)$$

By the uniform convergence of the series, given any $\varepsilon > 0$, there exists a positive integer m such that for all x in [a, b],

$$| S_n (x) - f(x) | < \varepsilon/(b - a) \; \forall \; n \geq m \quad ...(2)$$

$$\text{Now } \left| \sum_{1}^{n} \int_a^b f_r \, dt - \int_a^b f \, dt \right| = \left| \int_a^b \sum_{1}^{n} \int_a^b f_r \, dt - \int_a^b f \, dt \right|, \text{ by (1)}$$

$$= \left| \int_a^b (S_n - f) \, dt \right| \leq \int_a^b |S_n - f| \, dt$$

$$< \frac{\varepsilon}{(b-a)} (b - a) = \varepsilon \quad \forall \; n \geq m, \text{ using (2).}$$

$$\therefore \lim_{n \to \infty} \sum_{r=1}^{n} \int_a^b f_r dt = \int_a^b f dt.$$

$$\text{Hence, } \sum_{r=1}^{\infty} \int_a^b f_r dt = \int_a^b f dt.$$

Theorem 9:

If a series Σf_n converges uniformly to f on [a, b] and each f_n is integrable on [a, b], then f is integrable on [a, b]. Further

$$\int_a^b f dt = \sum_{n=1}^{\infty} \int_a^b f_n dt.$$

In this case, we say that the series is integrable term by term.

Remark 1:

A sequence (series) may converge to an integrable limit without being uniformly convergent.

Remark 2:

If $f_n \to f$ (point-wise) and f is not integrable (where each f_n is integrable), then the convergence cannot be uniform.

Further, if

$$\lim_{n\to\infty} \int_a^b f_n dt \neq \int_a^b fdt$$

(where each f_n is integrable), then also the convergence cannot be uniform.

The similar results are also true for a series Σf_n.

UNIFORM CONVERGENCE AND INTEGRABILITY

Theorem 10:

If a sequence $\{f_n\}$ converges uniformly to f on $[a, b]$, and each f_n is integrable on $[a, b]$, then f is integrable o n $[a, b]$ and

$$\int_a^b fdt \lim_{n\to\infty} \int_a^b f_n dt.$$

Proof:

Let $\varepsilon > 0$ be given. Since $f_n \to f$ uniformly on [a, b], there exists a positive integer m such that for all $x \in [a, b]$.

$$| f_n(x) - f(x) | < \frac{\varepsilon}{3(b-a)} \quad \forall\, n \geq m. \qquad ...(1)$$

For n = m, we obtain

$$| f_m(x) - f(x) | < \frac{\varepsilon}{3(b-a)} \qquad ...(2)$$

Since f_m is integrable on [a, b], there exists a partition P of [a, b] such that

$$U(P, f_m) - L(P, f_m) < \tfrac{1}{3}\varepsilon \qquad ...(3)$$

From (2), we get

$$f(x) - \frac{\varepsilon}{3(b-a)} < f_m(x) < f(x) + \frac{\varepsilon}{3(b-a)} \qquad ...(4)$$

$$\Rightarrow \quad f(x) < f_m(x) + \frac{\varepsilon}{3(b-a)}$$

$$\Rightarrow \quad U(P, f) < U(P, f_m) + \frac{\varepsilon}{3(b-a)}.(b-a) \qquad ...(5)$$

Again, from (4), $f(x) > f_m(x) - \frac{\varepsilon}{3(b-a)}$

$\Rightarrow \quad L(P, f) > L(P, f_m) - \frac{\varepsilon}{3(b-a)}(b-a)$ ...(6)

From (5) and (6), we get

$$U(P, f) - L P, f) < U(P, f_m) - L(P, f_m) + \tfrac{1}{3}\varepsilon$$

$$< \tfrac{1}{3}\varepsilon + \tfrac{2}{3}\varepsilon = \varepsilon, \text{ by (3)}.$$

Hence, f is integrable on [a, b].

Now, we prove the second part.

Since $f_n \to f$ uniformly on [a, b], there exists a positive integer N such that for all $x \in [a, b]$,

$$|f_n(x) - f(x)| < \varepsilon/(b-a) \quad \forall\, n \geq N. \quad ...(7)$$

Consider $\left|\int_a^b f_n dt - \int_a^b f_n dt\right| = \left|\int_a^b (f_n - f)\, dt\right|$

$$\leq \int_a^b |f_n - f|\, dt < \frac{\varepsilon}{(b-a)}(b-a) = \varepsilon, \text{ by (7)}$$

$\therefore \quad \left|\int_a^b f_n\, dt - \int_a^b f_n\, dt\right| < \varepsilon \;\; \forall\, n \geq N.$

Hence, $\lim_{n\to\infty} \int_a^b f_n dt = \int_a^b f dt.$

Hence, the theorem.

UNIFORM CONVERGENCE AND DIFFERENTIATION

The series $\sum \frac{\sin nx}{n^2}$ converges uniformly for all values of x, and every term, without exception, is continuous and differentiable. The series of differentials, $\sum \frac{\cos nx}{n}$, however, diverges at x = 0. The situation, therefore, seems to be different in the case of differentiation, and accordingly the theorem on term by term differentiation must be of a different stamp.

Theorem 11:

(A) *Let $\{f_n\}$ be a sequence of differentiable functions an [a, b] such that it converges at least at one point $x_0 \in [a, b]$. If the sequence of differentials $\{f_n'\}$ converges uniformly to G on [a, b] then the given sequence $\{f_n\}$ converges uniformly on [a, b] to f, and $f'(x) = G(x)$.*

(B) Let Σf_n be a series of differentiable functions of [a, b] and such that it converges at least at one point $x_0 \in [a, b]$. If the series of differentials $\Sigma f_n'$ converges uniformly to G on [a, b], then the given series Σf_n converges uniformly on [a, b] to f, and $f'(x) = G(x)$.

(The existence of f' is part of the conclusion.)

Proof:

(A) Let $\varepsilon > 0$ be any number.

By the convergence of $\{f_n(x_0)\}$ and uniform convergence of $\{f_n'\}$, for $\varepsilon > 0$, we can choose a positive integer N such that for all $x \in [a, b]$

$$| f_{n+p}(x_0) - f_n(x_0) | < \varepsilon/2, \qquad \forall\, n \geq N,\ P \geq 1 \qquad ...(1)$$

$$| f'_{n+p}(x) - f_n'(x) | < \varepsilon/2(b-a), \qquad \forall\, n \geq N,\ p \geq 1 \qquad ...(2)$$

Applying Lagrange's mean value theorem to the function $(f_{n+p} - f_n)$ for any two points x and t of [a, b], we get for $x < \xi < t$, for all $n \geq N$, $p \geq 1$

$$| f_{n+p}(x) - f_n(x) - f_{n+p}(t) + f_n(t) | = | x - t | \, | f'_{n+p}(\xi) - f_n'(\xi) |$$

$$< \frac{|x-t|\varepsilon}{2(b-a)} \qquad ...(3)$$

$$< \varepsilon/2 \qquad ...(3a)$$

and

$$| f_{n+p}(x) - f_n(x) | \leq | f_{n+p}(x) - f_n(x) - f_{n+p}(x_0) + f_n(x_0) |$$
$$+ | f_{n+p}(x_0) - f_n(x_0) |$$
$$< \tfrac{1}{2}\varepsilon + \tfrac{1}{2}\varepsilon = \varepsilon \qquad \text{[using (1), (3a)\}}$$

$\Rightarrow$The sequence $\{f_n\}$ uniformly converges on [a, b].

Let it converge to f, say.

For a fixed x on [a, b] and for $t \in [a, b]$, $t \neq x$, let us define

$$\phi_n(t) = \frac{f_n(t) - f_n(x)}{t - x}, \; n = 1, 2, 3, ... \qquad ...(4)$$

$$\phi(t) = \frac{f(t) - f(x)}{t - x} \qquad ...(5)$$

Since each f_n is differentiable, therefore for each n

$$\lim_{t \to x} \phi_n(t) = f'_n(x) \qquad (6)$$

$$\therefore | \phi_{n+p}(t) - \phi_n(t) | = \frac{1}{|t-x|} | f_{n+p}(t) - f_{n+p}(x) - f_n(t) + f_n(x) |$$

$$= \frac{1}{|t-x|} | \{f_{n+p}(t) - f_n(t)\} - \{f_{n+p}(x) - f_n(x)\} |$$

$$< \frac{\varepsilon}{2(b-a)}, \quad \forall\, n \geq N,\ p \geq 1 \qquad \text{(using (3))}$$

so that $\{\phi_n(t)\}$ converges uniformly on [a, b], for $t \neq x$.

Since $\{f_n\}$ also converges uniformly to f, therefore from (4).

$$\lim_{n\to\infty} \phi_n(t) = \lim_{n\to\infty} \frac{f_n(t) - f_n(x)}{t-x} = \frac{f(t) - f(x)}{t-x} = \phi(t)$$

Thus, $\{\phi_n(t)\}$ converges uniformly to $\phi(t)$ on [a, b], for $t \in [a, b]$, $t \neq x$.

Applying Theorem 7(A) to the uniformly convergent sequence $\{\phi_n(t)\}$ and using (6), we get

$$\lim_{t\to x} \phi_n(t) = \lim_{n\to\infty} f_n{}'(x) = G(x)$$

$$\Rightarrow \lim_{t\to x} \phi_n(t) = \text{exists,}$$

and therefore, (5) implies that f is differentiable and

$$\lim_{t\to x} \phi_n(t) = f'(x)$$

Hence

$$f'(x) = G(x) = \lim_{n\to\infty} f'(x)$$

This completes the proof of the theorem.

If in addition to the above hypothesis of the theorem, continuity of the functions $f_n{}'$ is also assumed, then a much shorter proof of the theorem exists.

We now prove a simple version of the theorem for series; that for sequence may be written down on the same lines.

WEIERSTRASS'S APPROXIMATION THEOREM

Theorem:

If f is a real valued continuous function defined on a closed interval [a, b], then there exists a sequence of real polynomials $\{P_n(x)\}$ which converges uniformly to f(x) on [a, b].

Proof:

Without any loss of generality, we may suppose that [a, b] = [0, 1].

Let $g(x) = f(x) - f(0) - x\,[f(1) - f(0)]$, $0 \leq x \leq 1$.

Here g(0) = 0 = g)1) and further f(x) – g(x) is a polynomial of degree one. If g(x) can be expressed as the limit of a uniformly convergent sequence of polynomials, then the same is true of f(x), since f(x) = [f(x) – g(x)] + g(x) and f(x) – g(x) is a polynomial.

So we may assume that f(0) = f(1) = 0.

Since f is defined on [0, 1], we may take

f(x) = 0 outside [0, 1].

We write $q_n(x) = c_n(1 - x^2)^n$, $n \in N$; ...(1)

where C_n is such that

$$\int_{-1}^{1} Q_n(x)\,dx = 1, n \in N \quad ....(2)$$

$$\text{or } 1 = \int_{-1}^{1} C_n(1-x^2)^n\,dx = 2C_n \int_0^1 (1-x^2)^n dx$$

$$\geq 2C_n \int_0^{1/\sqrt{n}} (1-x^2)^n dx$$

$$\geq 2C_n \int_0^{1/\sqrt{n}} (1-nx^2)dx \quad [\because (1-x^2)^n > (1-nx^2) \text{ for } x \in [0, 1]]$$

$$\therefore 1 \geq 2\,C_n \left| x - \frac{1}{3}nx^2 \right|_0^{1/\sqrt{n}} = \frac{4C_n}{3\sqrt{n}} > \frac{C_n}{\sqrt{n}}$$

Thus, $C_n < \sqrt{n}$. Using in (1), we get

$Q_n(x) < \sqrt{n}\,(1 - x^2)^n$

or $Q_n(x) \geq \sqrt{n}\,(1 - \delta^2)^n$, where $\delta \leq |x| \leq 1$. ...(3)

Let $P_n(x) = \int_{-1}^{1} f(x+t)\,Q_n(t)\,dt, 0 \leq x \leq 1$

or $P_n(x) = \int_{-1}^{-x} + \int_{-x}^{1-x} + \int_{1-x}^{1}$ ...(5)

Now, $-1 \leq t \leq -x \Rightarrow -1 + x \leq t + x \leq 0 \Rightarrow x + t$ lies outside [0, 1] and so f(x + t) = 0.

Again $1 - x \leq t \leq 1 \Rightarrow 1 \leq x + t \leq 1 + x \Rightarrow x + t$ lies outside [0, 1] and so f(x + t) = 0.

Thus the first and third integrals on the R.H.S. of (5) vanish.

Consequently,

$P_n(x) = \int_{-x}^{1-x} f(x+t)\,Q_n(t)\,dt$, from (5)

or $P_n(x) = \int_0^1 f(t)\,Q_n(t-x)\,dt$,

which is a real polynomial of degree 2n in x, by (1).

We now proceed to show that the sequence $\{P_n(x)\}$ converges uniformly to f on [0, 1].

Since f is continuous on the closed interval [0, 1], f is bounded and uniformly continuous on [0, 1]. Thus, there exists some number M > 0 such that

$$| f(x) | \leq M \quad \forall \quad x \in [0, 1] \qquad ...(6)$$

Since f is uniformly continuous on [0, 1], for any $\varepsilon > 0$, there exists some δ, $0 < \delta < 1$ such that for any two points x_1, x_2 in [0, 1],

$$| f(x_1) - f(x_2) | < \tfrac{1}{2} \varepsilon, \text{ when } | x_1 - x_2 | < \delta \leq 1. \qquad ...(7)$$

For $0 \leq x \leq 1$, we have, by (4),

$$| P_n(x) - f(x) | = \left| \int_{-1}^{1} f(x+t)\, Q_n(t)\, dt - f(x) \right|$$

$$= \left| \int_{-1}^{1} \{f(x+t) - f(x)\} Q_n(t)\, dt \right|, \text{ by (2)}$$

$$\leq \int_{-1}^{1} |f(x+t) - f(x)|\, Q_n(t)\, dt$$

$$= \int_{-1}^{-\delta} + \int_{-\delta}^{\delta} + \int_{\delta}^{1}$$

Using (6) and (7), we get

$$| P_n(x) - f(x) | \leq 2M \int_{-1}^{-\delta} Q_n(t)\, dt + \tfrac{1}{2}\varepsilon \int_{-\delta}^{\delta} Q_n(t)\, dt + 2M \int_{\delta}^{1} Q_n(t)\, dt$$

$[\because | f(x + t) - f(x) | \leq | f(x+t) | + | f(x) | \leq M + M = 2M$, by (6)]

$$| P_n(x) - f(x) | \leq 2M \sqrt{n}\, (1 - \delta^2) \left[\int_{-1}^{-\delta} dt + \int_{\delta}^{1} dt \right] + \frac{1}{2} \varepsilon \text{ by (3)}$$

$$= 2M \sqrt{n}\, (1 - \delta^2)^n\, [(1 - \delta) + (1 - \delta)] + \tfrac{1}{2} \varepsilon$$

$$< 4M \sqrt{n}\, (1 - \delta^n) + \tfrac{1}{2} \varepsilon \qquad (\because 0 < \delta < 1).$$

Now, $0 < \delta < 1 \Rightarrow 0 < (1 - \delta^2) < 1 \Rightarrow \lim_{n\to\infty} (1 - \delta^2)^n = 0.$

Thus, there exists a positive integer N such that

$$(1 - \delta^2)^n < \frac{\varepsilon}{8M\sqrt{n}} \quad \forall\, n \geq N.$$

$\therefore | P_n(x) - f(x) | < \tfrac{1}{2} \varepsilon + \tfrac{1}{2} \varepsilon = \varepsilon \quad \forall\, n \geq N,$

where N is independent of

Hence, $\{P_n(x)\}$ converges uniformly to f(x) on [0, 1].

Corollary:

For any interval [–a, a], there is a sequence of real polynomials $\{P_n(x)\}$ such that $P_n(0) = 0$, and that

$\lim_{n\to\infty} p_n(x) = |x|$, *uniformly on [–a, a].*

Proof:

Since | x | is a continuous function on [–a, a], so there exists a sequence $\{Q_n(x)\}$ of real polynomials such that

$$\lim_{n\to\infty} Q_n(x) = |x| \quad \forall x \in [-a, a].$$

In particular, $\lim_{n\to\infty} Q_n(0) = 0.$ $(\because 0 \in [-a, a])$

Let $P_n(x) = Q_n(x) - Q_n(0) \quad \forall n \in N, x \in [-a, a].$

Hence $\lim_{n\to\infty} P_n(x) = |x| - 0 = |x| \quad \forall x \in [-a, a].$

UNIFORM CONVERGENCE AND DERIVABILITY

Theorem 1:

Suppose a series Σf_n of derivable functions converges point-wise to f on [a, b], each f'_n is continuous on [a, b] and the series $\Sigma f'_n$ converges uniformly to a function G on [a, b]. Then

$$f'(x) = G(x) \quad \forall x \in [a, b]$$

or $$\frac{d}{dx}\left\{\sum_{n=1}^{\infty} f_n(x)\right\} = \sum_{n=1}^{\infty} \frac{d}{dx} f_n(x).$$

Further the given series $\Sigma f_n(x)$ converges uniformly to f on [a, b].

Proof:

Since the series $\Sigma f'_n$ of continuous functions converges uniformly to G on [a, b], therefore, its sum function G is continuous on [a, b]. Consequently $\int_a^x G(t)\,dt$ is differentiable and

$$\frac{d}{dx}\int_a^x G(t)\,dt = G(x) \ \forall x \in [a, b] \quad \text{...(1)}$$

We are given that

$$f(x) = \sum_{n=1}^{\infty} f_n(x), \qquad \forall x \in [a, b] \quad \text{...(2)}$$

Since each function f'_n being continuous is integrable on [a, b],

So we have

$$\int_a^x f'_n(t)\,dt = f_n(x) - f_n(a) \qquad \forall\ n \geq 1,\ x \in [a, b] \qquad ...(3)$$

From (2) and (3), we obtain

$$\sum_{n=1}^{\infty} \int_a^x f'_n(t)\delta = f(x) - f(a)\ \forall\ x \in [a, b] \qquad ...(4)$$

Since the series $\Sigma f'_n$ of continuous functions converges uniformly to G on [a, b], therefore we have

$$\int_a^x G(t)\,dt = \sum_{n=1}^{\infty} \int_a^x f'_n(t)\,dt,\ \forall\ x\ \in [a, b] \qquad \text{[Theorem 9]}$$

$$= f(x) - f(a) \qquad \forall\ x \in [a, b], \text{ by } (4)$$

$$\therefore \frac{d}{dx}\int_a^x G(t)\,dt = f'(x) \qquad \forall\ x \in [a, b]$$

Hence $f'(x) = G(x)$ $\qquad \forall\ x \in [a, b]$, by (1)

or $\quad \dfrac{d}{dx}\left\{\sum_{n=1}^{\infty} f_n(x)\right\} = \sum_{n=1}^{\infty} f_n'(x) \qquad \forall\ x \in [a, b]$

(ii) Finally we show that Σf_n converges uniformly to f on [a, b]

For any $x_0 \in [a, b]$, $\Sigma f_n(x_0)$ to $f(x_0)$ is point-wise convergent.

By the convergence of $\Sigma f_n(x_0)$ to $f(x_0)$ and uniform convergence of $\Sigma f_n'$ to f', for $\varepsilon > 0$, we can choose a positive integer m such that for all $x \in [a, b]$,

$$| S_n(x_0) - f(x_0) | < ½\,\varepsilon \quad \forall\ n \geq m, \qquad ...(5)$$

and $\quad |S_n'(x) - f'(x)| < \dfrac{\varepsilon}{2(b-a)} \qquad \forall\ n \geq m, \qquad ...(6)$

where $\quad S_n(x) = \sum_{r=1}^{n} f_r(x)$.

Applying Lagrange's mean value theorem to the function $[\delta - (x) - f(x)]$ for any two points x_0, x of [a, b], there exists some $x \in\ ]x_0, x[$ such that $\forall\ n \geq m$,

$$| \{S_n(x) - f(x)\} - \{S_n(x_0) - f(x_0)\} | - | x - x_0 |\, | S_n'(c) - f'(c) |$$

$$< \frac{\varepsilon|x - x_0|}{2(b-a)} < \frac{\varepsilon}{2}, \text{ [by (6)]} \qquad ...(7)$$

Now,

$$| S_n(x) - f(x) | = | \{s_n(x) - f(X)\} - \{S_n(x_0) - f(x_0)\} + \{S_n(x_0) - f(x_0)\}|$$

$$< ½\,\varepsilon + ½\,\varepsilon = \varepsilon \qquad \forall\ n \geq m \text{ and } \forall\ x \in [a, b].$$

(using (5) and (7))

Hence the series Σf_n converges uniformly to f on [a, b].

Theorem 2:

Suppose a sequence $\{f_n\}$ of differentiable functions converges point-wise to f on [a, b] and that each fn' is continuous on [a, b]. If the sequence $\{f_n'\}$ converges uniformly to G on [a, b], then the given sequence $\{f_n\}$ converges uniformly to f on [a, b], and f' (x) = G (x) or

$$f'(x) = \lim_{n\to\infty} f_n'(x).$$

SOLVED EXAMPLES

Example 1:

Test the series $\sum_{n=1}^{\infty} \frac{\sin nx}{n^p}$ *for uniform convergence in any finite interval.*

Solution:

(i) When P > 1, $\frac{\sin nx}{n^p} \le \frac{1}{n^p}$ ...(1) as $|\sin x| \le 1$.

So if u_n (x) = The *n*th term of the given series = $\frac{\sin nx}{n^p}$ and $M_n = \frac{1}{n^p}$

we have from (1) $|u_n(x)| \le M_n$.

But $\Sigma M_n = \sum \frac{1}{n^p}$ is convergent when p > 1.

Hence by Weierstrass's M-test it follows that $\sum \frac{\sin nx}{n^p}$ is uniformly convergent in any finite interval.

(ii) When $0 < p \le 1$ we have by taking $u_n(x) = \sin nx$ and v_n (x) $\frac{1}{n^p}$.

$$S_n(x) = \sum_{r=1}^{n} u_r(x) = \sin x + \sin 2x + \sin 3x \ldots + \sin nx$$

$$= \frac{\sin\left(x + \frac{n-1}{2}x\right)\sin\frac{nx}{2}}{\sin\frac{x}{2}} = \frac{\sin\frac{n+1}{2}x \sin\frac{nx}{2}}{\sin\frac{nx}{2}}$$

$$\text{Hence } |S_n(x)| = \frac{\left|\sin\frac{n+1}{2}x\right| . \left|\sin\frac{nx}{2}\right|}{\left|\sin\frac{nx}{2}\right|} \le \frac{1}{\left|\sin\frac{x}{2}\right|}$$

or $| S_n(x) | \le | \text{cosec } x/2 |$ ...(1)

But cosec x/2 becomes infinitely great if

$x = 2m\pi, m = 0, 1, 2, 3, ...$

Excluding these points from any finite interval we see that cosec x/2 is bounded for all the remaining values of that interval.

Also the sequence $\{v_n (x)\} = \left\{\frac{1}{n^p}\right\}$ is monotonic decreasing sequence converging to zero for $0 < p \le 1$.

Hence by Dirichlet's test the series $\sum_{n=1}^{\infty} \frac{\sin nx}{n^p}$ is uniformly convergent in every interval from which the points $x = 2m\pi$, $m = 0, 1, 2$ .... are excluded.

(iii) When p = 0 the given series is reduce to the series $\sum_{n=1}^{\infty} \sin nx$.

Example 2:

Examine for term by term integration

$$S_n(x) = nx\, e^{-nx^2}.$$

Solution:

$$S(x) = \lim_{n\to\infty} S_n(x) = \lim_{n\to\infty} \frac{nx}{e^{nx^2}}$$

$$= \lim_{n\to\infty} \frac{x}{e^{nx^2}.x^2} = 0$$

$S(x) = 0$

$\therefore \int_0^1 s(x)\, dx = 0.$

Also

$$\int_0^1 S_n(x)\, dx = \int_0^1 nx\, e^{-nx^2} dx = \left[-\tfrac{1}{2}e^{-nx^2}\right]_0^1$$

$$= \left[\tfrac{1}{2} - \tfrac{1}{2}e^{-n}\right]$$

Thus, $\lim_{n\to\infty} \int_0^1 S_n(x)\, dx = \lim_{n\to\infty} \left[\tfrac{1}{2} - \tfrac{1}{2}.1/e^n\right] = \tfrac{1}{2}$

$\therefore \int_0^1 S(x)\, dx \neq \lim_{n\to\infty} \int_0^1 S_n(x)\, dx.$

Thus, the series is not term by term integrable.

Example 3(a):

Test for uniform convergence and term by term integration the series for which nth partial sum $S_n(x)$ = nxe^{-ne^2} *the limits of integration between 0 to 1.*

Solution:

$$\text{Here } S(x) = \lim_{n\to\infty} [S_n(x)] \lim_{n\to\infty} \left[\frac{nx}{e^{nx^2}}\right]$$

$$= \lim_{n\to\infty} \left[\frac{nx}{1+nx^2+(n^2x^4)/2!+.......}\right] = 0$$

for all finite values of x.

Then in the interval $0 \le x \le 1$ we have

$$\int_0^1 S(x)\,dx = \int_0^1 0\,dx = 0 \qquad ...(1)$$

$$\text{Also } \lim_{n\to\infty} \int_0^1 S_n(x)\,dx = \lim_{n\to\infty} \int_0^1 nx\,e^{-nx^2}\,dx$$

$$= \lim_{n\to\infty} \left[\frac{e^{-nx^2}}{-2}\right]_0^1$$

$$= \lim_{n\to\infty} \left[\frac{1-e^{-n}}{2}\right]$$

$$= ½.$$

Since $\lim_{n\to\infty} \int_0^1 S_n(x)\,dx \ne \int_0^1 \lim_{n\to\infty} [S_n(x)]\,dx$, the series does not admit of term by term integration over the interval $0 \le x \le 1$. In fact $x = 0$ is a point of non-uniform convergence.

Note that the term by term integration is justified over the range (k, 1) where $0 < k < 1$. For we have

$$\lim_{n\to\infty} \int_k^1 S_n(x)\,dx = \lim_{n\to\infty} \int_k^1 nx\,e^{-nx^2}\,dx$$

$$= \lim_{n\to\infty} \left[\frac{e^{-nx^2}}{-2}\right]_k^1$$

$$= \lim_{n\to\infty} \frac{1}{2}\left[e^{-nk^2} - e^{-n}\right]$$

$$= 0$$

Thus, $\int_k^1 S(x)\,dx = \lim_{n\to\infty} \int_k^1 S_n(x)\,dx$ so that the series admits of term by term integration in (k, 1) where $0 < k < 1$.

Example 3(b):

Prove that $\int_0^1 \log \frac{1}{1-x}\,dx = 1$ *by expanding in powers of x and integrating term by term.*

Solution:

We have $\log \frac{1}{1-x} = -\log(1-x)$

$$= x + \frac{x^2}{2} + \frac{x^3}{3} + ... + \frac{x^n}{n} + ... \qquad ...(1)$$

Then $\lim_{n\to\infty} \int_0^1 S_n(x)\,dx = \lim_{n\to\infty} \left[\int_0^1 \left\{ x + \frac{x^2}{2} + \frac{x^3}{3} + ... + \frac{x^n}{n} \right\} dx \right]$

$$= \lim_{n\to\infty} \left[\int_0^1 x\,dx + \int_0^1 \frac{x^2}{2}dx + \int_0^1 \frac{x^3}{3}dx + ... + \int_0^1 \frac{x^n}{n}dx \right]$$

$$= \lim_{n\to\infty} \left[\frac{1}{1.2} + \frac{1}{2.3} + \frac{1}{3.4} + ... + \frac{1}{n(n+1)} \right]$$

$$= \lim_{n\to\infty} \left[\left(1-\tfrac{1}{2}\right) + \left(\tfrac{1}{2}-\tfrac{1}{3}\right) + \left(\tfrac{1}{3}-\tfrac{1}{4}\right) + ... + \left(\frac{1}{n} - \frac{1}{n+1}\right) \right] \qquad ...(2)$$

$$= \lim_{n\to\infty} \left[\left(1 - \frac{1}{n+1}\right) \right] = 1$$

and $S(x) = \log \frac{1}{1-x}$ which is ∞ at $x = 1$

Hence, $\int_0^1 S(x)\,dx = \int_0^1 \log \frac{1}{1-x}\,dx$

$$= \lim_{\epsilon\to\infty} \int_0^{1-\epsilon} \log \frac{1}{x+1}\,dx$$

$$= \lim_{\epsilon\to\infty} \left[\left\{ x \log \frac{1}{1-x} \right\}_0^{1-\epsilon} - \int_0^{1-\epsilon} \frac{x}{1-x}\,dx \right]$$

$$= \lim_{\in\to\infty}\left[(1-\in)\log 1/\in + \{x+\log(1-x)\}_0^{1-\in}\right]$$

$$= \lim_{\in\to\infty}\left[(1-\in)\log 1/\in + \{1-\in+\log\in\}\right]$$

$$= \lim_{\in\to\infty}\left[-(1-\in)\log\in + 1-\in+\log\in\right]$$

$$= \lim_{\in\to\infty}[\in\log\in + -\in] \qquad ...(3)$$

But $\lim_{\in\to\infty}[\in\log\in]\lim_{\in\to\infty}\left[\frac{\log\in}{1/\in}\right]$ $\left(\text{from }\frac{\infty}{\infty}\right)$

$$= \lim_{\in\to\infty}\left[\frac{1/\in}{-1/\in^2}\right] = \lim_{\varepsilon\to\infty}(-\in) = 0$$

$$\therefore \int_0^1 \log\frac{1}{1-x}\,dx = 1.$$

Since $\int_0^1 S(x)\,dx \lim_{n\to\infty}\int_0^1 S_n(x)\,dx = 1$, the term by term integration is justified i n (0, 1) and $\int_0^1 \log\frac{1}{1-x}\,dx = 1$.

Note that the sum function $\log\frac{1}{1-x}$ has infinite discontinuity at x = 1 so that the series is not uniformly convergent in any neighbourhood of x = 1 yet it admits of term by term integration.

Example 3(c):

Examine whether the series for which $S_n(x)$ $\frac{1}{n+n^3x^2}$ is differentiable term by term.

Solution:

We have $S(x) = \lim_{n\to\infty}[S_n(x)]$

$$= \lim_{n\to\infty}\left[\frac{1}{n+n^3x^2}\right] = 0 \qquad \text{for all values of x.}$$

$\therefore$ $S'(x) = 0$

Also $\lim_{n\to\infty}[S'_n(x)]\lim_{n\to\infty}\left[\frac{-1}{(n+n^3x^2)^2\,.\,2n^3x}\right] = 0$ for all values of x.

$\therefore \quad S'(x) = \lim_{n\to\infty} [S'_n(x)]$.

Hence, the term by term differentiation holds.

Note that the series of derivatives $\Sigma\, u_n'(x)$ is uniformly convergent for every real value of x since the sequence of $\{S'_n(x)\}$ is uniformly convergent as can be seen from the following working.

$$\text{Here } |S'_n(x) - S'(x)| = \left|\frac{-1}{2n^5x(1+n^2x^2)^2} = 0\right|$$

$$= \frac{1}{2n^5x(1+n^2x^2)^2}$$ which is < any preassigned ε_1 however small it may be, by choosing n independent of x and large enough for every x.

Hence the sequence $\{S'_n(x)\}$ is uniformly convergent for all real values of x and so the series $\Sigma\, u'_n(x)$ is uniformly convergent for every real value of x.

Example 4:

Show that the series

$$\frac{x}{x+1} + \frac{x}{(x+1)(2x+1)} + \frac{x}{(2x+1)(3x+1)} + \ldots$$

is uniformly convergent in (k, ∝; where k is any positive number. Show also that the series is non-uniformly convergent near the point x = 0.

Solution:

$$\text{Here } S_n(x) = \sum_{n=1}^{n} u_n(x)$$

where $U_n(x)$ = nth, term of the series.

$$= \frac{x}{x+1} + \frac{x}{(x+1)(2x+1)} + \frac{x}{(2x+1)(3x+1)} + \ldots + \frac{x}{\{(n-1)x+1\}\,(nx+1)}$$

$$= \left(1 - \frac{1}{x+1}\right) + \left(\frac{1}{x+1} - \frac{1}{2x+1}\right) + \left(\frac{1}{2x+1} - \frac{1}{3x+1}\right) + \ldots + \left(\frac{1}{(n-1)x+1} - \frac{1}{nx+1}\right)$$

$$= 1 - \frac{1}{nx+1} = \frac{nx}{nx+1}$$

$$\therefore\ s(x) = \lim_{n\to\infty}[S_n(x)] = \lim_{n\to\infty}\left(\frac{nx}{nx+1}\right) = \begin{cases} 1 \text{ when } x > 0 \\ 0 \text{ when } x = 0 \end{cases}$$

Then for $x > 0$ and for a given ε we have

$$|S_n(x) - S(x)| \le \varepsilon$$

$$\text{if } \left|\frac{nx}{nx+1} - 1\right| \le \varepsilon$$

$$\text{or} \quad \text{if } \left|\frac{-1}{nx+1}\right| \le \varepsilon$$

$$\text{or} \quad \text{if } \frac{1}{nx+1} \le \varepsilon \text{ or if } nx + 1 \ge 1/\varepsilon$$

$$\text{or} \quad \text{if } n \ge \frac{1-\varepsilon}{x\varepsilon} \qquad \text{...(1)}$$

Also (1) shows that if $x\to 0$, $n\to\infty$ so that it is not possible to choose N such that $|S_n(x) - S(x)| \le \varepsilon$ for every $n \ge N$ and for every x in $(0, \infty)$.

So the convergence is non-uniform near $x = 0$.

However, if we consider the interval (k, ∞) where k is any fixed number > 0 however large, then the maximum value of $\frac{1-\epsilon}{x\epsilon}$ is $\frac{1-\epsilon}{k\epsilon}$ so that we take N = any integer greater than $\frac{(1-\epsilon)}{k\epsilon}$ then $|S_n(x) - S(x)| \le \epsilon$ for every $n \ge N$ and for every x in (k, ∞).

Thus, the series is uniformly convergent in (k, ∞) where $k > 0$ and is non-uniformly convergent in $(0, \infty)$.

Hence $x = 0$ is a point of non-uniform convergence of the series.

Example 5:

Show that the series $\sum \frac{x}{n(n+1)}$ *is non-uniformly convergent in $(0, \infty)$.*

Solution:

$$S_n(x) = x\left[\frac{1}{1.2} + \frac{1}{2.3} + \frac{1}{3.4} + \dots + \frac{1}{n(n+1)}\right]$$

$$= x\left[\left(1-\frac{1}{2}\right) + \left(\frac{1}{2}-\frac{1}{3}\right) + \left(\frac{1}{3}-\frac{1}{4}\right) + \dots + \left(\frac{1}{n}-\frac{1}{n+1}\right)\right]$$

$$= x\left[1-\frac{1}{n+1}\right] = \frac{nx}{n+1}$$

Now, $s(x) = \lim_{n\to\infty} S_n(x) \lim_{n\to\infty} \frac{nx}{n+1} = x.$

We have $| S_n(x) - s(x) | \le \in \quad (\in < 0)$

$$\Rightarrow \left| \frac{nx}{n+1} - x \right| \le \in$$

$$\Rightarrow \left| \frac{-x}{n+1} \right| \le \in$$

$$\Rightarrow \frac{x}{n+1} \le \in$$

$\Rightarrow n + 1 \ge x/\in \quad$ thus $n \ge x/\in - 1.$

Obviously, if the interval is $(0, \infty)$, then no upper bound of N exists i.e. the given series is non-uniformly convergent in this interval, but if we take the interval (0, c) where c is any finite quantity however large, then series is uniformly convergent for in this case

N = integral part of $(c/\in - 1)$.

Example 6(a):

If f is continuous on [0, 1] such that

$$\int_0^1 x^n f(x)\, dx = 0, \text{ for } n = 0, 1, 2, 3, ... \qquad ...(A)$$

then show that f(x) = 0 on[0, 1].

Solution:

For any real polynomial,

$h(x) = a_0 + a_1x + a_2x^2 + ... + a_nx^n$

$$\int_0^1 h(x) f(x) = 0, \quad \text{[using (A)]} \qquad ...(1)$$

Since f is continuous on [0. 1], so by Weierstrass's approximation theorem, there exists a sequence $\{P_n(x)\}$ of real polynomials such that

$\lim_{n\to\infty} P_n(x) = f(x)$ uniformly $\forall\ x \in [0, 1]$

$\Rightarrow \lim_{n\to\infty} P_n(x) = f(x) = f^2(x)$ uniformly $\forall\ x \in [0, 1]$

Since f is continuous on [0, 1], therefore

$$\int_0^1 f^2\,dx = \lim_{n\to\infty} \int_0^1 P_n f\,dx.$$

$$= 0, \text{ by (1)}$$

$\therefore f^2 = 0$ on [0, 1]. Hence f = 0 on [0, 1].

Example 6(b):

State Weierstrass approximation theorem and assuming its validity for the interval [0, 1], prove it for any closed and bounded interval [a, b].

Solution:

Suppose f is continuous in [a, b]. The function defined by

$$g(x) = f[a + (b - a)x\} - f(a) - x(f\{b) - f(a)\} \qquad ...(1)$$

is continuous on [0, 1]. By Weierstrass approximation theorem,

$P_n(x) \to g(x)$ uniformly on [0, 1].

Replacing a + (b − a) x by x, and using (1),

$$P_n\left(\frac{x-a}{b-a}\right) \to f(x) - f(a) - \frac{x-a}{b-a}\{f(b) - f(a)\}$$

Hence, $P_n(x) \equiv P_n\left(\frac{x-a}{b-a}\right) + f(a) - \left(\frac{x-a}{b-a}\right)\{f(b) - f(a)\}$

converges uniformly to f(x) on [a, b].

Example 6(c):

Show that $\sum \frac{(-1)^n\, x^{2n}}{n^p\,(1+x^{2n})}$ *converges absolutely and uniformly for all real values of x if p > 1.*

Solution:

$$|f_n(x)| = \left|\frac{(-1)^n\, x^{2n}}{n^p(1+x^{2n})}\right| = \frac{x^{2n}}{n^p(1+x^{2n})}.$$

Clearly $x^{2n} < 1 + x^{2n} \quad \forall\, x \in R$ and $\forall\, n \in N$.

$\therefore \quad \frac{x^{2n}}{1+x^{2n}} < 1 \ \forall\, x \in R$ and $\forall\, n \in N$.

Thus, $\quad |f_n(x)| < \frac{1}{n^p} = M_n \quad \forall\, x \in R, \quad \forall\, n \in N.$

Since $\quad \sum M_n = \sum \frac{1}{n^p}$ is convergent $\qquad (\because p > 1).$

by μ-test, the given series is uniformly and absolutely convergent for all real values of x if p > 1.

Example 6(d):

Prove that the series

$$(1 - x) + x(1 - x) + x^2(1 - x) + \ldots .$$

is absolutely convergent in the interval $-c \le x < 1$ *where* $0 < c < 1$ *but that it is not uniformly convergent in this interval.*

Solution:

The given series is absolutely convergent in the given interval if the series

$$|1 - x| \; |x| \, |1 - x| + |x^2| \, |1 - x| + \ldots \qquad \ldots(1)$$

is convergent for every value of x in that interval.

But the series (1) is a geometric series with $|x|$ (which is < 1) as its common ratio.

Thus, the sum to n terms of the series (1)

$$= \frac{|1-x| \, [1-|x|^n]}{1-|x|} \text{ which tends to } \frac{|1-x|}{1-|x|}$$

for every x in $-c \le x < 1$ as $n \to \infty$

This shows that the given series is absolutely convergent in the interval $-c \le x < 1$ where $0 < c < 1$.

To test the series for uniform convergence, we have

$S_n(x)$ = Sum to n terms of the original series

$$= \frac{(1-x)(1-x^n)}{1-x} = 1 - x^n \qquad \ldots(1)$$

$$\therefore S(x) = \lim_{n\to\infty} S_n(x) = \begin{cases} 1 \text{ when } -c \le x < 1 \text{ and } 0 < c < 1 \\ 0 \text{ when } x = 1 \end{cases}$$

Then for $-c \le x < 1$ and for a given $\in$ we have

$|S_n(x) - S(x)| \le \in$ if $|(1 - x^n) - 1| \le \in$

$\Rightarrow$if $|x^n| \le \in$ or if $|1/x|^n \ge 1/\in$

$\Rightarrow$if $n \log |1/x| \ge \log 1/\in$

$$\Rightarrow \text{ if } n \ge \frac{\log 1/\in}{\log 1/|x|} \qquad \ldots(2)$$

Now (2) shows that if $x \to 1$, $n \to \infty$ so that it is not possible to choose a positive integer N such that $|S_n(x) - S(x)| < \varepsilon$ for every $n \ge N$ and for every x in $-c \le x < 1$.

So the convergence is non-uniform in the interval $-c \leq x < 1$.

Note that $x = 1$ is a point of non-uniform convergence in the interval $-c \leq x < 1$.

Example 7:

Show that the series

$$\sum_{n=0}^{\infty} \frac{(-1)^n x^{2^{n+1}}}{(2n+1)!} = x - \frac{x^2}{3!} + \frac{x^5}{5!} - \ldots$$

is uniformly convergent in every interval (a, b).

Solution:

Let M be any positive number greater than each of | a | and | b | so that for every x in (a, b) we have

$$\left| \frac{(-1)^n x^{2^{n+1}}}{(2n+1)!} \right| \leq \frac{M^{2^{n+1}}}{(2n+1)!}$$

Now consider the series $M + \frac{M^3}{2!} + \frac{M^5}{5!} + \ldots$ ...(1)

Since $e^M = 1 + m \frac{M^2}{2!} + \frac{M^3}{3!} + \ldots$

and $e^{-M} = 1 - M + \frac{M^2}{2!} + \frac{M^3}{3!} + \ldots$

we have $\frac{1}{2}(e^M - e^{-M}) = M + \frac{M^3}{3!} + \frac{M^5}{5!} + \ldots$ ...(2)

(2) shows that the series (1) is convergent whatever the positive constant M may be.

Hence, by Weierstrass's M-test the given series

$$\sum_{n=0}^{\infty} \frac{(-1)^n x^{2n+1}}{(2n+1)!} \text{ is uniformly convergent in (a, b)}$$

Note that the above example justified the definition

$$\sin x = x \frac{x^3}{3!} + \frac{x^5}{5!} - \ldots + \frac{(-1)^n x^{2n+1}}{(2n+1)!} + \ldots$$

Example 8:

Show that the series $\sum_{n=1}^{\infty} x^n(1-x^n)$ *is not uniformly convergent in the interval* $0 \le x \le 1$.

Solution:

The given series is $x(1-x) + x^2(1-x^2) + x^3(1-x^3) + ...$

Thus $S_n(x) = (x + x^2 + x^3 + ... + x^n) - (x^2 + x^4 + x^6 + ... + x^{2n})$

$$= \frac{x(1-x^n)}{1-x} - \frac{x^2(1-x^{2n})}{1-x^2}$$

$$\therefore\ S(x) = \lim_{n\to\infty}[S_n(x)] = \begin{cases} \dfrac{x}{1-x} - \dfrac{x^2}{1-x^2} & \text{when } 0 < x < 1 \\ 0 & \text{when } x = 0 \text{ or } x = 1 \end{cases}$$

Then for $0 < x < 1$ and for a given $\in$ we have

$$| S_n(x) - S(x) | \le \in$$

$$\text{if} \left| \frac{x(1-x^n)}{1-x} - \frac{x^2(1-2^{2n})}{1-x^2} - \frac{x}{1-x} + \frac{x^2}{1-x^2} \right| \le \in$$

$$\Rightarrow \text{if} \left| \frac{x^{2n+2}}{1-x^2} - \frac{x^{n+1}}{1-x} \right| \le \in$$

$$\Rightarrow \text{if} \left| \frac{x^{2n+2} - (x+1)x^{n+1}}{1-x^2} \right| \le \in$$

$\Rightarrow$if $(x + 1)x^{n+1} - x^{2n+2} \le \in (1 - x^2)$ $\quad$ {for $x < 1$ $(x + 1)x^{n+1} < x^{2n+2}$

$\Rightarrow$if $-(x + 1)x^{n+1} + x^{2n+2} \ge -\in (1 - x^2)$

$\Rightarrow$if $x^{2n+2} - (x + 1)x^{n+1} + (1 - x^2) \in \ge 0$

$$\Rightarrow \text{if } x^{n+1} \ge \frac{(x+1) \pm \sqrt{[(x+1) - 4 \in (1-x^2)]}}{2}$$

$$\Rightarrow \text{if } (n+1) \ge \frac{\log\left[\dfrac{(x+1) \pm \sqrt{[(x+1) - 4 \in (1-x^2)]}}{2}\right]}{\log x} \qquad ...(1)$$

Now (1) show that when $x \to 1$, $n \to \infty$ so that it is not possible to find m such that $| S_n(x) - S(x) | \le \in$ for every $n \ge N$ and for every x in $0 < x \le 1$.

Here x = 1 is a point of non-uniform convergence of the series so that the series is non-uniformly convergent in $0 \le x \le 1$.

Example 9:

Prove that the series

$$\frac{1}{a} - \frac{2a}{a^2 - 1^2}\cos x + \frac{2a}{a^2 - 2^2}\cos^2 x - .....$$

is uniformly convergent in any finite interval.

Solution:

Neglecting the first term we have

$$|u_n(x)| = \left|\frac{(-1)^{n-1}2a}{a^2 - n^2}\cos nx\right| \le \left|\frac{2a}{a^2 - n^2}\right| \quad ...(1)$$

But we have $n^2 - a^2 > \frac{1}{2}n^2$ if $\frac{1}{2}n^2 > a^2$

$\Rightarrow$ if $n > |a|\sqrt{2}$.

So if m is a positive integer just greater than $|a|\sqrt{2}$ then $|a^2 - n^2| =$

$|n^2 - a^2| < \frac{1}{2}n^2$ for every $n \ge m$ ...(2)

From (1) and (2) we have

$$|u_n(x)| \le \frac{2|a|}{|a^2 - n^2|} < \frac{2|a|}{\frac{1}{2}n^2}\ \frac{4|a|}{n^2} \quad \text{for } n \ge m \quad ...(3)$$

But $4|a|\sum 1/n^2$ is obviously convergent.

Hence the given series is uniformly convergent for $x \ge 0$ by Welerstrass's M-test.

Example 10(a):

Show that the series $\cos x + \frac{1}{2}\cos 2x + \frac{1}{3}\cos 3x + ...$ *converge uniformly in* $0 < a \le x \le b < 2\pi$.

Solution:

Let $u_n(x) = \cos nx$ and $v_n(x) = 1/n$.

Then $S_n(x) = \sum_{r=1}^{n} u_r(x) = \cos x + \cos 2x + \cos 3x + ... + \cos nx$

$$= \frac{\cos\left\{x + \frac{1}{2}(n-1)x\right\}\sin nx/2}{\sin x/2}$$

$$= \frac{\cos\frac{1}{2}(n+1)x \sin nx/2}{\sin x/2}$$

$$\therefore |S_n(x)| = \frac{|\cos\frac{1}{2}(n+1)x|\,|\sin nx/2|}{|\sin x/2|} \leq \frac{1}{|\sin x/2|}$$

$\leq |\text{cosec } x/2|$

$< k$ where k is the least upper bound of cosec x/2 in the open interval $(0, 2\pi)$ for cosec x/2 is bounded for all values of x in $0 < a \leq x \leq b < 2\pi$,

Also the sequence $\{v_n(x)\} = \{1/n\}$ is monotonic decreasing converging uniformly to zero in $0 < a \leq x \leq b < 2\pi$.

Hence by Dirichlet's test the given series.

$\sum_1^{\infty} 1/n \cos nx$ is uniformly convergent in.

$0 < a \leq x \leq b < 2\pi$.

Example 10(b):

Show that the series $\sum_1^{\infty} \frac{x}{n(1+nx^2)}$ *converges uniformly for all real x.*

Solution:

Let $f_n(x) = \frac{x}{n(1+nx^2)}$

$$\therefore f_n'(x) = \frac{(1+nx^2) - x(2x^2)}{n(1+nx^2)} = \frac{1-nx^2}{n(1+nx^2)^2}$$

Now, $f_n'(x) = 0$

$\Rightarrow x = \pm 1/\sqrt{n}$.

It can be verified that $f_n'(x) < 0$ for $x = 1/\sqrt{n}$.

$$\therefore \text{Maximum value of } f_n(x) = \frac{1/\sqrt{n}}{n(1+1)} = \frac{1}{2n^{2/2}}$$

Thus, $f_n(x) < \frac{1}{2n^{3/2}} < \frac{1}{n^{3/2}} \quad \forall x \in R$
and $\forall n \in N$.

Since $\sum M_n = \sum \frac{1}{n^{3/2}}$ converges $\left(\because \frac{3}{2} > 1\right)$, so the given series is uniformly convergent for all real x.

Example 10(c):

Show that the series $\sum \frac{\cos n\theta}{n^p}, \sum \frac{\sin n\theta}{n^p}$ *converge uniformly for all values of* $p > 0$ *in* $[\alpha, 2\pi - \alpha]$, *where* $0 < \alpha < \pi$.

Solution:

Let p > 1 so that

$$\left|\frac{\cos n\theta}{n^p}\right| \leq \frac{1}{n^p} = M_n.$$

Since $\sum M_n = \sum \frac{1}{n^p}$ (p > 1) is convergent, so by Weierstrass's M-test, $\sum \frac{\cos n\theta}{n^p}$ is uniformly convergent $\forall\theta$.

For the case $0 < p \leq 1$, we take $b_n = \frac{1}{p}$ so that b_n is positive, monotonic decreasing function of n and converges uniformly to zero. Let $M_n(\theta) = \cos n\theta$.

$$\left|\sum_{r=1}^{n} u_r(\theta)\right| \leq \left|\operatorname{cosec} \frac{1}{2}\alpha\right| \; \forall\, n, \;\; \forall\theta \in [\alpha, 2\pi, -\alpha]$$

Hence, by Dirichlet's test, $\sum \frac{\cos n\theta}{n^p}$

converges uniformly in $[\alpha, 2\pi, -\alpha]$ for $0 < p \leq 1$.

Similarly, we can do the second part.

Example 10(d):

Show that the following series are uniformly convergent in $[-a, a]$, $0 < a < 1$:

(i) $\sum_{n=0}^{\infty} x^n$, (ii) $\sum_{n=0}^{\infty} (n+1)\, x^n$, (iii) $\sum_{n=0}^{\infty} n^2 x^n$,

Solution:

(i) $|x^n| = |x|^n \leq a^n \quad \forall\, x \in [-\alpha, \alpha]$ i.e., $|x| < \alpha$.

Let $M_n = \alpha^n$ so that $M_{n+1} = \alpha^{n+1}$

Now $\lim_{n\to\infty} \frac{M_{n+1}}{M_n} = a$, where $0 < \alpha < 1$.

By D' Alembert's Ratio Test, $\sum M_n$ is convergent.

Hence, by M-test, Σx^n is convergent $\forall$ x $\in [-\alpha, \alpha]$, where $0<\alpha<1$.

(ii) $| (n + 1)x^n | = (n + 1) | x |^n < \alpha^n (n + 1)$ $(\because | x | < \alpha)$

Let $M_n = \alpha^n (n + 1)$ so that $M_{n+1} = a^{n+1} (n + 2)$.

Now, $\lim_{n\to\infty} \frac{M_{n+1}}{M_n} = \lim_{n\to\infty} \frac{\alpha(n+2)}{(n+1)} = a < 1.$

By D' Alembert's Ratio Test, ΣM_n converges.

Hence, by M-test $\Sigma(n+1) x^n$ is uniform convergent in $[-\alpha, \alpha]$, where $0 < \alpha < 1$.

(iii) $| n^2.x^n | = n^2 | x |^n < \alpha^n n^2$.

Let $M_n = \alpha^n n^2$ so that $M_{n+1} = \alpha^{n+2} (n + 1)^2$.

Now, $\lim_{n\to\infty} \frac{M_{n+1}}{M_n} = \lim_{n\to\infty} \alpha\left(\frac{n+1}{n}\right)^2 = \alpha < 1.$

By D' Alembert's Ratio Test, Σm_n converges.

Hence, by M-test n^2x^n) x^n is uniform convergent in $[-\alpha, \alpha]$, where $0 < \alpha < 1$.

Example 11:

Show that the sequence $\{f_n(x)\} = \{x^n\}$ is not uniformly convergent on [0, 1].

Solution:

Let $f_n(x) = x^n$.

For x = 1, $f_n(x) = 1$ and so that point-wise limit is 1.

For $x \in [0, 1[$, $\lim_{n\to\infty} f_n(x) = \lim_{n\to\infty} x^n = 0$.

Thus, the point-wise limit function of $\{x^n\}$ is

$$f(x) = \begin{cases} 0 \text{ in } [0, 1[, \\ 1 \text{ for } x = 1. \end{cases}$$

Thus f is not continuous on [0, 1] and so the convergence is not uniform on [0, 1].

Example 12:

Show that the series $\sum_{n=1}^{\infty} \frac{x}{1+n^2x^2}$ is not uniformly convergent on [0, 1], while it is so on [½, 1].

Or

Show that $\sum_{n=1}^{\infty} \frac{x}{1+n^2x^2}$ *is uniformly convergent in [δ, 1] for any δ > 0 but is not uniformly convergent in [0, 1].*

Solution:

Let $f_n(x) = \frac{x}{1+n^2x^2}$, $x \in [0, 1]$, $\delta > 0$.

$\therefore |f_n(x)| = \left|\frac{x}{1+n^2x^2}\right| \le \frac{x}{1+n^2\delta^2}$

Let $M_n = \frac{x}{1+n^2\delta^2}\left(\sim \frac{1}{n^2}\right)$, so that ΣM_n is a convergent series and by M-test, $\Sigma f_n(x)$ is uniformly convergent in [δ, 1].

(ii) Let, if possible, $\Sigma f_n(x)$ be uniformly convergent on [0, 1].

By Cauchy's general principle of convergence, for δ = 1/6, there exists a positive integer m such that

$$| S_{n+p}(x) - S_n(n) | < \frac{1}{6} \quad \forall\, n \ge m,\ p \ge 1,\ \forall\, x \in [0, 1]$$

Example 13:

Show that $\int_1^{\infty} \frac{\sin x}{x^4}\,dx$ is absolutely convergent.

Solution:

The integral $\int_1^{\infty} \frac{\sin x}{x^4}\,dx$ will absolutely convergent if $\int_1^{\infty} \left|\frac{\sin x}{x^4}\right| dx$ is convergent.

Let $f(x) = \left|\frac{\sin x}{x^4}\right|$, then f(x) is bounded in the interval (1, ∞) we have

$$f(x) = \left|\frac{\sin x}{x^4}\right| \le 1/x^4 \sin x\ |\sin x| = 1$$

∴ By Comparison Test $\int_1^{\infty} f(x)\,dx$ is convergent if $\int_1^{\infty} \frac{1}{x^4}\,dx$ is convergent but the comparison integral $\int_1^{\infty} \frac{1}{x^4}\,dx$ is convergent because $\int_1^{\infty}$ Here n = 4 which is > 1

Hence $\int_1^{\infty} f(x)\,dx$ is convergent and so the given integral is absolutely convergent.

Taking p = n = m, we obtain

$$| S_{2m}(x) - S_m(x) | < \frac{1}{6} \qquad \forall\ x \in [0, 1]$$

or $\dfrac{x}{1+(2m)^2 x^2} + \dfrac{x}{1+(2m-1)^2 x^2} + ... + \dfrac{x}{1+(m+1)^2 x^2} < \dfrac{1}{6}$

or $\dfrac{mx}{1+(2m)^2 x^2} < \dfrac{1}{6} \quad \forall\ x \in [0, 1]$

Taking x = 1/m, the above inequality becomes

$\dfrac{1}{5} < \dfrac{1}{6}$, which is impossible.

Hence $\Sigma f_n(x)$ is not uniformly convergent on [0, 1].

Example 14(a):

Show that the series

$$\frac{1}{a} - \frac{2a}{a^2 - 1^2} \cos\theta + \frac{2a}{a^2 - 2^2} \cos 2\theta \ ......$$

is uniformly convergent w.r.t. θ in any finite interval.

Solution:

We know

$$| a^2 - n^2 | = | n^2 - a^2 | > ½\, n^2, \qquad ...(1)$$

when n exceeds a certain integer N. Using (1), we get

$$\left| \frac{(-1)^n 2a \cos n\theta}{a^2 - n^2} \right| \geq 2\ |a|.\,1\,.\,\frac{2}{n^2}$$

$$= \frac{4|a|}{n^2}, \qquad \forall\ n > N \text{ and } \forall\ \theta$$

Since $\sum M_n = \sum \dfrac{1}{n^2}$ is convergent, therefore

by Weierstrass's M-test, the given series is uniformly convergent in any finite interval.

Example 14(b):

Show that the sequence $\{f_n\}$, where

$$f_n(x) = \begin{cases} n^2 x, \ 0 \leq \ \leq 1/n \\ -n^2 x + 2n, 1/n \leq x \leq 2/n \\ 0, 2/n \leq x \leq 1 \end{cases}$$

is not uniformly convergent on [0, 1].

Solution:

We have $\lim_{n\to\infty} f_n(x) = 0 \quad \forall\, x \in [0, 1]$.

Thus the given sequence converges point-wise to f, where $f(x) = 0$ $\forall\, x \in [0, 1]$. Further each f_n and f are continuous on [0, 1]. Now

$$\int_0^1 f_n dx = \int_0^{1/n} n^2 x\, dx + \int_{1/n}^{2/n} (-n^2 x + 2n)\, dx + \int_{2/n}^1 0\, dx$$

$$= \frac{1}{2} - \frac{1}{2} n^2 \left(\frac{4}{n^2} - \frac{1}{n^2}\right) + 2n\left(\frac{2}{n} - \frac{1}{n}\right) + 0$$

$$= \frac{1}{2} - \frac{3}{2} + 2 = 1.$$

Also $\int_0^1 f\, dx = 0$

Thus, $\lim_{n\to\infty} \int_0^1 f\, dx \neq \int_0^1 f\, dx.$

Hence, the given sequence is not uniformly convergent on [0, 1].

Example 14(c):

Show that the sequence $\{f_n\}$, $f_n(x) = nx\,(1 - x^2)^n$, converges but not uniformly to f, where $f(x) = 0$, for $0 \le x \le 1$, and that

$$\lim_{n\to\infty} \int_0^1 f_n\, dx \neq \int_0^1 f\, dx.$$

Solution:

We have $f_n(x) = 0 \quad \forall\, n$ if $x = 0$ or 1.

Now $0 < x < 1 \Rightarrow 0 < 1 - x^2 < 1$.

$$\text{Now } \lim_{n\to\infty} \frac{f_{n+1}(x)}{f_n(x)} = \lim_{n\to\infty} \frac{(n+1)\, x(1-x^2)^{n+1}}{nx(1-x^2)^n}$$

$$= \lim_{n\to\infty} \left(1 + \frac{1}{n}\right)(1-x^2) \text{ for } 0 < x < 1$$

$$= (1 - x^2) < 1 \quad (\because\ 0 < x < 1)$$

$\therefore \lim_{n\to\infty} f_n(x) = 0$ for $0 < x < 1$.

Thus, the point-wise limit of $\{f_n\}$ is $f(x) = 0 \quad \forall\, x \in [0, 1]$

Now $\int_0^1 f_n(x)\, dx = \int_0^1 nx\,(1-x^2)^n\, dx$

$$= -\frac{n}{2(n+1)}\left|(1-x^2)^{n+1}\right|_0^1 = \frac{n}{2(n+1)}$$

$$\therefore \lim_{n\to\infty} \int_0^1 f_n(x)\,dx = \frac{1}{2} \text{ and } \int_0^1 f(x)\,dx = 0.$$

Since $\lim_{n\to\infty} \int_0^1 f_n(x)\,dx \neq \int_0^1 f(x)\,dx$, $\{f_n\}$ is not uniformly convergent to f on [0, 1]

Example 15:

Show that the sequence $\{f_n\}$, $f_n(x) = x/(1+nx^2)$ converges uniformly to a function f on [0, 1], and that the equation, $f'(x) = \lim_{n\to\infty} fn'(x)$ is true if $x \neq 0$ and false if $x = 0$. Why so?

Solution:

The sequence $\{f_n\}$ converges uniformly to $f(x) = 0 \quad \forall\, x \in R$.

For $x \neq 0$, $f_n'(x) = \dfrac{1-nx^2}{(1+nx^2)^2} \to 0$ as $n \to \infty$,

so that $\lim_{n\to\infty} f_n'(x)$ is true for $x \neq 0$.

For $x = 0$, $f_n'(0) = \lim_{x\to 0} \dfrac{f_n(x) - f_n(0)}{x-0} = \lim_{x\to 0} \dfrac{1}{1+nx^2} = 1$

and $\quad f'(0) = 0$. Thus

$\lim_{x\to\infty} f_n'(0) = 1 \neq f'(0)$ and so the given equation is false for $x = 0$. It is so because the sequence $\{f_n')$ is not uniformly convergent in any interval containing zero.

Example 16:

If $f(x) = \sum_{n=1}^{\infty} \dfrac{\sin nx}{n^2}$, *then prove that*

$$\int_0^{\pi} f(x)\,dx = 2\sum_{n=1}^{\infty} \frac{1}{(2n-1)^4}$$

Solution:

We have $\left|\dfrac{\sin nx}{n^3}\right| \leq \dfrac{1}{n^3} \;\forall\, x.$

By Weierstrass's M-test, the given series is uniformly convergent for all x, in particular $0 \leq x \leq p$, and can be integrated term by term. Thus

$$\int_0^{\pi} f(x)\,dx = \int_0^{\pi}\left(\sum_{n=-1}^{\infty}\frac{\sin nx}{n^3}\right)dx = \sum_{n=1}^{\infty}\int_0^{\pi}\frac{\sin nx}{n^3}\,dx$$

$$= \sum_{n=1}^{\infty}\left(\frac{1-\cos n\pi}{n^4}\right) = 2\left(\frac{1}{1^4}+\frac{1}{3^4}+\frac{1}{5^4}+\ldots\right)$$

$$= 2\sum_{n=1}^{\infty}\frac{1}{(2n-1)^4}.$$

Example 17:

Show that converges $\sum_{n=1}^{\infty}\frac{1}{1+n^2+n^4x^2}$ *uniformly for all values of x.*

Solution:

It is clear that

$$f_n(x) = \frac{1}{1+n^2+x^4n^2} < \frac{1}{n^2} \qquad \forall\, x \in R \text{ and } \forall\, n \in N.$$

Since $\sum M_n = \sum \frac{1}{n^2}$ is convergent, by Weierstrass's M-test, Σf_n is uniformly convergent for all $x \in R$.

Example 18(a):

Given $f(x) = \sum_{n=1}^{\infty}\frac{1}{n^2+n^4x^2}$, *justify the validity of*

$$f'(x) = -2x\sum_{n=1}^{\infty}\frac{1}{n^2(1+nx^2)^2},$$

or

Show that the series $\sum_{n=1}^{\infty}\frac{1}{n^2+n^4x^2}$ *is uniformly convergent for all x.*

Solution:

It is clear that

$$f_n(x) = \frac{1}{n^2+n^4x^2} \le \frac{1}{n^2} \qquad \forall\, x \in R \text{ and } \forall\, x \in N.$$

Since $\sum M_n = \sum \frac{1}{n^2}$ is convergent, by Weiestrass's M-test, Σf_n is uniformly convergent for all $x \in R$.

We have $$f_n'(x) = \frac{2xn^4}{(n^2+n^4x^2)^2} = -\frac{2x}{n^2(1+nx^2)^2} \qquad \ldots(1)$$

Let $g_n(x) = f_n'(x)$ so that $g_n'(x) = \dfrac{2 + 6x^2n}{1+nx^2}$.

Now $g_n'(x) = 0 \Rightarrow -2 + 6x^2n = 0 \Rightarrow x^2 = 1/3n$. Putting in (1), the maximum value of

$$|g_n(x)| = \frac{2}{\sqrt{(3n)}\, n^2(1+\frac{1}{2})^2} = \frac{3\sqrt{3}}{8n^{5/2}}$$

$$\therefore |g_n(x)| < \frac{1}{n^{5/2}} \ \forall\ n.$$

Since $\sum M_n = \sum \dfrac{1}{n^{5/2}}$ is convergent, the series $\Sigma\ g_n = \Sigma\ f_n'$ is uniformly convergent for all $x \in R$ and consequently term by term differentiation of the series $\Sigma\ f_n$ is justified. Hence

$$f'(x) = \sum_{n=1}^{\infty} f_n'(x) = -2x \sum_{n=1}^{\infty} \frac{1}{n^2(1+nx^2)^2}, \text{ using (1).}$$

Example 18(b):

Show that the function represented by $\displaystyle\sum_1^{\infty} \frac{\sin nx}{n^3}$ *is differentiable for every x and its derivative is* $\displaystyle\sum_1^{\infty} \frac{\cos nx}{n^2}$.

Solution:

Let $f_n(x) = \dfrac{\sin nx}{n^3}$ so that $f_n'(x) = \dfrac{\cos nx}{n^2}$

Since $|f_n'(x)| = \left|\dfrac{\cos nx}{n^2}\right| \le \dfrac{1}{n^2} \ \forall\ x$

and $\sum M_n = \sum \dfrac{1}{n^2}$

is convergent, so by Weierstrass's M-test, the series $\Sigma\ f_n'$ is uniformly convergent for all values of x and consequently the series $\Sigma\ f_n$ can be differentiated term by term. Hence

$$f'(x) = \sum_1^{\infty} f_n'(x) = \sum_1^{\infty} \frac{\cos nx}{n^2}.$$

Example 18(c):

Show that the sequence $\{f_n\}$*, where* $f_n(x) = x - x^n/n$*, converges uniformly on [0, 1]. Show also that the s sequence* $\{f_n'\}$ *of differentials does not converge uniformly on [0, 1].*

Solution:

$$\lim_{n\to\infty} f_n(x) \lim_{n\to\infty}\left(x - \frac{x^n}{n}\right) = x \quad \forall\ x \in [0, 1]$$

Thus, the point-wise limit of $\{f_n\}$ is f, $f(x) = x\ \ \forall\ x \in [0, 1]$.

$$\text{Now,}\ \ |\ f_n(x) - f(x)\ | = \left|x - \frac{x^n}{n} - x\right| \leq \frac{1}{n} \qquad \forall\ x \in [0, 1].$$

$$\therefore M_n = \sup\ \{\ |\ f_n(x) - f(x)\ |\ : 0 \leq x \leq 1\ \} = \frac{1}{n} \to 0 \text{ as } n \to \infty.$$

Hence, the sequence $\{f_n\}$ converges uniformly to f on [0, 1], where $f(x) = \forall\ [0, 1]$. We have $f'(x) = 1\ \ \forall\ x \in [0, 1]$.

Thus $f'(1) = 1$. Now $f_n'(x) = 1 - x^{n-1}\ \ \ \forall\ x \in [0, 1]$ and so $f_n'(1) = 0$. We see that $\lim_{n\to\infty} f_n'(1) \neq f'(1)$

Hence, the sequence $\{f_n'\}$ does not converge uniformly on [0, 1].

Example 19:

Show that the sequence $\{f_n\}$, *where*

$$f_n(x) = \frac{1}{n^2}\ \log\ (1 + n^2x^2)$$

is uniformly convergent on [0, 1].

Solution:

We have $f_n'(x) = \dfrac{2nx}{1+n^2x^2}$. Let $g_n(x) = f_n'()\ \ \forall\ n$.

$$\text{Now}\ \lim_{n\to\infty} g_n(x) = \lim_{n\to\infty} \frac{2mx}{1+n^3x^2} = \lim_{n\to\infty} \frac{2x/n^2}{x^2+(1/n)^3} = 0,$$

$\forall\ x \in [0, 1]$, so that the point-wise limit is $g(x) = 0\ \ \forall\ x \in [0, 1]$.

Let $y = g_n(x) - g(x) = \dfrac{2mx}{1+n3x^2}$, so that

$$\frac{dy}{dx} = \frac{2n(1-n^2x^2)}{(1+n^3x^2)^2} \text{ and so } \frac{dy}{dx} = 0 \Rightarrow x = \frac{1}{n^{3/2}}$$

It can be verified that for this value of x, $y_3 < 0$.

$$\text{Now, } M_n = \sup\ \{\ |\ g_n(x) - g(x)\ |\ : 0 \leq x \leq 1\} = \frac{2n}{n^{3/2}\ (1+1)} = \frac{1}{\sqrt{n}}$$

$\to 0$ as $n \to \infty$.

Hence, $\{g_n\} = \{f_n'\}$ converges uniformly to g, where $g(x) = 0 \ \forall \ x \in [0, 1]$. Also each function g_n is continuous on [0, 1].

We see that

$$\lim_{n\to\infty} f_n(x) = \lim_{n\to\infty} \frac{\log(1+n^2x^2)}{n^2} \left(\frac{\infty}{\infty}\right)$$

$$= \lim_{n\to\infty} \frac{3n^2x^2}{(1+n^3x^2)} \lim_{n\to\infty} \frac{3x^2}{2(3n^2x^2)} = 0$$

$$\therefore \lim_{n\to\infty} f_n(x) = 0 \quad \forall \ x \in [0, 1].$$

Thus, $\{f_n\}$ is a sequence of differentiable functions which converges point-wise to zero o n [0, 1]. The converges, $\{f_n\}$ converges uniformly to zero on [0, 1].

Example 20:

Prove that each of the series

$$\sum \frac{1}{n} \cos nx, \sum \frac{1}{n} \sin nx$$

converges uniformly in $[\alpha, 2\pi - \alpha]$, where α is any fixed positive number less than π.

Solution:

(i) Let $b_n = \frac{1}{n}$ so that b_n is positive, monotonic decreasing function of n and converges uniformly to zero.

Let $u_n(x) = \cos nx$. Then

$$\sum_{r=1}^{n} u_r(x) = \cos x + \cos 2x + \ldots.. + \cos nx$$

$$= \frac{\cos\left\{\frac{1}{2}(n+1)x\right\} \sin\left\{\frac{1}{2}nx\right\}}{\sin\left(\frac{1}{2}x\right)}$$

$$\therefore \left|\sum_{r=1}^{n} u_r(x)\right| = \frac{|\cos\left\{\frac{1}{2}(n+1)x\right\}|\,|\sin \frac{1}{2}nx|}{|\sin(\frac{1}{2}x)|}$$

$$\le \frac{1}{|\sin(\frac{1}{2}x)|} |\operatorname{cosec}(\tfrac{1}{2}x)| \qquad \forall \ n \text{ and } \forall \ x \in [\alpha, 2\pi - \alpha],$$

where cosec $\frac{1}{2}$x) is bounded $\forall \ x \in [\alpha, 2\pi - \alpha]$.

Hence, by Dirichlet's test, $\sum \frac{1}{n} \cos nx$ converges uniformly in $[\alpha, 2\pi - \alpha]$.

(ii) Let $u_n(x) = \sin nx$. Then

$$\left|\sum_{r=1}^{n} u_r(x)\right| = \frac{|\sin\{\frac{1}{2}(n+1)x\}|\,|\sin\frac{1}{2}nx|}{|\sin(\frac{1}{2}x)|}$$

$$< |\operatorname{cosec}(½x)| \ \forall\ n \text{ and } \forall\ x \in [\alpha, 2\pi - \alpha].$$

Hence, by Dirichlet's test, $\sum \frac{1}{n} \sin nx$ converges uniformly in $[\alpha, 2\pi - \alpha]$.

Example 21:

Show that the series $\sum \frac{(-1)^n}{n} |x|^n$ *is uniformly convergent in* $-1 \le x \le 1$.

Solution:

Let $b_n(x) = |x|^n$ so that $b_n(x)$ is positive, monotonically decreasing and bounded in $-1 \le x \le 1$ (i.e., $|x| \le 1$).

Let $u_n(x) = \frac{(-1)^n}{n}$ so that the series is uniformly convergent

Hence, by Abel's test,

$$\sum b_n(x)\ u_n(x) = \sum \frac{(-1)^n}{x} |x|^n$$

is uniformly convergent in $-1 \le x \le 1$.

Example 22(a):

Discuss the uniform convergence of

$$1 + \frac{e^{-2x}}{2^2 - 1} - \frac{e^{-4x}}{4^2 - 1} + \frac{e^{-6x}}{6^2 - 1} - \ldots.$$

for all real $x \ge 0$.

Solution:

Neglecting the first term, we get

$$f_n(x) = \frac{(-1)^{n+1}\ e^{-2nx}}{(2n)^2 - 1}.$$

Let $b_n(x) = e^{-2nx}$ and $u_n(x) = \dfrac{(-1)^{n+1}}{(2n)^2 - 1}$

Now, $b_n(x)$ is a monotonic decreasing function of n $\forall\ x \geq 0$.

$$\text{and} \qquad \lim_{n\to\infty} b_n(x) = \lim_{n\to\infty} \frac{1}{e^{2nx}} = 0 \qquad \forall\ x \geq 0.$$

$$\text{Further, } \sum u_n(x) = \sum \frac{(-1)^{n+1}}{(2n)^{2-1}}$$

$$= \frac{1}{2^2-1} - = \frac{1}{4^2-1} + = \frac{1}{6^2-1} - \ldots\ldots$$

is an alternating series such that $u_{n+1} < u_n\ \forall$ n and $\lim\limits_{n\to\infty} u_n = 0$.

Using Leibnitz's test. $\Sigma\, u_n(x)$ is convergent.

Hence by Dirichlet's test, the series $\Sigma b_n(x)\, u_n(x)$ i.e., the given series is uniformly convergent for all $x \geq 0$.

Example 22(b):

Prove that the series $\sum (-1)^n \dfrac{x^2+n}{n^2}$ *converges uniformly in every bounded interval, but does not converge absolutely for any value of x.*

Solution:

Let $\Sigma u_n = \Sigma(-1)^n$, which oscillates finitely.

$$\text{Let } b_n(x) = \frac{x^2+n}{n^2} < \frac{b^2+n}{n^2} \qquad \forall\ x \in [a, b]$$

where [a, b] is any bounded interval.

It is clear that $b_n(x)$ is a positive, monotonic decreasing function of n for each x in [a, b], and converges uniformly to zero for $a \leq x \leq b$.

Hence, by Dirichlet's test, the series $\sum (-1)^n \dfrac{x^2+n}{n^2}$ converges, uniformly on [a, b].

$$\text{Again } \left| (-1)^n \frac{x^2+n}{n^2} \right| = \frac{x^2+n}{n^2} \sim \frac{1}{n}, \text{ which diverges.}$$

Hence the given series is not absolutely convergent for any value of x.

Example 23:

Show that the series

$$\frac{2x}{1+x^2}+\frac{4x^2}{1+x^4}+\frac{8x^7}{1+x^8}+.....$$

is uniformly convergent in $-1 < x < 1$.

Solution:

We have $u_n(x)$ = The nth, term of the series

$$=\frac{2^n x^{2^n-1}}{1+x^{2n}}$$

Then $|u_n(x)| = \left|\frac{2^n x^{2^n-1}}{1+x^{2n}}\right| \le 2^n, k^{2^n-1}$ ...(1)

for $|x| \le k < 1$

Thus to test the convergence of the series $\sum 2^n, k^{2^n-1}$ let $v_n = 2^n, k^{2^n-1}$

$$\therefore \lim_{n\to\infty}\left(\frac{v_n}{v_{n+1}}\right) = \lim_{n\to\infty}\frac{2^n, k^{2^n-1}}{2^{n+1}k^{2^{n+1}-1}}$$

$$\lim_{n\to\infty}\left[\frac{1}{2.k^{2n}}\right]$$

which tends to ∞ since $k < 1$

Hence by D' Alembert's ratio test, the series $\Sigma\, v_n$ is convergent.

So by Weierstrass's M-test the given series is uniformly convergent for $|x| \le k$ 1.

Example 24:

Show that the sequence $\{f_n\}$, where

$$f_n(x) = \frac{n^2x}{1+n^2x^2}$$

is not uniformly convergent on [0, 1].

Solution:

$$\text{Here } f(x) = \lim_{n\to\infty}\frac{n^2x}{1+n^2x^2} = \frac{x/n}{(1/n^3)+x^2} = 0\ \forall\ x \in R.$$

$$\text{Let}\quad y = f_n(x) - f(x) = \frac{n^2x}{1+n^3x^2}.$$

$$\therefore \frac{dy}{dx} = \frac{(1+n^3x^2) \cdot n^2 - n^2x\,(2nx)}{(1+n^3x^2)^2} = \frac{n^2(1-n^3x^2)}{(1+n^3x^2)^2}$$

It can be verified that $y_2 < 0$ for $x = 1/n^{3/2}$.

Putting $x = 1/n^{3/2}$ in the value of y, we obtain

$M_n = \max |y| = \frac{1}{2}\sqrt{n} \to \infty$ as $n \to \infty$.

Hence the given sequence is not uniformly convergent on [0. 1]

Example 25:

Show that the sequence $\{f_n\}$, where

$$f_n(x) = \frac{nx}{1+n^2x^2}$$

is not uniformly convergent on any interval containing zero.

Solution:

Here $f(x) \lim_{n\to\infty} \frac{nx}{1+n^2x^2} = \lim_{n\to\infty} \frac{x/n}{1/n^2+x^2} = 0.$

$\therefore f(x) = 0 \quad \forall \quad x \in r.$

Consider any interval [a, b] containing 0.

Let $y = f_n(x) - f(x) = \frac{nx}{1+n^2x^2}$.

$$\therefore \frac{dy}{dx} = \frac{(1+n^2x^2) \cdot n - nx\,(2n^2x)}{(1+n^2x^2)^2} = \frac{n(1-n^2x^2)}{(1+n^2x^2)^2}$$

Now $\frac{dy}{dx} = 0 \;\Rightarrow\; x = \frac{1}{n} \in [a, b].$

It can be verified that $y_2 < 0$ for $x = 1/n$.

So $M_n = \max. |y| = \frac{1}{2}$.

which does not tend to zero is $n \to \infty$. Hence the given sequence is uniformly convergent in any interval containing zero.

Example 26:

Show that the sequence $\{f_n\}$, where $f_n(x) = e^{-nx}$ is point-wise but not uniformly convergent in $[0, \infty[$. Also show that the convergence is uniform in $[k, \infty[$, k being any positive number.

Solution:

For $x \neq 0$, $f(x) = \lim_{n\to\infty} f_n(x) = \lim_{n\to\infty} \left(\frac{1}{e^{nx}}\right) = 0$

For $x = 0$, $f_n(x) = 1 \quad \forall\, n$ so that the point-wise limit is

$$f(x) = \begin{cases} 0, & x \neq 0, \\ 1, & x = 0, \end{cases}$$

We choose ε, $0 < \varepsilon < 1$. For $x \in]0, \infty[$, we have

$| f_n(x) - f(x) | = | e^{-nx} - 0 | = 1/e^{nx} < \varepsilon$.

if $e^{nx} > 1/\varepsilon \Rightarrow nx > \log (1/\varepsilon) \Rightarrow n > \frac{1}{x} \log (1/\varepsilon)$.

For $x = 0$, $| f_n(x) - f(x) | = | 1 - 1 | = 0 < \varepsilon \quad \forall\ n \geq 1$.

Let m be a positive integer such that

$$m = \max. \left\{1, \frac{1}{x} \log (1/\varepsilon)\right\}. \qquad ...(1)$$

Then $m \to \infty$ as $x \to 0+$, so that there does not exist any positive integer m such that $\forall\ n \geq m$ and $\forall\ x \in [0, \infty[$,

$| f_n(x) - f(x) | < \varepsilon$.

Hence, the convergence is not uniform in $[0, \infty[$,

Let k be any positive number. Then by (1),

$m = \frac{1}{k} \log (1/\varepsilon)$. Thus $| f_n(x) - f(x) | < \varepsilon \ \ \forall\ n \geq m, \ \ \forall\ x \in [k, \infty[$.

Hence, $\{f_n(x)\}$ is uniformly convergent in $[k, \infty[, k > 0$.

Ex.: *Show that the sequence* (e^{-nx}) *is uniformly convergent in any interval* $[a, b]$*, where a and b are positive numbers but only point-wise in* $[a, b]$*.*

Example 27:

Show that the series $\sum \{\log(n + 1)\}^{-x} \cos nx$ *is uniformly convergent in* $[\theta_1, \theta_2]$*, where* $0 < \theta_1 \leq x \leq \theta_2 < 2\pi$*.*

Solution:

When $x \in [\theta_1, \theta_2]$, $\{\log)n + 1)\}^{-x}$ is a positive monotonic decreasing function of n. Also since $\{\log (n + 1)\}^{-x} \leq \{\log (n + 1)\}^{-\theta^1}$, the function $\{\log (n + 1)\}^{-x}$ tends uniformly to zero as $n \to \infty$. Moreover, as in Example 14,

$$\left| \sum_{r=1}^{n} \cos rx \right| \leq \frac{1}{\sin(x/2)} \leq \max\left(\frac{1}{\sin(\theta_1/2)} \cdot \frac{1}{\sin(\theta_2/2)}\right),$$

both of which are independent of x and n.

Thus by Dirichlet's test the series Σ {log (n + 1)}$^{-x}$ cos nx is uniformly convergent in $[\theta_1, \theta_2]$.

Example 28:

Prove that the series $\sum \frac{\cos n\theta}{n^p}$ *and* $\sum \frac{\sin n\theta}{n^p}$ *converge uniformly for all values of* $p > 0$ *in an interval* $[\alpha, 2\pi - \alpha[$, *where* $0 < \alpha < \pi$.

Solution:

When p > 1, Weierstrass's M-test at once proves that both the series converge uniformly for all real values of θ.

When $0 < p \le 1$, both the series converge uniformly in any interval $[\alpha, 2\pi - \alpha]$, $\alpha > 0$. This can be proved by taking $b_n = (1/n^p)$ and $u_n = \cos n\theta$ (or sin nθ) in Dirichiet's test.

$(1/n^p)$ is positive monotonic decreasing and tending uniformly to zero for $0 < p \le 1$, and

$$\left|\sum_{r=1}^{n} u_r\right| = \left|\sum_{r=1}^{n} \cos r\theta\right| = |\cos\theta + \cos 2\theta + \ldots + \cos n\theta|$$

$$= \left|\frac{\cos ((n+1)/2)\,\theta \sin (n/2)\theta}{\sin (\theta/2)}\right| \le \text{cosec}\,(\alpha/2), \quad \forall\, n,$$

for $\theta \in [\alpha, 2\pi - \alpha]$.

Thus, all the conditions are fulfilled and the series Σ (cos nθ/n^p) and similarly Σ (sin nθ/n^p) converge uniformly on $[\alpha, 2\pi - \alpha]$ where $0 < \alpha < \pi$.

Ex.: Examine for uniform convergence on R, the sequence $\{f_n(x)\}$, $f_n(x) = \dfrac{n^2x^2}{1 + n^5x^4}$.

Example 29:

Show that the sequence (f_n), *where* $f_n(x) = nx\, e^{-nx^2}$ *is not uniformly convergent on [0, 1].*

Solution:

Here $f(x) = \lim\limits_{n\to\infty} f_n(x)$

$$= \lim_{n\to\infty} \frac{nx}{e^{nx^2}} \left(\frac{\infty}{\infty}\right)$$

$$= \lim_{n\to\infty} \frac{x}{x^2 e^{nx^2}} = 0 \qquad " x \in [0.\ 1]$$

Let $= f_n(x) - f(x) = nxe^{nx^2}$ $(\because f(x) = 0)$

$$\therefore \frac{dy}{dx} = n\left[e^{-nx^2} + x(-2nx)e^{-nx^2}\right]$$

Now $\frac{dy}{dx} = 0 \Rightarrow 1-2nx^2 = 0 \Rightarrow x = \frac{1}{\sqrt{2n}} \in [0, 1]$

if can be verified that $d^2y/dx^2 < 0$ for this value of x.

$$\therefore M_n = \max. |y| = n. \frac{1}{\sqrt{2n}} e^{-1/2} = \sqrt{\frac{n}{2e}} \to \infty \text{ as } n \to \infty.$$

Thus, M_n does not tend to 0 as $n \to \infty$ and so the given sequence is not uniformly convergent in [0, 1].

Example 30(a):

Test the following sequence for uniform c convergence:

(i) $\left\{\frac{\sin nx}{\sqrt{n}}\right\}, 0 \le x \le 2\pi$ (ii) $\left\{\frac{nx}{1+n^3x^2}\right\}, 0 \le x \le 1.$

Solution:

(i) We have $\left|\frac{\sin nx}{\sqrt{n}}\right| \le \frac{1}{\sqrt{n}} \quad \forall x \in [0, 2\pi].$

$$\therefore \lim_{n\to\infty} \frac{\sin nx}{\sqrt{n}} = 0 \quad \forall x \in [0, 2\pi].$$

Now, $|f_n(x) - f(x)| = \left|\frac{\sin nx}{\sqrt{n}} - 0\right| \le \frac{1}{\sqrt{n}} \quad \forall x \in [0, 2\pi]$

$$\therefore M_n = \sup \{|f_n(x) - f(x)| : 0 \le x \le 2p\} = \frac{1}{\sqrt{n}}$$

Since $M_n = \frac{1}{\sqrt{n}} \to 0$ as $n \to \infty$, so the given sequence is uniformly convergent on $[0, 2\pi]$

(ii) Here $f(x) = \lim_{n\to\infty} \frac{nx}{1+n^3x^2}$

$$= \lim_{n\to\infty} \frac{(n/x^2)}{x^2 + (1/n^3)} = 0 \quad \forall x \in [0, 1].$$

Let $$y = f_n(x) - f(x) = \frac{nx}{1+n^3x^2}$$

$$\therefore \quad \frac{dy}{dx} = \frac{(1+n^3x^2).n - nx(2n^3x)}{(1+n^3x^2)^2} = \frac{1-n^4x^2}{(1+n^3x^2)^2}$$

Now, $$\frac{dy}{dx} = 0 \quad \Rightarrow \quad x = \frac{1}{n^2} \in [0, 1]$$

It can be verified that $y_8 < 0$ for $x = 1/\pi^2$.

So M_n Max. $| y | = \frac{1}{(1+n)} \to 0$ as $n \to \infty$.

Hence, the given sequence is uniformly convergent o n [0, 1].

Example 30(b):

Test for uniform convergence the series

$$\sum \frac{x}{n(1+nx^{-})}.$$

Solution:

Here $u_n(x)$ = The *n*th. term of the series

$$= \frac{x}{n(1+nx^2)}$$

The $u_n(x)$ is maximum if $\frac{d}{dx}\, u_n(x) = 0.$

or if $$\frac{n(1+nx^2) - 2n^2x^2}{n^2(1+nx^2)^2} = 0$$

or if $$x = \pm \frac{1}{\sqrt{(n)}}.$$

So the maximum value f u_n (x) $$= \frac{1/\sqrt{n}}{n[1+n.1/n]}$$

$$= \frac{1}{2n^{3/2}} = M_n \text{ say}$$

since $x = \frac{1}{\sqrt{n}}$ makes $u_n(x)$ a maximum.

But $\Sigma\, M_n = \Sigma \frac{1}{\sqrt{n}}$ is convergent.

Hence, it follows from Weierstrass's M-test that the given series is uniformly convergent for all values of x.

Example 30(c):

Show that the series $\sum_{n=0}^{\infty}(1-x)x^n$ *is not uniformly convergent on [0, 1].*

Solution:

The given series is

$(1 - x) + (1 - x)x + (1 - x)x^2 + \ldots\ldots + f_n(x) = (1 - x)x^n.$

For x = 1, the sum of the series = (x) = 0.

For $x \neq 1$, the given series is a geometric series with common ratio x, so that the sum function is

$$f(x) = \frac{a}{1-r} = \frac{1-x}{1-x} = 1. \text{ Thus}$$

$$f(x) = \begin{cases} 1, & x \neq 0 \\ 0, & x = 0 \end{cases}$$

Thus, sum function is discontinuous on [0, 1] and so the given series is not uniformly convergent on [0, 1].

Example 31:

Show that the series

$$\frac{x}{x+1} + \frac{x}{(x+1)(2x+1)} + \frac{x}{(2x+1)(3x+1)} + \ldots\ldots.$$

in uniformly convergent in] k, ∞ [, where k is a positive number. Show also that the series i s not uniformly convergent in [0, ∞[.

Solution:

We have

$$f_n(x) = \frac{x}{(nx+1)[(n-1)x+1]} = \frac{1}{(n-1)x+1} - \frac{1}{nx+1}$$

Now, $S_n(x) = f_1(x) + f_2(x) + \ldots. + f_n(x)$

$$= \left(1 + \frac{1}{x+1}\right) + \left(\frac{1}{x+1} - \frac{1}{2x+1}\right) + \ldots\ldots\ldots$$

$$+ \left(\frac{1}{(n-1)x+1} - \frac{1}{nx+1}\right)$$

$\therefore S_n(x) = 1 - \dfrac{1}{nx+1}$. The sum function is given by

$$f(x) = \lim_{n\to\infty} S_n(x) = \begin{cases} 1, & \text{if } x \neq 0, \\ 0, & \text{if } x = 0. \end{cases}$$

It is clear that the sum function f is discontinuous on [0, 1] and so the given series is not uniformly convergent in [0, ∞[.

Now we show that Σf_n is uniformly convergent on]k, ∞ [k > 0. Let ε > 0 be given. For x ≠ 0

$$| S_n(x) - f(x) | = \left| 1 - \frac{1}{nx+1} - 1 \right| = \frac{1}{nx+1} < \varepsilon,$$

if $n > \frac{1}{x}\left(\frac{1}{\varepsilon} - 1\right)$.

Now $\frac{1}{x}\left(\frac{1}{\varepsilon} - 1\right)$ decreases with x and has a maximum value

$$m_0 = \frac{1}{k}\left(\frac{1}{\varepsilon} - 1\right) \infty \ x, \infty k$$

Here m_0 is independent of x.

Thus, ∀ x ∈]k, ∞[, there exists a positive integer m(> m_0) such that

$| S_n(x) - f(x) | < e \quad \forall\, n \geq m.$

Hence, the given series is uniformly convergent on]k, ∞[, k > 0.

Example 32:

Show that the sequence $\{f_n\}$, $f_n(x) = nx\, e^{-nx^2}$ converges but not uniformly to f, where f(x) = 0 for all x ∈ [0, 1], and that

$$\lim_{n\to\infty} \int_0^1 f_n \, dx \neq \int_0^1 f \, dx.$$

Solution:

Here $f(x) = \lim_{n\to\infty} nx\, e^{-nx^2} = 0 \ \forall\, x \in [0, 1]$.

$\therefore \int_0^1 f(x)\,dx = 0.$ Now $\int_0^1 f(n)\,dx = \int_0^1 nx\, e^{-nx^2}\, dx$

$$= -\frac{1}{2}\left| e^{-nx^2} \right|_0^1 = \frac{1}{2}(1 - e^{-n})$$

$$\therefore \lim_{n\to\infty} \int_0^1 f_n dx = \lim_{n\to\infty} \frac{1}{2}(1 - e^{-n}) = \frac{1}{2} \neq \int_0^1 f dx.$$

Since $\lim_{n\to\infty} \int_0^1 f_n \, dx \neq \int_0^1 f dx$, $\{f_n\}$ is not uniformly convergent to f on [0,1].

Example 33:

Show that the series $\sum_{n=0}^{\infty} \frac{x^2}{(1+x^2)^n}$ *is not uniformly convergent for* $x \geq 0$.

Solution:

The given series is

$$x^2 + \frac{x^2}{(1+x^2)} + \frac{x^2}{(1+x^2)} + \frac{x^2}{(1+x^2)} + \ldots$$

For x = 0, $f_n(x) = 0 \; \forall$ n so that the sum of the series is f(x) = 0.

For $x \neq 0$, the given series is a geometric series with common ratio $r = /(1 + x^2)$, so that the sum function is

$$f(x) = \frac{a}{1-r} = \frac{x^2}{1-(1/1+x^2)} = 1 + x^2.$$ Thus

$$f(x) = \begin{cases} 1 + x^2, & x \neq 0, \\ 0, & x = 0. \end{cases}$$

Clearly f is discontinuous on $[0, \infty[$ and so convergence is not uniform on $[0, \infty]$.

Example 34:

Let $\{f_n\}$ *be defined by* $f_n(x) = nx/(1+ n^2x^2)$ " $x \in [0, 1]$. *Show that* $\{f_n\}$ *is not uniformly convergent on [0, 1], although*

$$\lim_{n\to\infty} \int_0^1 f_n(x)\,dx = \int_0^1 f(x)\,dx = \int_0^1 \lim f_n(x)dx,$$

f(x) being the point-wise limit of $\{f_n\}$.

Solution:

Note that each $f_n(x)$ is continuous on [0, 1] and thus integrable on [0, 1]. $\{f_n\}$ is not uniformly convergent on [0, 1], but the point-wise limit of $\{f_n\}$ is $f(x) = 0 \; \forall \; x \in [0, 1]$. Now

$$\int_0^1 f_n(x)\,dx = \int_0^1 \frac{nx}{1+n^2x^2}\,dx = \frac{1}{2n} \log(1+n^2)$$

$$\therefore \lim_{n\to\infty} \int_0^1 f_n(x)\,dx = \lim_{n\to\infty} \frac{\log(1+n^2)}{2n}$$

$$= \lim_{n\to\infty} \frac{1}{1+n^2} = 0$$ (D Hospital's Rule)

$$= 0 = \int_0^1 f(x)\,dx.$$

EXERCISES

1. *Discuss the uniform convergence of the series*

$$x^2 + \frac{x^2}{1+x^2} + \frac{x^2}{(1+x^2)^2} + \frac{x^2}{(1+x^2)^2} + ... \text{ in } [0, 1].$$

2. *Show that* $\sum \frac{\sin(x^2 + nx^2)}{n(n+1)}$ *is uniformly convergent for all real x.*

3. *Show that* $\sum \frac{\sin n\theta}{n^p}$, $\sum \frac{\cos n\theta}{n^p}$ *are uniformly convergent for all real values of x if p > 1.*

4. *Show that the uniform sum of a series of continuous functions is continuous. Establish by on example that the point-wise limit of a sequence of continuous functions may be a continuous functions.*

5. *Show that the series* $\sum \frac{1}{n}$ *sin nx converges uniformly in* $[0, 2\pi]$.

6. *Decide whether or not the sequence* $\{f_n'\}$ *converges uniformly on* $[0, 1]$ *where* $f_n(x) = x - \frac{x^n}{n}$.

7. *Prove that the series* $\sum \frac{x^4}{n(n+1)}$ *is uniformly convergent in 1 – 1.*

8. *Show that* $\sum_{n=1}^{\infty} \frac{\cos nx}{n^2}$ *is uniformly convergent for all real values of x.*

9. *Show that the series* $\sum \frac{-(1)^{n-1}}{n}$ *is uniformly convergent in [0, 1].*

10. *Examine for uniform convergence on R, the sequence* $\{f_n\}$, $f_n(x) = \log (1 + n^3x^2)/2n^2$.

11. *Show that the sequence* $\{f_n\}$ *defined on [0, 1] by*

$$f_n(x) = \begin{cases} n(1-nx), & 0<x<1/n \\ 0, & \text{otherwise} \end{cases}$$

converges point-wise but not uniformly to f, where $f(x) = 0 \ \forall x \in [0, 1]$, *and that*

$$\lim_{n\to\infty} \int_0^1 f_n dx \neq \int_0^1 f dx.$$

12. *Show that the sequence* $\{f_n'\}$ *where* $f_n(x) = \dfrac{x}{1+x^2x}$ *is not uniformly convergent on [0, 1].*

13. *Show that* $\sum\limits_{n=1}^{n} (\sin x)^n$ *converges uniformly for* $|x| \leq h,\ 0 < h < 1$.

14. *Show that the series* $\sum\limits_{n=1}^{n} (-1)^{n-1} x^4$ *is uniformly convergent for* $|x| \leq$ *1/2.*

15. *Show that the series*

    $x^4 + x^4/(1 + x^4) + x^4/(1 + x^4)^2 + x^4/(1 + x^4)^2 + ...$

    is not uniformly convergent on [0, 1].

4

Fourier, Power, Double Series and Series of Arbitrary Terms

INTRODUCTION OF FOURIER SERIES

A series of the form

$$\frac{1}{2} a_n + (a_1 \cos x + b_1 \sin x) + (a_2 \cos 2x + b_2 \sin 2x) + \ldots.$$

or more compactly

$$\frac{1}{2} a_0 + \sum_{n=1}^{\infty} (a_n \cos nx + b_n \sin nx)$$

is called a *trigonometric series,* and the constants a_0, a_n, b_n ($n = 1, 2, 3 \ldots$) are called its *coefficients.* If the series converges for all values of x (in fact, in an interval $c \leq x \leq c + 2\pi$) then, in view of the periodicity of the trigonometric functions, its sum is a function f defined for all values of x and periodic with period 2π. To save tedious repetition, it may be mentioned once for all that the *functions considered in this chapter are all assumed to be periodic.* We now proceed to obtain a relationship between the function f and the coefficients a_n, b_n, which was conjectured by *Euler.* The following results of integral calculus form the basis of all subsequent work. Throughout we use m, n to *denote positive integers or zero*:

$$\int_{-\pi}^{\pi} \cos mx \cos nx \, dx = \begin{cases} 0, & \text{for } m \neq n \\ \pi, & \text{for } m \;\; n > 0 \\ 2\pi, & \text{for } m = n = 0 \end{cases}$$

$$\int_{-\pi}^{\pi} \sin mx \sin nx \, dx = \begin{cases} 0, & \text{for } m \neq n \text{ and } m = n = 0 \\ \pi, & \text{for } m = n > 0 \end{cases}$$

$$\int_{-\pi}^{\pi} \cos mx \sin nx \, dx = 0 .$$

Fourier Series

If the numbers a_0, a_1,...., b_1, b_2, ..., b_n, ... are derived from a function f by means of Euler-Fourier formulas:

$$a_n = \frac{1}{n}\int_{-\pi}^{\pi} f(x)\cos nx \, dx \ (n=0, 1, 2, ...),$$

$$b_n = \frac{1}{n}\int_{-\pi}^{\pi} f(x)\sin nx \, dx \ (n = 1, 2, ...), \qquad ...(1)$$

then the series

$$\tfrac{1}{2} a_0 + \sum_{n=1}^{\infty}(a_n \cos nx + b_n \sin nx) \qquad ...(2)$$

is called the *Fourier series of* f or the *Fourier series generated by f,* and the coefficients a_n, b_n defined by (1), the *Fourier coefficients** of f.

PERIODIC FUNCTION

We know that a period function of period λ is such that $f(x \pm \lambda) = f(x)$, for all x.

Theorem 1:

For a periodic function of period 2p, prove that

(i) $\int_{\alpha}^{\beta} f \, dx = \int_{\alpha+2\pi}^{\beta+2\pi} f \, dx$, (ii) $\int_{-\pi}^{\pi} f \, dx = \int_{\alpha}^{\alpha+2\pi} f \, dx$,

(iii) $\int_{-\pi}^{\pi} f(x) \, dx = \int_{-\pi}^{\pi} f(\gamma + x) \, dx$.

α, β γ being any numbers whatsoever.

Proof:

(i) For a periodic function of period 2π, we know

$f(t - 2\pi) f(t)$

Hence, putting $t = x + 2\pi$, for all a and b, we get

$$\int_{\alpha}^{\beta} f(x) \, dx = \int_{\alpha+2\pi}^{\beta+2\pi} f(t-2\pi)dt = \int_{\alpha+2\pi}^{\alpha+2\pi} f(t)dt = \int_{\alpha+2\pi}^{\beta+2\pi} f(x) \, dx$$

(ii) $$\int_{\alpha}^{\alpha+2\pi} fdx = \int_{\alpha}^{-\pi} fdx + \int_{-\pi}^{\pi} f dx + \int_{\pi}^{\alpha+2\pi} f dx$$

$$= \int_{\alpha}^{-\pi} fdx + \int_{-\pi}^{\pi} fdx + \int_{-\pi}^{\alpha} fdx \qquad \text{[using (i)]}$$

$$= \int_{-\pi}^{\pi} f dx$$

(iii) Let $\gamma + x = t$

$$\therefore \quad \int_{-\pi}^{\pi} f(\gamma + x)\, dx = \int_{\gamma-\pi}^{\gamma+\pi} f(t)\, dt$$

$$= \int_{\gamma-\pi}^{-\pi} f dt + \int_{-\pi}^{\pi} f dt + \int_{\pi}^{\gamma+\pi} f dt$$

$$= \int_{\gamma+\pi}^{\pi} f dt + \int_{-\pi} f dt + \int_{\pi}^{\gamma+\pi} f dt \qquad \text{[using (i)]}$$

$$= \int_{-\pi}^{\pi} f dt = \int_{-\pi}^{\pi} f dx.$$

These results, in fact mean that the integral of a periodic function over any interval whose length is equal to its period always has the same value.

Theorem 2:

(Dirichlet's Conditions) *: Let a function f be bounded, periodic with period 2π, and integrable in [–π, π] and let it be possible to divide [–π, π] into a finite number of open sub-intervals in each of which f is monotonic (i.e., f is piecewise monotonic in [–π, π]). Then*

$$\frac{1}{2}a_0 + \sum_{n=1}^{\infty}(a_n \cos nx + b_n \sin nx)$$

$$= \begin{cases} \frac{1}{2}[f(\xi - 0) + f(\xi + 0), & \text{for } -\pi < \xi < \pi \\ \frac{1}{2}[f(\pi - 0) + f(-\pi + 0) & \text{for } \xi - \pm\pi. \end{cases}$$

(Here $f(\xi + 0)$ and $f(\xi - 0)$ are the right and left hand limits of $f(x)$ at $(x = \xi)$.

The proof is omitted.

Corollary:

Suppose f is continuous at $x = \xi$ so that

$$f(\xi - 0) = f(\xi + 0) = f(\xi)$$

$$\Rightarrow \frac{1}{2}[f(\xi - 0) + f(\xi + 0)] = f(\xi).$$

If follows that if f satisfies the conditions of the above theorem, then

$$\frac{1}{2}a_0 + \sum_{n=1}^{\infty}(a_n \cos nx + b_n \sin nx)$$

$$= \begin{cases} f(x), \text{ if } x \text{ is a point of continuity,} \\ \frac{1}{2}[f(x-0)+f(x+0), \text{ if } x \text{ is a point of discontinuity.} \end{cases}$$

Ex.: *State the conditions under which $\phi(x)$ can be expressed as a Fourier series in $[-\pi, \pi]$.*

Remark:

While computing the Fourier coefficients, we can save a lot of time and labour by using certain properties of even and odd functions: A function f(x) is called even if $f(-x) = f(x)$ for all x in the domain of f. A function g(x) is called odd if $g(-x) = -g(x)$ for all x in the domain of g.

We know that if f is even and g is odd, then

$$\int_{-a}^{a} f(x)\,dx = 2\int_{0}^{a} f(x)\,dx, \quad \int_{-a}^{a} g(x)\,dx = 0.$$

Theorem 3:

If f is bounded and integrable on $[-\pi, \pi]$ and if a_n, b_n are its Fourier coefficients, then $\sum_{n=1}^{\infty}(a_n^2 + b_n^2)$ converges.

The integral

$$\int_{-\pi}^{\pi}\left(f - \sum_{n=1}^{m}(a_n \cos nx + b_n \sin nx)\right)^2 dx \geq 0$$

as its integrand is never negative, now

$$\int_{-\pi}^{\pi}\left(f - \sum_{n=1}^{m}(n_n \cos nx + b_n \sin nx)\right)^2 dx$$

$$= \int_{-\pi}^{\pi} f^2 dx + \int_{-\pi}^{\pi}[\sum_{n=1}^{m}(a_n \cos nx + b_n \sin nx)]^2 dx$$

$$-2\Sigma\left[a_n \int_{-\pi}^{\pi} f \cos nx\,dx\right] - 2\Sigma\left[b_n \int_{-\pi}^{\pi} f \sin nx\,dx\right]$$

$$= \int_{-\pi}^{\pi} f^2 dx + \pi\Sigma a_n^2 + \pi\Sigma b_n^2 - 2\pi\Sigma a_n^2 - 2\pi\Sigma b_n^2$$

$$= \int_{-\pi}^{\pi} f^2 dx - \pi\sum_{n=1}^{m}\left(a_n^2 + b_n^2\right)$$

Since this is non-negative, we have

$$\sum_{n=1}^{m}\left(a_n^2 + b_n^2\right) \leq \frac{1}{\pi}\int_{-\pi}^{\pi} f^2 dx,$$

taking limit as $m \to \infty$, we obtain

$$\sum_{n=1}^{m}\left(a_n^2 + b_n^2\right) \le \frac{1}{\pi}\int_{-\pi}^{\pi} f^2 dx \qquad \textit{(Bessel's Inequality)}$$

Hence, the series $\sum_{n=1}^{m}\left(a_n^2 + b_n^2\right)$ is convergent, since f, and so f^2 is integrable on $[-\pi, \pi]$.

Corollary:

Since $\sum\left(a_n^2 + b_n^2\right)$ *converges, its general term* $\left(a_n^2 + b_n^2\right) \to 0$, *and so,* $a_n \to 0$, $b_n \to 0$. *Thus the Fourier coefficients* a_n *and* b_n *of an integrable (and bounded) function form a null sequence, i.e., when* $n \to \infty$,

$$a_n = \frac{1}{\pi}\int_{-\pi}^{\pi} f \cos nx\, dx \to 0, \text{ and } b_n = \frac{1}{\pi}\int_{-\pi}^{\pi} f \sin nx\, dx \to 0.$$

Theorem 4:

If a function f is bounded periodic with period 2π *and integrable on* $[-\pi, \pi]$, *and piecewise monotonic on* $[-\pi, \pi]$, *then*

$$\tfrac{1}{2}a_0 + \sum_{n=1}^{\infty}(a_n \cos n\xi + b_n \sin n\xi) = \begin{cases} \tfrac{1}{2}[f(\xi-) + f(\xi+)], & \text{for } -\pi < \xi < \pi, \\ \tfrac{1}{2}[f(\pi-) + f(-\pi+)], & \text{for } \xi = \pm\pi \end{cases}$$

where a_n, b_n *are Fourier coefficients of f.*

Solution:

Let $\frac{1}{2}a_0 + \sum_{n=1}^{\infty}(a_n \cos nx + b_n \sin nx)$ be the Fourier series of f, and ξ a point of $[-\pi, \pi]$.

The mth partial sum at the point ξ,

$$\tfrac{1}{2}a_0 + \sum_{n=1}^{\infty}(a_n \cos n\xi + b_n \sin n\xi)$$

$$= \frac{1}{2\pi}\int_{-\pi}^{\pi} f\, dx + \sum_{n=1}^{m}\frac{1}{\pi}\int_{-\pi}^{\pi} f[\cos nx \cos n\xi + \sin nx \sin n\xi]\, dx$$

$$= \frac{1}{2\pi}\int_{-\pi}^{\pi} f\,[1 + 2\sum_{n=1}^{m}\cos n(x-\xi)]\, dx$$

$$= \frac{1}{2\pi}\int_{-\pi}^{\pi} f(x+\xi)\,[1 + 2\sum_{n=1}^{m}\cos nx)]\, dx,$$

$$= \frac{1}{2\pi}\int_{-\pi}^{\pi} f(x+\xi)\,\frac{\sin(m+\frac{1}{2})x}{\sin\frac{1}{2}x}\,dx$$

$$= \frac{1}{2\pi}\int_{-\pi}^{0} f(x+\xi)\,\frac{\sin(m+\frac{1}{2})x}{\sin\frac{1}{2}x}\,dx$$

$$= \frac{1}{2\pi}\int_{0}^{\pi} f(x+\xi)\,\frac{\sin(m+\frac{1}{2})x}{\sin\frac{1}{2}x}\,dx$$

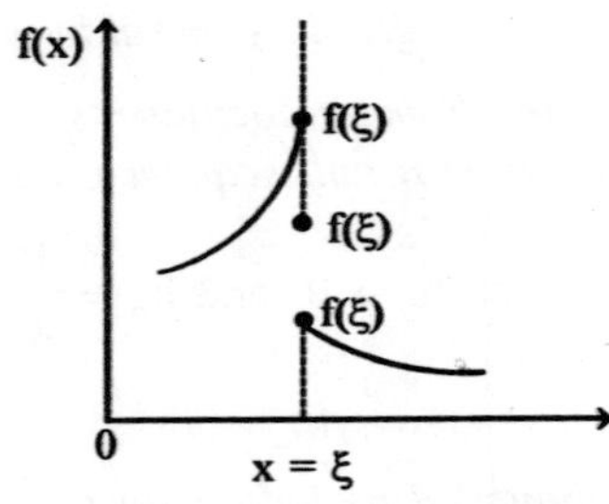

Discontinuity at ξ.

$$= \frac{1}{\pi}\int_{0}^{\pi/2} f(-2t+\xi)\,\frac{\sin(2m+1)t}{\sin t}\,dt$$

$$+ \frac{1}{\pi}\int_{0}^{\pi/2} f(2t+\xi)\,\frac{\sin(2m+1)t}{\sin t}\,dt$$

Proceeding to limits when $m \to \infty$.

$$\frac{1}{2}a_0 + \sum_{n-1}^{\infty}(a_n\ \cos n\xi + b_n\ \sin n\xi) = \frac{1}{\pi}\left[\frac{1}{2}\pi f(\xi-) + \frac{1}{2}\pi f(\xi+)\right]$$

$$= \frac{f(\xi-)+f(\xi+)}{2},\ \xi \in [-\pi, \pi]$$

Thus, the Fourier series of a (periodic) function f which is bounded, integrable and piecewise monotonic on $[-\pi, \pi]$, converges to ½ $[f(\xi-)+f(\xi+)]$ at a point ξ, $-\pi < \xi < \pi$, and (using periodicity of f) to ½$[f(\pi-) + f(-\pi+)]$ at the ends, $\pm\,\pi$.

Corollary 1:

At a point ζ of continuity of f, the sum of the Fourier series of f,

$$\tfrac{1}{2}a_0 + \sum_{n=1}^{\infty}(a_n\ \cos n\zeta + b_n\ \sin n\zeta) = \tfrac{1}{2}[f(\zeta-) + f(\zeta+1)]$$

$$= \tfrac{1}{2}[f(\zeta) + f(\zeta)] = f(\zeta)$$

Thus, for a function f satisfying the conditions of the theorem, *the sum of the Fourier series is actually f(x) at all points x where f is continuous;* while at points of (finite) discontinuity the sum of the series is 1/2 (f(x–) + f(x+)].

Corollary 2:

Since a function of bounded variation can be expressed as a difference of two monotone increasing functions (Jordan theorem), it may be easily shown that a periodic function of bounded variation can be expressed as a Fourier Series.

Even Function: If f is an even function, i.e., $f(-x) = f(x), \forall x$ then f cos nx is an even and f sin nx an odd function and therefore

$$a_n = \frac{1}{\pi}\int_{-\pi}^{\pi} f \cos nx \, dx = \frac{1}{\pi}\left[\int_{-\pi}^{0} f \cos nx \, dx + \int_{0}^{\pi} f \cos nx \, dx\right]$$

$$= \frac{2}{\pi}\int_{0}^{\pi} f \cos nx \, dx \qquad ...(1)$$

$$b_n = \frac{1}{\pi}\int_{-\pi}^{\pi} f \sin nx \, dx = 0 .$$

So, the Fourier Series of an even function consists of terms of cosines only and the coefficients a_n may be computed from (1).

Also, since for an even function.

$$f(0+) = f(0-) = f(0)$$

and

$$f(-\pi + 0) - f(\pi - 0)$$

the sum of the series of f(0) at 0 or ± (even multiple of π), and is f(π–) at ± π or ± (odd multiple of π).

Odd Function: If f is an odd function, i.e., $f(-x) = -f(x), \forall x$, then f cos nx is an odd and f sin nx and even function and therefore

$$a_n = \frac{1}{\pi}\int_{-\pi}^{\pi} f \cos nx \, dx = 0$$

$$b_n = \frac{1}{\pi}\int_{-\pi}^{\pi} f \sin nx \, dx = \frac{2}{\pi}\int_{0}^{\pi} f \sin nx \, dx \qquad ...(2)$$

So, the Fourier Series of an odd function consist of sine terms only and the coefficient b_n may be calculated from (2). Also for an odd function

$$f(0-) = 0 = f(0+)$$

and $f(-\pi + 0) = -f(\pi - 0)$

the sum of the series is 0 at x = 0 (or any multiple of π).

INTRODUCTION TO POWER SERIES

A series of the form $\sum_{n=0}^{\infty} a_n z^n$ or $\sum_{n=0}^{\infty} a_n (z-a)^n$, whose terms are variable is called a power series where z is a complex variable and a_n, are complex constants.

The second form car be reduced to the first from merely by changing origin. So the consideration of the series of the second form is always reduced to that of the first form

The form $\sum_{n=0}^{\infty} a_n z^n$ being simpler, we have discussion of power series by considering the form $\sum_{n=0}^{\infty} a_n z^n$

SOME USEFUL TESTS FOR CONVERGENCE

The following tests whose proofs are trivial should be recollected

Comparison Test : If $\sum|\sqrt{n}|$ converges and $|u_n|$ Σ $|v_n|$ Then $\Sigma|u_n|$ converges absolutely.

Ratio Test : The series Σu_n converges (absolutely) or diverges according as $\lim_{n\to\infty}\left|\frac{u_{n-1}}{u_n}\right| < 1$ or > 1. If this limit equal to 1 the test fails.

Root Test : (n^{th} root test) the series Σu_n converges (absolutely) or diverge according as $\lim_{n\to\infty} \sqrt[n]{(|u_n|)} < 1$ or > 1

If this limit is equal to 1, the test fails.

Series absolutely convergent : The series Σu_n is convergent (absolutely) the series $|\Sigma u_n|$ is convergent.

Theorem 1:

If the power series $\Sigma a_n x^n$ is such that $a_n \neq 0$ for all n and $\lim_{n\to\infty}\left|\frac{a_{n+1}}{a_n}\right| = \frac{1}{R}$, *then $\Sigma a_n x^n$ is convergent (absolutely convergent) for $|x| < R$ and divergent for $|x| > R$..*

Proof:

Let $u_n = a_n x^n$ so that $u_{n+1} = a_{n+1} x^{n+1}$.

$$\therefore \lim_{n\to\infty}\left|\frac{u_{n+1}}{u_n}\right| = \lim_{n\to\infty}\left|\frac{a_{n+1}\,x}{a_n}\right| = \frac{|x|}{R}.$$

By D' Alembert's ratio test, $\Sigma a_n x^n$ converges absolutely, if

$\frac{|x|}{R} < 1$ i.e., $|x| < R$.

Further $\Sigma a_n x^n$ diverges if $|x| > R$. Hence the theorem.

Remark 1:

By Cauchy's second theorem on limits, if $\{a_n\}$ is sequence of positive constants, then

$$\lim_{n\to\infty}(a_n)^{1/n} = \lim_{n\to\infty}\left(\frac{a_{n+1}}{a_n}\right),$$

provided the limit on the R.H.S. exists. It follows that :

If $\lim_{n\to\infty}|a_n|^{1/n} = \frac{1}{R}$, *Then the series $\Sigma a_n x^n$ is convergent (absolutely convergent) for $|x| < R$ and divergent for $|x| > R$..*

In view of the above discussion, the *radius of convergence R of a power series $\Sigma a_n x^n$ may be determined by the following :*

$$R = \lim_{n\to\infty}\left|\frac{a_n}{a_{n+1}}\right| = 1 \div \lim_{n\to\infty}|a_n|^{1/n}$$

$R = \infty$, when $\lim |a_n|^{1/n} = \infty$.

$R = 0$, when $\lim |a_n|^{1/n} = \infty$.

Remark 2:

Every power series converges absolutely within its interval of convergence.

Theorem 2:

If a power series $\Sigma a_n x^n$ converges for $|x| < R$, then it converges uniformly on $[-R + \varepsilon, R - \varepsilon]$, for each $\varepsilon > 0$.

Proof:

Let $x_0 = R - \varepsilon < R$ so that $\Sigma a_n x_0^{\ n}$ converges absolutely.

For $|x| \leq x_0$, $|a_n x^n| \leq |a_n|\ x_0^n$.

Since $\Sigma a_n x_0^n$ converges absolutely, therefore, by Weierstrass's M-test, the series $\Sigma a_n x^n$ converges uniformly on

$[-x_0, x_0] = [-R + \varepsilon, R - \varepsilon]$. Hence the theorem.

Ex.:

Show that a power series converges uniformly on any closed and bounded subinterval of its interval of convergence.

Theorem 3:

If a power series $\Sigma a_n x^n$ converges for $x = x_0$, then

(i) *it is absolutely convergent in the interval $|x| < |x_0|$*

(ii) *it is uniformly convergent in the interval $|x| \leq |x_1|$ where $|x_1| \leq |x_0|$.*

Proof:

(i) Since $\Sigma a_n x_0^n$ converges, so $\lim\limits_{n\to\infty} a_n x_0^n = 0$.

Consequently there exists a positive integer *m* such that

$$\left|a_n x_0^n = 0\right| < 1\ \forall\ n \geq m \text{ or } |a_n| < \frac{1}{|x_0|^n}\ \forall\ n \geq m \qquad \text{...(1)}$$

$$\Rightarrow |a_n x^n| = |a_n|\,|x|^n < \frac{|x|^n}{|x_0|^n}\ \forall\ n \geq m, \text{ by (1)}$$

$$\Rightarrow |a_n x^n| < \left\{\frac{|x|}{|x_0|}\right\}^n \quad \forall\ n \geq m.$$

The series on the R.H.S. converges for $|x| < |x_0|$ (being a geometric series with common ratio < (1), it follows by the comparison test that $\Sigma |a_n x^n|$ is convergent for $|x| < |x_0|$.

Hence $\Sigma a_n x^n$ is absolutely convergent for $|x| < |x_0|$.

(ii) Let $M_n = \dfrac{|x_1|^n}{|x_0|^n}$. Then ΣM_n converges, since $|x_1| < |x_0|$ (being a geometric series with common ratio <1).

Using (1), we obtain

$$|a_n x^n| = |a_n|\,|x|^n < \frac{|x|^n}{|x_0|^n} \leq \frac{|x_1|^n}{|x_0|^n},$$

for $| x | \leq | x_1 |$ and $n \geq m$

$\therefore | a_n x^n | < M_n \ \forall \ n \geq m$, where ΣM_n converges.

By the Weierstrass M-Test, $\Sigma a_n x^n$ is uniformly convergent in $| x | \leq |x_1|$.

Note : *If a power series whose radius of convergence is 1, then it converges absolutely in]– 1, 1[and uniformly in [–k, k], |k| < 1.*

Theorem 4: (Cauchy's Product of Two Power Series)

Two power series $\sum_{n=0}^{\infty} a_n x^n$ *and* $\sum_{n=0}^{\infty} b_n x^n$ *can be multiplied to obtain the power series* $\sum_{n=0}^{\infty} c_n x^n$, *where*

$$c_n = a_0 b_n + a_1 b_{n-1} + a_2 b_{n-2} + ... + a_n b_0,$$

the result being valid for each x within the common interval of convergence.

The proof of these theorems are omitted.

SOME IMPORTANT POWER SERIES

Consider the following power series :

$$1 + \frac{x}{1!} + \frac{x^2}{2!} + \frac{x^3}{3!} + ... + \frac{x^n}{n!} + ...$$

$$1 - \frac{x^2}{2!} + \frac{x^4}{4!} - ... + (-1)^n \frac{x^{2n}}{(2n)!} + ...$$

$$x - \frac{x^3}{3!} + \frac{x^5}{5!} - ... + (-1)^n \frac{x^{2n+1}}{(2n+1)!} + ...$$

For the first series, $\lim_{n\to\infty} \left| \frac{a_n}{a_{n+1}} \right| = \lim_{n\to\infty} \frac{(n+1)!}{n!} = \lim_{n\to\infty} (n+1) = \infty$.

Thus, the radius of convergence is $R = \infty$ so that the series is everywhere convergent (absolutely and uniformly). Similarly we can verify that the other two power series are also everywhere convergent (absolutely and uniformly). In this section we shall study their exclusive properties and show that the sums of these power series are actually the functions e^x, cos x and sin x respectively.

Theorem 5: (Abel's Theorem)

If a power series converges at an end point of its interval of convergence, then the power series is uniformly convergent in the interval which includes this end point.

Remarks:

1. *If a power series with interval of convergence]–R, R[converges at both the end-points – R and R, then the power series is uniformly convergent in [–R, R].*
2. *If a power series converges at the end-point x = R of the interval of convergence]–R, R[, then it is uniformly convergent in]–R, R] or [–R + ε, R], where ε > 0. In particular, the series is uniformly convergent in [0, R].*

Theorem 6: (Abel's Limit Theorem)

But $\Delta a_n x^n$ be a power series with finite radius of convergence R that $r(x) = \Sigma a_n x^n$, $-R\Delta x<R$. If the series $\Sigma a_n R^n$ converges, be

$$\lim_{x\to R-} f(x) = \Sigma a_n R^n.$$

Remark 3:

The case R = 1 is of much practical use :

Let $\Sigma a_n x^n$ *be a power series with unit radius of convergence such that* $f(x) = \Sigma a_n x^n$, $-1 \Delta x < 1$. *If the series* $\Sigma a_n x^n$ *converges for x = 1 i.e., if* Σa_n converges, then

$$\lim_{x\to 1-} f(x) = \Sigma a_n.$$

Theorem 7:

Suppose a power series $\Sigma a_n x^n$ converges for $|x| < R$ and that $f(x) = \Sigma a_n x^n$. Then

(i) *f(x) is continuous in]–R, R[.*

(ii) *$\Sigma a_n x^n$ can be integrated term by term on]–R, R[and the integrated series has the same radius of convergence as $\Sigma a_n x^n$.*

(iii) *$\Sigma a_n x^n$ can be differentiated term and term on]–R, [R and $f'(x) = \Sigma n a_n x^{n-1}$, $|x| < R$. Further the differentiated series has the same radius of convergence as $\Sigma a_n x^n$.*

Proof:

(i) Since each term of the power series $\Sigma a_n x^n$ is continuous in]–R, R[and the series is uniformly convergent on]–R, R[, so the sum f(x) of the given series is continuous in]–R, R[. (See Theorem 6, p. 165).

(ii) Since each term of $\Sigma a_n x^n$ is continuous in]–R, R[and $\Sigma a_n x^n$ converges uniformly to f(x) in]–R, R[, so $\Sigma a_n x^n$ can be integrated term by termin] –R, R[and the integrated series is $\Sigma \frac{a_n}{n+1} x^{n+1}$. Further

$$\lim_{n\to\infty}\left|\frac{u_n}{u_{n+1}}\right| = \lim_{n\to\infty}\left(\frac{n+2}{n+1}\frac{a_n}{a_{n+1}}\right) = \lim_{n\to\infty}\frac{a_n}{a_{n+1}} = R,$$

which shows that $\Sigma a_n x^n$ and $\Sigma\frac{a_n x^{n+1}}{n+1}$ have the same radius of convergence.

(iii) We know each term of $\Sigma a_n x^n$ is continuous, has continuous derivative in $]-R, R[$ and $\Sigma a_n x^n$ is uniformly convergent in $]-R, R[$. Thus the series obtained by differentiating the power series $\Sigma a_n x^n$ is $\Sigma n a_n x^{n-1}$. Further

$$\lim_{n\to\infty}\left|\frac{u_n}{u_{n+1}}\right| = \lim_{n\to\infty}\left|\frac{n\,a_n}{(n+1)\,a_{n+1}}\right| = \lim_{n\to\infty}\left|\frac{a_n}{a_{n+1}}\right| = R.$$

Hence the differentiated series $\Sigma n\, a_n\, x^{n-1}$ is also a power series which has the same radius of convergence as $\Sigma\, a_n\, x^n$.

Note : If *$f(x) = \Sigma a_n x^n$, then $f(x)$ is called the sum function of the series.*

Theorem 8:

Define cosine and sine functions as sums of power series. Prove that

(i) $S(x + y) = S(x)C(y) + C(x)S(y)$,

(ii) $C(x + y) = C(x)C(y) - S(x)S(y)$,

where C, S denote cosine and since respectively.

Proof:

The *cosine function* is defined as

$$C(x) = 1-\frac{x^2}{2!}+\frac{x^4}{4!}-...+(-1)^n\frac{x^{2n}}{(2n)!}+... \quad \forall\ x \in \Re \qquad ...(1)$$

The *sine function* is defined as

$$S(x) = x-\frac{x3}{3!}+\frac{x^5}{5!}-...+(-1)^n\frac{x^{2n+1}}{(2n+1)!}+... \qquad ...(2)$$

$\forall\ x \in \Re$

These two power series are uniformly convergent $\forall\ x \in \Re$ and consequently, the functions C(x) and S(x) are continuous $\forall\ x \in \Re$.

Differentiating term by term the convergent series (2).

$$S'(x) = 1-\frac{x^2}{2!}+\frac{x^4}{4!}+......$$

which gives us the uniformaly convergent series (1),

Thus S(x) is variable, $\forall x\in\Re$ and $S'(x) = C(x)$ ...(3)

Similarly $C'(x) = -S(x)$. ...(4)

From (1) and (2); $S(0) = 0$ and $C(0) = 1$. ...(5)

Let y be any arbitrary but fixed real number. We write

$$f(x) = S(x + y) - S(x)C(y) - C(x)S(y), \quad ...(6)$$

$$g(x) = C(x + y) - C(x)C(y) + S(x) S(y). \quad ...(7)$$

Differentiating w.r.t. x, we obtain

$$f'(x) = C(x + y) - C(x)C(y) + S(x)S(y) = g(x), \quad ...(6)$$

$$g'(x) = -S(x + y) - S(x)C(y) + C(x) S(y) = -f(x). \quad ...(7)$$

[Using (3), (4)]

Now $\frac{d}{dx}[f^2(x) + g^2(x)] = 2f(x) f'(x) + 2g(x) g'(x)$

$= 2f(x) g(x) - 2g(x) f(x) = 0 \quad \forall x$

If follows that $f^2(x) + g^2(z)$ is a constant $\quad \forall x$

$\Rightarrow f^2(x) + g^2(x) = f^2(0) + g^2(0) \quad \forall x$

$= [S(y) - S(0) C(y) - C(0) S(y)]^2$

$+ [C(y) - C(0) C(y) + S(0) S(y)]^2$

$= 0 \quad$ [Using (5)],

$\therefore f^2(x) + g^2(x) = 0 \ \forall x \in \Re$ implies the $f(x) = 0$ and $g(x) = 0$.

Hence (6) and (7) yield

$$S(x + y) = S(x) C(y) + C(x) S(y),$$

$$C(x + y) = C(x) C(y) - C(x) S(y).$$

Cor. 1. $S(-x) = -S(x); C(-x) = C(x) \quad \forall x \in \Re$.

We have $S(-x) = -x - \frac{(-x)^8}{3!} + \frac{(-x)^5}{5!} - \ldots\ldots$

$$= -\left[x - \frac{x^3}{3!} + \frac{x^5}{5!} - \ldots\ldots\right] = -S(x).$$

Similarly we can prove the second part.

Cor. 2. $S(x - y) = S(x) C(y) - C(x) S(y)$,

$C(x - y) = C(x) C(y) + S(x) S(y)$.

Replacing y by $-y$ in (i) and (ii)

$$S(x-y) = S(x) C(-y) + C(x) S(-y),$$

$$C(x-y) = C(x) C(-y) - S(x) S(-y).$$

The desired results follow by Cor. 1.

Cor. 3. $C^2(x) + S^2(x) = 1$ or $\cos^2 x + \sin^2 x = 1 \quad \forall x \in \mathfrak{R}$.

We have $1 = C(0) = C(x - x)$

$= C(x)\, C(-x) - S(x)\, S(-x)$

$= C^2(x) + S^2(x)$, by Cor. 1.

Hence $\cos^2 x + \sin^2 x = 1\ \forall x \in \mathfrak{R}$.

Cor. 4. $|C(x)| \le 1, |S(x)| \le 1, \quad \forall x \in \mathfrak{R}$.

OR

$|\cos x| \le 1, |\sin x| \le 1 : \forall x \in \mathfrak{R}$.

This follows by Cor. 3.

Cor. 5. *$\sin 2x = 2 \sin x \cos x$, $\cos 2x = \cos^2 x - \sin^2 x$.*

Replacing y by x in (i) and (ii) of Theorem 8, we obtain

$S(2x) = 2S(x)\, C(x),\ C(2x) = C^2(x) - S^2(x)$

Hence the results follow.

Remark:

It may be noted that the above properties of C(x) and S(x) are similar to those of the trigonometric functions cos x and sin x respectively which the students do at the elementary, stage. Thus we have

$C(x) \equiv \cos x$ and $S(x) \equiv \sin x$.

Theorem 9:

Let f be a non-zero function continuous on R^+ such that $f(xy) = f(x) + f(y)\ \forall x, y > 0$. ...(1)

Show that $f(x) \log_a x$, $x \in \mathfrak{R}^+$ ($a > 0$, $a \ne 1$).

Proof:

For $x > 0$, let us take $t = \log_e x \Leftrightarrow x = e^t$, $t \in \mathfrak{R}$. ...(2)

Let $g(t) = f(e^t)\ \forall t \in \mathfrak{R}$. Then g is continuous on $\mathfrak{R}$ and

$g(t + s) = f(e^{t+s}) = f(e^t e^3) = f(e^t) + f(e^s)$ by (1)

or $g(t + s) = g(t) + g(s)$.

Here $t, s \in \mathfrak{R}$. $g(t) = ct$, $\forall t \in \mathfrak{R}$, c being a constant.

Thus $f(x) = c \log_e x\ \forall x \in \mathfrak{R}^+$, where $c \ne 0$ as f is non-zero.

Let $a = e^{1/e}$. Then $f(x) = \log_a x\ \forall x \in \mathfrak{R}^+$ ($a > 0$, $a \ne 1$).

Ex.:

See the exponential function as the sum of a power series and determine its range. Hence define the logarithmic function and deduce that it is continuous.

[**Hint.** It can be verified that the exponential function E : R → R^+ is strictly increasing on R i.e., E is one-one and onto. Thus E has inverse function, say L : R^+ → R. L is called the *logarithmic function,* defined as

E {L (y)} = y (y > 0) or L {E (x)} = x (x ∈ R).

In other words, E (x) = y ⇒ I. (y) = x or e^x = y ⇒ $\log_e y = x$.

Since the function E is differentiable, so its inverse function L is false differentiable and hence continuous.

Theorem 10:

Define exponential function E(x) as the sum of power series. Show that the domain is the set of all real number Prove that

E (x + y) = E (x) E (y) for all x, y ∈ ℜ.

If e denotes E(1), prove that E (x) = e^2 for all real x.

Proof:

The exponential function E(x) is defined as the sum of the power series

$$E(x) = 1+\frac{x}{1!}+\frac{x^2}{2!}+...+\frac{x^n}{n!}+..... \qquad ...(1)$$

$$\text{We have } R = \lim_{n\to\infty}\left|\frac{a_n}{a_{n+1}}\right| = \lim_{n\to\infty}\frac{(n+1)!}{n!} = \lim_{n\to\infty}(n+1) = \infty.$$

Thus, the radius of convergence of the above power series is R = ∞ that the power series is everywhere convergent i.e., E(x) is defined ∀ x ∈ ℜ. This shows that the domain of the exponential function is ℜ.

It is clear that the function E(x), defined by (1), is continuous and has derivatives of all orders, for each x ∈ ℜ.

Differentiating (1), we get

E'(x) = E(x) and so E'(x) = E(x), E'(x) = E(x) etc.

In fact $E^{(n)}$ (x) = E(x) ∀ n ∈ N.

By Taylor's Theorem, for all real values t and x

$$E(t) = E(x) + \frac{E'(x)}{1!}(t-x)+\frac{E''(x)}{2!}(t-x)^2+.......$$

$$\text{or } E(t) = E(x) + \frac{E(x)}{1!}(t-x)+\frac{E(x)}{2!}(t-x)^2+.......$$

Replacing t by x + y, we get

$$E(x+y) = E(x)\left\{1+\frac{y}{1!}+\frac{y^2}{2!}+........\right\} = E(x)E(y)$$

Thus, E (x + y) = E(x) E (y) ∀ x, y ∈ ℜ. ...(2)

From (1), $E(1) = 1 + \frac{1}{1!} + \frac{1}{2!} + ... + \frac{1}{3!} + ...$

The above series on the right converges to a number which lies between 2 and 3.

We denote this number by e so that E(1) = e.

The property (2) can be extended as

$E(x_1 + x_2 + ... + x_n) = E(x_1) E(x_2)...E(x_n)$ ...(3)

for any positive integer n and $\forall\ x \in \Re$.

Taking $x_1 = x_2 = ... = x_n = x$ in (3), we obtain

$E(nx) = \{E(x)\}^n \quad \forall\ x \in \Re, \forall\ n \in N.$ ...(4)

For x = 1, we obtain

$E(n) = e^n \quad \forall\ n \in N$ $(\because E(1) = e)$ ...(5)

Taking x = m/n (m and n being positive integers) in (4),

$E(m) = \{E(m/n)\}^n$ or $e^m = \{E(m/n)\}^n$, by (5)

$\therefore E(m/n) = (e^m)^{1/n} = e^{m/n}$.

Thus, $E(q) = e^n$ for all positive rationals q. ...(6)

Let x be any *positive irrational* number. Then there always exists a sequence $\{x_n\}$ of positive rational numbers such that $x_n \to x$.

Using (6), $E(x_n) = e^{xn}\ \forall\ n \in N$.

Since $x_n \to x$ as $n \to \infty$ and E is a continuous function $\forall\ x \in \Re$, therefore

$E(x) = e^n$ for all positive irrationals. ...(7)

From (6) and (7), we conclude that

$E(x) = e^x$ for all positive reals. ...(8)

Taking y = – x in (2), we get

$E(0) = E(x)E(-x)$ or $1 = E(x)\ E(-x)\ (x \geq 0)$

$[\because E(0) = 1$, by (1)].

or $E(-x) = \frac{1}{E(x)} = \frac{1}{e^x} = e^{-x}$, using (8)

Hence, $E(x) = e^x \quad \forall\ x \in \Re$.

Theorem 11:

Let f be a continuous function on $\Re$ such that

$f(x + y) = f(x) + f(y)\ \forall\ x, y \in \Re.$ ...(1)

Show that $f(x) = e^x$ for some constant c and $\forall\ x \in \Re$.

Proof:

Putting y = x in (1), f(2x) = 2 f(x).

In fact f(nx) = nf(x) ∀ x ∈ ℜ, ∀ n ∈ N.

Replacing x by x/n, we get

f(x) = nf(x/n) or f(x/n) = (1/n) = (1/n) f(x) ∀ x ∈ ℜ.

For any integers p and p (q > 0), it follows that

$$f\left(\frac{p}{q}x\right) = \frac{p}{q} f(x) . \forall x \in \Re.$$

Thus, f(rx) = r f(x), for any rational number r. ...(2)

Let us write f(1) = c. Taking x = 1 in (2), we get

f(r) = cr, for every rational number r. ...(3)

Let α be any irrational number. Then there exists a sequence $\{r_n\}$ of rational numbers such that $r_n \to a$

$\Rightarrow f(r_n) \to f(a)$, continuity of f

$\Rightarrow f(a) = \lim f(r_n) = \lim cr_n = c \lim r_n = ca$, using (3).

Hence f(x) = cx ∀ x ∈ ℜ.

Theorem 12:

Define cos x as a power series and determine its domain. Prove that there exists a least positive number π such that $\cos\left(\frac{1}{2}\pi\right) = 0$.

Proof:

cos x is defined as the sum of the power series

$$1 - \frac{x^2}{2!} + \frac{x^4}{4!} - ... + (-1)^n \frac{x^{2n}}{(2n)!} + ...$$

Its radius of convergence is

$$R = \lim_{n\to\infty} \left|\frac{a_n}{a_{n+1}}\right| = \lim_{n\to\infty} \left|\frac{(2n+2)!}{(2n+1)!}\right| = \lim_{n\to\infty} (2n+2) = \infty .$$

Thus, this power series is convergent for ∀ x ∈ ℜ and so the domain of cos x is ℜ.

$$\cos x = 1 - \frac{x^2}{2!} + \frac{x^4}{4!} - \frac{x^6}{6!} + \frac{x^8}{8!} + ... \quad \forall x \in \Re \qquad ...(1)$$

∴ cos 0 = 1 > 0 and

$$\cos 2 = 1 - \frac{2^2}{2!} + \frac{2^4}{4!} - \frac{2^6}{6!} + \frac{2^8}{8!} + \ldots\ldots\ldots$$

$$= 1 - \frac{2^2}{2!}\left(1 - \frac{2^2}{3.4}\right) - \frac{2^6}{6!}\left(1 - \frac{2^2}{7.8}\right) - \ldots\ldots.$$

Since the brackets are all positive, therefore

$$\cos 2 < 1 - \frac{2^2}{2!}\left(1 - \frac{2^2}{3.4}\right) = -\frac{1}{3} \text{ and so } \cos 2 < 0.$$

Thus, we see that cos x is a continuous function in the closed interval [0, 2], where cos 0 and cos 2 are of opposite signs. By Intermediate value theorem, there exists at least one number $\alpha \in]0, 2[$ such that $\cos \alpha = 0$. We now proceed to show that α is unique.

Let, if possible, $\beta \in]0, 2[$ satisfy $\cos \beta = 0$ so that

$$\cos \alpha = \cos \beta = 0,\ 0 < \alpha, \beta < 2.$$

By Rolle's theorem, there exists some number λ, $\alpha < \lambda < b$, such that $-\sin \lambda = 0$.

$$\therefore \sin \lambda = 0,\ 0 < \lambda < 2. \qquad \ldots(2)$$

By definition,

$$\sin x = x - \frac{x^3}{3!} + \frac{x^5}{5!} - \frac{x^7}{7!} + \ldots\ldots \quad \forall\, x \in \Re$$

$$\therefore \quad \sin \lambda = \frac{\lambda}{1!}\left(1 - \frac{\lambda^2}{2.3}\right) + \frac{\lambda^5}{5!}\left(1 - \frac{\lambda^2}{6.7}\right) + \ldots\ldots > 0, \text{ for } 0 < \lambda < 2.$$

This contradicts (2).

Thus, there exists one and only one that of the equation cos x = 0 lying between 0 and 2. i.e., $\cos \alpha = 0,\ 0 < \alpha < 2$.

We denote α as $\frac{1}{2}\pi$. Hence $\frac{1}{2}\pi$ is the *least positive root* of the equation $\cos x = 0$ i.e., $\cos \frac{1}{2}\pi = 0$.

Also, therefore, we have $\cos x > 0$ when $0 \le x < \frac{1}{2}\pi$.

Theorem 13:

Let f be a non-zero continuous function on $\Re$ such that

$$f(x + y) = f(x)\, f(y)\ \forall\, x, y \in \Re. \qquad \ldots(1)$$

Show that $f(x) = e^{cx}$, for some constant $c \neq 0$.

Proof:

It is given that $f(x) \neq 0\ \forall\, x \in \Re$. From (1)

$$f(x) = f\left(\frac{1}{2}x\right) f\left(\frac{1}{2}x\right)$$

$$\text{or } f(x) = \left\{f\left(\frac{1}{2}x\right)\right\}^2 > 0 \ \forall \ x \in \Re.$$

Let $g(x) = \log_e f(x) \ \forall \ x \in \Re$. Then by (1),

$$g(x + y) = g(x) + g(y) \ \forall \ x, y \in \Re. \quad ...(2)$$

Now g, being the composition of two continuous functions f(x) and $\log_e$ x, is itself continuous on R and g satisfies (2).

It follows that

$$g(x) = cx \ \forall \ x \in \Re, \text{ c being a constant}$$

$$\Rightarrow \log_e f(x) = cx$$

$$\Rightarrow f(x) = e^{cx} \ \forall \ x \in \Re.$$

ABSOLUTE CONVERGENCE OF POWER SERIES $\Sigma A_N Z^N$

The power series $\Sigma a_n z^n$ is said to be absolutely convergent if series $\Sigma |a_n z^n|$ is convergent

If the series $\sum_{n=0}^{\infty} a_n z^n$ convergent but the series $\sum_{n=0}^{\infty} |a_n z|^n$ not converge,

then the series $\sum_{n=0}^{\infty} a_n z^n$ is called conditionally convergent.

RADIUS OF CONVERGENT OF POWER SERIES

Applying Cauchy's not test to the series $\Sigma |a_n| |z|^n$ we find that $\Sigma a_n z^n$ is convergent or divergent according as limit of nth root of the nth term is less or greater than unity Thus.

$\Sigma |a_n| |z|^n$ convergent if $\lim_{n\to\infty} |a_n|^{1/n} |z| < |$ i.e. if 1/R.

$|z| < |$ where $\lim|a_n|^{1/n} = 1/R$ or if $|z| < R$ and divergent if $|z| > R$

Thus, for every power series $\Sigma a_x z^n$. There exist a non-negative real number R such that for every $|z| < R$ The series is absolutely convergent, and for every $|z| > R$ the series is not convergent.

Such a number R is called the Radius of convergence of the power series $\Sigma a_x z^n$. For if we draw a circle whose centre is origin and radius equal to R, then for all values of z with in this circle power series $\Sigma a_x z^n$ is absolutely convergent and for all values of z outside this circle the power series $\Sigma a_x z^n$ is divergent.

SOLVED EXAMPLES

Example 1:

Show that

(i) $\log(1-x) = -x - \frac{x^2}{2} - \frac{x^3}{3} - \ldots, \ -1 \le x < 1.$

(ii) $\log 2 = 1 - \frac{1}{2} + \frac{1}{3} - \frac{1}{4} + \ldots\ldots$

(iii) $\frac{1}{2}[\log(1-x)]^2 = \frac{x^2}{2} + \left(1+\frac{1}{2}\right)\frac{x^3}{3} + \left(1+\frac{1}{2}+\frac{1}{3}\right)\frac{x^4}{4} + \ldots, \ -1 < x < 1.$

Solution:

(i) We know that

$$(1-x)^{-1} = 1 + x + x^2 + x^3 + \ldots, \ |x| < 1. \qquad \ldots(1)$$

Its radius of convergence is 1. So the above power series converges absolutely for $|x| < 1$ and uniformly for [– k, k\, $|k| < 1$. The integrated series is also convergent absolutely for $|x| < 1$ and uniformly for $[-k, k]$, $|k| < 1$,

Integrating, we get

$$-\log(1-x) = x + \frac{x^2}{2} - \frac{x^3}{3} + \ldots + c$$

Putting $x = 0$, we get $c = 0$.

$$\therefore \log(1-x) = -x - \frac{x^2}{2} - \frac{x^3}{3} - \ldots\ldots, \ -1 < x < 1.$$

By Leibnitz Test, the power series on the R.H. S. converges at $x = -1$. So by Abel's Theorem.

$$\log(1-x) = -x - \frac{x^2}{2} - \frac{x^3}{3} - \ldots\ldots, \ -1 < x < 1. \qquad \ldots(2)$$

(ii) At $x = -1$, by Abel's Limit Theorem

$$\log 2 = \lim_{x \to -1-} \log(1-x) = 1 - \frac{1}{2} + \frac{1}{3} - \frac{1}{4} + \ldots$$

(iii) both the series given on the R.H.S. of (1) and (2) are absolutely convergent in]–1, 1[, so their Cauchy's product will converge to $(1-x)^{-1}\log(1-x)$. Thus

$$(1-x)^{-1}\log(1-x) = -\left[x + \left(1+\frac{1}{2}\right)x^2 + \left(1+\frac{1}{2}+\frac{1}{2}\right)x^2 + \ldots\right], \quad -1 < x < 1$$

Integrating, we get

$$\frac{1}{2}[\log(1-x)]^2 = \frac{x^2}{2} + \left(1+\frac{1}{2}\right)x^3 + \left(1+\frac{1}{2}+\frac{1}{3}\right)\frac{x^4}{4} + \ldots\ldots$$
$$-1 < x < 1,$$

where the constant of integration vanishes.

Since the series on the right converge at $x = -1$, so by Abel's theorem, the above expansion is valid for $-1 < x < 1$.

Example 2:

Find the radius of convergence and the exact interval of convergence of the following power series :

(i) $\sum \frac{(n+1)}{(n+2)(n+3)} x^n,$

(ii) $\sum \frac{(n!)^2 x^{2n}}{(2n)!}$.

Solution:

(i) $a_n = \frac{(n+1)}{(n+2)(n+3)},\ a_{n+1} = \frac{(n+2)}{(n+3)(n+4)}$

$$\therefore\ R = \lim_{n\to\infty}\left|\frac{a_n}{a_{n+1}}\right| = \frac{(n+1)(n+4)}{!(n+2)^2} = 1.$$

Thus, the given series converges for $|x| < 1$ and diverges for $|x| > 1$. We now check at the end points $x = 1, -1$.

For $x = 1$, the nth term of the series is

$$u_n = \frac{(n+1)}{(n+2)(n+3)} \sim \frac{n}{n^2} = \frac{1}{n},\ \text{which diverges.}$$

However, for $x = -1$, the given series is an alternating series for which $u_{n+1} < u_n\ \forall\ n$ and $\lim_{n\to\infty} u_n = 0$. Thus, by Leibnitz's test, the series converges for $x = -1$. Hence the interval of convergence is $[-1, 1[$.

(ii) We have $a_n = \frac{(n!)^2}{(2n)!},\ a_{n+1} = \frac{[(n+1)!]^2}{(2n+2)!}$.

Then $\frac{a_n}{a_{n+1}} = \frac{(n!)^2}{\{(n+1)!\}^2}\ \frac{(2n+2)!}{(2n)!} = \frac{(2n+2)(2n+1)}{(n+1)^2}$

$$R = \lim_{n\to\infty}\left|\frac{a_n}{a_{n+1}}\right| = \lim_{n\to\infty}\frac{\left(2+\frac{2}{n}\right)\left(2+\frac{1}{n}\right)}{(1+1/n)^2} = 4.$$

It follows that the given series converges for $|x| < 4$ diverges for $|x| > 4$. Hence the interval of convergence is]–4, 4[.

Example 3:

Show that the series $\Sigma x^n/n^2$ is uniformly convergent in [–1, 1].

Solution:

We have $a_n = \frac{1}{n^2}$, $a_{n+1} = \frac{1}{(n+1)^2}$ so that

$$\lim_{n\to\infty}\left|\frac{a_n}{a_{n+1}}\right| = \lim_{n\to\infty}\frac{(n+1)^2}{n^2} = \lim_{n\to\infty}\left(1+\frac{1}{n}\right) = 1$$

The radius of convergence is R = 1 and so the given series is uniformly convergent in]–1, 1[.

- For x = 1, given series is $\sum\frac{1}{n^2}$, which is convergent.

For x = –1, given series becomes

$$-1+\frac{1}{2^2}-\frac{1}{3^2}+\frac{1}{4^2}-\ldots\ldots$$

This is an alternating series for which $u_{n+1} \leq u_n$ $\forall$ n and

$$\lim_{n\to\infty} u_n = \lim_{n\to\infty}\frac{1}{n^2} = 0.$$

So by Leibnitz's Test, the series to convergent or x = –1. Hence by Abel's Theorem, given power series is uniformly convergent in [–1, 1].

Example 4:

Show that the following series

$$x+\frac{1}{2}\frac{x^2}{3}+\frac{1.3}{2.4}\frac{x^5}{5}+\frac{1.3.5}{2.4.6}\frac{x^7}{7}+\ldots\ldots$$

is uniformly convergent in [–1, 1].

Solution:

Ignoring the first term, we have

$$a_n = \frac{1.3.5\ldots\ldots(2n-1)}{2.4.6\ldots.2n\ (2n+1)},\ a_{n+1} = \frac{1.3.5\ldots.(2n-1)\ (2n+1)}{2.4.6\ldots..2n(2n+2)\ (2n+3)}$$

$$\therefore \lim_{n\to\infty} \frac{a_n}{a_{n+1}} = \lim_{n\to\infty} \frac{(2n+2)(2n+3)}{(2n+1)^2} = 1.$$

Since the radius of convergence is R = 1, given power series is uniformly convergent in]–1, 1[. For x = 1, we see that the D'Alembert's Ratio Test fails. Now

$$\lim_{n\to\infty} \left(\frac{a_n}{a_{n+1}} - 1\right) = \lim_{n\to\infty} \frac{n(6n+5)}{(2n+1)^2} = \frac{6}{4} > 1.$$

By Raabe's Test, the series is convergent for x = 1.

Again for x = – 1, the series is convergent.

($\because$ Σa_n is convergent $\Rightarrow -\Sigma a_n$ is convergent).

Hence, the given power series is uniformly convergent in [–1, 1] (Abel's Theorem).

Example 5:

Show that

$$\frac{1}{2}[\log(1+x)]^2 = \frac{x^2}{2} - \frac{x^3}{3}\left(1+\frac{1}{2}\right) + \frac{x^4}{4}\left(1+\frac{1}{2}+\frac{1}{3}\right) + \ldots, -1 < x < 1.$$

Solution:

By the above example,

$$\log(1+x) = x - \frac{x^2}{2} + \frac{x^3}{3} + \frac{x^4}{4} + \ldots, -1 < x < 1.$$

and $(1 + x)^{-1} = 1 - x + x^2 - x^3 + x^4 - \ldots\ldots, -1 < x < 1.$

Both the series are absolutely convergent in]–1, 1[, ($\because$ R = 1), so their Cauchy's product will converge to $(1 + x)^{-1} \log(1 + x)$. Thus

$$(1+x)^{-1}\log(1+x) = x - x^2(1 + ½) + x^3(1 + ½ + ½) - \ldots\ldots, \quad -1 < x < 1$$

Integrating, we get

$$\frac{1}{2}[\log(1+x)]^2 = \frac{x^2}{2} - \frac{x^3}{3}\left(1+\frac{1}{2}\right) + \frac{x^4}{4}\left(1+\frac{1}{2}+\frac{1}{3}\right) - \ldots\ldots, -1 < x < 1$$

where the constant of integration vanishes.

Since the series on the right converges at x = 1, so by Abel's theorem the above expansion is valid for $-1 < x \le 1$.

Example 6:

Find the radius of convergence of $\sum_{n=1}^{\infty} nx^{n-1}$.

Solution:

Here $\lim_{n\to\infty}\left|\frac{a_n}{a_{n+1}}\right| = \lim_{n\to\infty}\frac{n}{n+1} = 1$.

Hence the radius of convergence = 1.

Example 7(a)

Find the radius of convergence of the series

$$x+\frac{x^2}{2^2}+\frac{2\,!}{3^3}x^2+\frac{3\,!}{4^6}x^6+......$$

Solution:

$$a_n=\frac{(n-1)\,!}{n^n},\ a_{n+1}=\frac{n\,!}{(n+1)^{n+1}}$$

$$\therefore\quad R=\lim_{n\to\infty}\left|\frac{a_n}{a_{n+1}}\right| = \lim_{n\to\infty}\frac{(n-1)\,!}{n\,!}\,\frac{(n+1)^{n+1}}{n^n}$$

$$=\lim_{n\to\infty}\left(\frac{n+1}{n}\right)^{n+1} = \lim_{n\to\infty}\left(1+\frac{1}{n}\right)^n\left(1+\frac{1}{n}\right)=e.$$

Hence the radius of convergence = e.

Example 7(b):

Show that the series

$$1+\frac{x}{2}+\frac{x^2}{3}+\frac{x^3}{4}+......$$

is uniformly convergent in [–1, k], 0 < k < 1.

Solution:

We have

$$\lim_{n\to\infty}\left|\frac{a_n}{a_{n+1}}\right| = \lim_{n\to\infty}\left(\frac{n+1}{n}\right) = \lim_{n\to\infty}\left(1+\frac{1}{n}\right)=1.$$

The radius of convergence is R = 1 and so the given series is uniformly convergent in]–1, 1[.

For x = 1, given series $\sum\frac{1}{n}$, which is divergent.

However for x = 1, given series becomes $1-\frac{1}{2}+\frac{1}{3}+\frac{1}{4}+.......$, which is convergent by Leibnitz's Test. Hence by Abel's Theorem, given power series is uniformly convergent in [–1, 1[or [–1, k], where 0 < k < 1.

Example 8:

Prove that

$$\int_0^x \frac{dt}{1+t^n} = x - \frac{x^{n+1}}{n+1} + \frac{x^{(2n+1)}}{2n+1} - \text{.........}, - < x < 1, n > 0.$$

Solution:

For n > 0, we have

$$(1 + t^n)^{-1} = 1 - t^n + t^{2n} - t^{2n} + \text{.....}$$

It is clear that the series on the right has its radius of convergence R = 1.

This series can be integrated term by term on]–1, 1[and the integrated series will also have the radius of convergence as unity. Thus

$$\int_0^x \frac{dt}{1+t^n} = x - \frac{x^{n+1}}{n+1} + \frac{x^{(2n+1)}}{2n+1} - \text{.........}, - < x < 1. \qquad \text{....(1)}$$

At x = 1, the series on the right is convergent, by Leibnitz's test.

Hence, (1) is true for $-1 < x \le 1$.

Example 9:

Show that

$$\log(1+x) = x - \frac{x^2}{2} + \frac{x^3}{3} - \frac{x^4}{4} + \text{......}, -1 < x \le 1$$

and deduce that

$$\log 2 = 1 - \frac{1}{2} + \frac{1}{3} - \frac{1}{4} + \text{......}$$

Solution:

We know

$$(1 + x)^{-1} = 1 - x + x^2 - x^3 + \text{........}, \quad -1 < x < 1. \qquad \text{...(1)}$$

Its radius of convergence is R = 1.

Thus the power series is absolutely convergent for $|x| < 1$ and uniformly convergent for [–k, k], $|k| < 1$.

The integrated series is also convergent absolutely in]–1, 1[and uniformly in [–k, k], $|k| < 1$.

Integrating (1), we get

$$\log(1+x) = c + x - \frac{x^2}{2} + \frac{x^3}{3} - \frac{x^4}{4} + \text{......}, 1 < x < 1.$$

Putting x = 0 gives c = 0. This

$$\log(1+x) = x - \frac{x^2}{2} + \frac{x^3}{3} - \frac{x^4}{4} + \ldots\ldots, 1 < x < 1.$$

The series on the R.H.S. is convergent for x = 1 (by Leibnitz's test). Hence by Abel's theorem,

$$\log(1+x) = x - \frac{x^2}{2} + \frac{x^3}{3} - \frac{x^4}{4} + \ldots\ldots, 1 < x < 1.$$

At x = 1, by Abel's limit theorem, we get

$$\log 2 = \lim_{x \to 1-} \log(1+x) = 1 - \frac{1}{2} + \frac{1}{3} - \frac{1}{4} + \ldots\ldots$$

Example 10:

Show that

(i) $\tan^{-1} x = x - \dfrac{x^3}{3} + \dfrac{x^5}{5} \dfrac{x^7}{7} + \ldots\ldots, -1 < x \le 1$

(ii) $\dfrac{\pi}{4} = 1 - \dfrac{1}{3} + \dfrac{1}{5} - \dfrac{1}{7} + \ldots\ldots$

(iii) $\dfrac{1}{2}(\tan^{-1} x)^2 = \dfrac{x^2}{2} - \dfrac{x^4}{4}\left(1 + \dfrac{1}{3}\right)$.

$$+ \frac{x^6}{6}\left(1 + \frac{1}{3} + \frac{1}{5}\right) + \ldots\ldots, -1 < x \le 1.$$

Solution:

We know

$$(1 + x^2)^{-1} = 1 - x^2 + x^4 - x^6 + \ldots\ldots, \quad |x| \le 1. \qquad \ldots(1)$$

Radius of convergence is

$$R = \lim_{n \to \infty} \left|\frac{a_n}{a_{n+1}}\right| = 1.$$

Thus the series is convergent absolutely for all x such that $|x| < 1$, and is uniformly convergent in $[-k, k]$, $|k| < 1$.

The integrated series is also convergent absolutely in $]-1, 1[$ and uniformly in $[-k, k]$, $|k| < 1$.

Integrating (1), we get

$$\tan^{-1} x = \left(x - \frac{x^3}{3} + \frac{x^5}{5} - \frac{x^7}{7} + \ldots\ldots\right) + C, \quad |x| < 1.$$

where C is constant of integration.

Putting x = 0, we get C = 0 and so

$$\tan^{-1} x = x - \frac{x^3}{3} + \frac{x^5}{5} - \frac{x^7}{7} + \ldots\ldots, \quad |x| < 1.$$

Now the series

$$x - \frac{x^3}{3} + \frac{x^5}{5} - \frac{x^7}{7} + \ldots\ldots \text{ is convergent at } x = \pm 1.$$

$\left[\because \text{ for } x = 1,\ 1 - \frac{1}{3} + \frac{1}{5} - \frac{1}{7} \ldots\ldots\right.$ is an alternating series which converges, by Leibnitz's test, Similarly it is convergent for $\left. x = -1\right]$.

By Abel's Theorem, it is uniformly convergent in [–1, 1].

$$\text{Hence, } \tan^{-1} x = x - \frac{x^3}{3} + \frac{x^5}{5} - \frac{x^7}{7} + \ldots, \ -1 < x \le 1.$$

(ii) Putting x = 1, by Abel's Limit Theorem,

$$\frac{\pi}{4} = \tan^{-1} 1 = \lim_{x \to 1-0} \tan^{-1} x = 1 - \frac{1}{3} + \frac{1}{5} - \frac{1}{7} + \ldots\ldots$$

(iii) We have

$$\tan^{-1} x = x - \frac{x^3}{3} + \frac{x^5}{5} - \frac{x^7}{7} + \ldots, \ -1 < x \le 1.$$

and $(1 + x^2)^{-1} \tan^{-1} x = x - x^2 + x^4 - x^6 + \ldots\ldots,\ 1 < x < 1.$

By part (i), both the series are absolutely convergent in]–1, 1[, so they can be multiplied and their Cauchy's product will be equal, to

$$(1 + x^2)^{-1} \tan^{-1} x = x - \left(1 + \frac{1}{3}\right)x^2 + \left(1 + \frac{1}{3} + \frac{1}{5}\right)x^2 \quad -1 < x < 1.$$

Integrating, we get

$$\frac{1}{2}(\tan^{-1} x)^2 = \frac{x^2}{2} - \frac{x^4}{4}\left(1 + \frac{1}{3}\right) + \frac{x^6}{6}\left(1 + \frac{1}{3} + \frac{1}{5}\right) - \ldots\ldots,$$

$$-1 < x < 1$$

Here the constant of integration vanishes on taking x = 0.

By Leibnitz's Test, the series on the R.H.S. converges for x = 1. Hence the above expansion is valid for – 1 < x < 1.

Example 11:

Find the radius of convergence of the series

$$\frac{1}{2}x + \frac{1.3}{2.5}x^2 + \frac{1.3.5}{2.5.8}x^2 + \ldots\ldots$$

Solution:

$$R = \lim_{n\to\infty} \left|\frac{a_n}{a_{n+1}}\right|$$

$$= \lim_{n\to\infty} \frac{1.3.5.....(2n-1)}{2.5.8.....(3n-1)}, \frac{2.5.8......(3n-1)(3n+2)}{1.3.5.....(2n-1)(2n+1)}$$

$$= \lim_{n\to\infty} \frac{3n+2}{2n+1} = \frac{3}{2}.$$

Example 12:

Show that

$$\sin^{-1} x = x + \frac{1}{2}\frac{x^3}{3} + \frac{1.3}{2.4}\frac{x^5}{5} + \frac{1.3.5}{2.4.6}\frac{x^7}{7} + ..., 1<x<1.$$

Deduce that

$$\frac{\pi}{2} = 1 + \frac{1}{2}.\frac{1}{3} + \frac{1.3}{2.4}.\frac{1}{5} + \frac{1.3.5}{2.4.6}.\frac{1}{7} + ...$$

Solution:

We know that

$$\frac{1}{\sqrt{1-x^2}} = (1 - x^2)^{-1/2}$$

$$= 1 + \frac{1}{2}x^2 + \frac{1.3}{2.4}x^4 + \frac{1.3.5}{2.4.6}x^6 + ..., |x| < 1.$$

The series on the R.H.S. is a power series with radius of convergence 1 and so it is absolutely convergent in]–1, 1[and uniformly convergent for [–k, k], | k | < 1. The integrated series is also convergent in]–1, 1[and uniformly convergent for [–k, k], | k | < 1. Integrating, we get

$$\sin^{-1} x = c + x + \frac{1}{2}.\frac{x^3}{3} + \frac{1.3}{2.4}.\frac{x^5}{5} + \frac{1.3.5}{2.4.6}.\frac{x^7}{7} + ...,|x| < 1$$

Putting x = 0, we get c = 0.

$$\therefore \sin^{-1} x + \frac{1}{2}.\frac{x^3}{3} + \frac{1.3}{2.4}.\frac{x^5}{5} + \frac{1.3.5}{2.4.6}.\frac{x^7}{7} + ...,-1< x <1 \quad ...(1)$$

For x = 1, the nth term of the series of the R.H.S. of (1) is

$$u_n = \frac{1.3.5..........(2n-1)}{2.4.6........2n(2n+1)}.$$

$$u_{n+1} = \frac{1.3.5..........(2n+1)}{2.4.6........(2n+2)(2n+3)}.$$

$$\therefore \frac{u_n}{u_{n+1}} = \frac{(2n+2)(2n+3)}{(2n+1)^2} . \; n\left(\frac{u_n}{u_{n+1}} - 1\right) = \frac{6n^2 + 5}{(2n+1)^2}$$

$$\lim_{n\to\infty} n\left(\frac{u_n}{u_{n+1}} - 1\right) = \frac{3}{2} > 1.$$

So by Raabe's Test, the series is convergent for x = 1. Hence

$$\sin^{-1} x = x + \frac{1}{2}.\frac{x^3}{3} + \frac{1.3}{2.4}\frac{x^5}{5} + \frac{1.3.5}{2.4.6}\frac{x^7}{7} + ..., - < x < 1.$$

(ii) At x = 1, by Abel's Limit Theorem

$$\frac{\pi}{2} = \lim_{x\to 1-} \sin^{-1} x = 1 + \frac{1}{2}.\frac{1}{3} + \frac{1.3}{2.4}.\frac{1}{5} + \frac{1.3.5}{2.4.6}\frac{1}{7} + ...$$

Example 13:

Prove that the power series

$$1 + \frac{a.b}{1.c}x + \frac{a(a+1)\,b(b+1)}{1.2c\,(c+1)}x^2 + ...$$

has unit radius of convergence.

Solution:

$$a_n = \frac{a(a+1)\,.....(a+n-1)\,b(b+1)\,.....(b+n-1)}{1.2......(n+1)\,c(c+1)......(c+n-1)}$$

(omitting the first term)

$$a_{n+1} = \frac{a(a+1)\,.....(a+n)\,b(b+1)\,.....(b+n)}{1.2......(n+1)\,c(c+1)......(c+n)}$$

$$R = \lim_{n\to\infty}\left|\frac{a_n}{a_{n+1}}\right| = \lim_{n\to\infty}\frac{(n+1)\,(c+n)}{(a+n)\,(b+n)}$$

$$= \lim_{n\to\infty}\frac{(1+1/n)\,(1+c/n)}{(1+a/n)(1+b/n)} = 1.$$

Example 14:

Find the domain of convergence of the series

(i) $\sum \frac{x^n}{n^2}$, (ii) $\sum \frac{(-1)^n (x-1)^n}{2^n (3n-1)}$.

Solution:

(i) $a_n = \frac{1}{n^3}$, $a_{n+1} = \frac{1}{(n+1)}$ so that

$$\lim_{n\to\infty} \frac{a_n}{a_{n+1}} = \lim_{n\to\infty} \left(\frac{n+1}{n}\right)^3 = 1. \text{ Thus } R = 1.$$

For x = 2, the given series is $\sum \frac{(-1)^{n+1}}{n}$, which is convergent by Leibnitz's Test. Hence the exact interval of convergence is]0, 2].

(iii) The given power series is about the point x = –2.

$$\lim_{n\to\infty} \left|\frac{a_n}{a_{n+1}}\right| = \lim_{n\to\infty} \frac{n!}{n^n} \frac{(n+1)^{n+1}}{(n+1)!} = \lim_{n\to\infty} \left(\frac{n+1}{n}\right)^n$$

$$= \lim_{n\to\infty} \left(1+\frac{1}{n}\right)^n = e.$$

∴ R = e and the interval of convergence is]–2–e, – 2 + e[.

Example 15(a):

Are the following series the Fourier Series?

(i) $\frac{\sin x}{\sqrt{1}} + \frac{\cos 2x}{\sqrt{2}} + \frac{\sin 3x}{\sqrt{3}} + \frac{\cos 4x}{\sqrt{4}} + \ldots\ldots$

(ii) $\frac{\cos x}{1^2} + \frac{\cos 2x}{2^2} + \frac{\cos 3x}{3^2} + \frac{\cos 4x}{4^2} + \ldots\ldots\ldots$

Solution:

(i) Suppose the given series is a Fourier Series

$$a_n = \frac{1}{\pi}\int_{-\pi}^{\pi} f(x)\cos nx = \frac{1}{\pi}\int_{-\pi}^{\pi} \sin nx \cos nx\, dx$$

$$= \begin{cases} 0, & \text{if n is odd.} \\ \frac{1}{\sqrt{n}}, & \text{if n is even} \end{cases}$$

$$\therefore \quad a_n = \begin{cases} \frac{1}{\sqrt{n}}, & \text{if n is odd.} \\ 0, & \text{if n is even} \end{cases}$$

Similarly, $b_n = \begin{cases} \frac{1}{\sqrt{n}}, & \text{if n is odd.} \\ 0, & \text{if n is even} \end{cases}$

Then $a_n^2 + b_n^2 = \frac{1}{n}$ in both the cases and so

$$\sum_{n=1}^{\infty}(a_n^{\ 2} + b_n^{\ 2}) = \sum_{n=1}^{\infty}\frac{1}{n}, \text{ which is divergent.}$$

This is a contradiction.

Hence the given series is not a Fourier Series.

(ii) The given series may be written as

$$\sum_{n=1}^{\infty}\left[\frac{\cos(2n-1)x}{(2n-1)^2} + \frac{\sin 2nx}{(2n)^2}\right]$$

We have

$$\left|\frac{\cos(2n-1)x}{(2n-1)^2} + \frac{\sin 2nx}{(2n)^2}\right| \le \frac{|\cos (2n-1)x|}{(2n-1)^2} + \frac{|\sin 2nx|}{(2n)^2}$$

$$\le \frac{1}{(2n-1)^2} + \frac{1}{(2n)^2} \le \frac{1}{n^2} + \frac{1}{n^2} = \frac{2}{n^2}.$$

Since $\sum M_n = \sum \frac{1}{n^2}$ is convergent, so by Weierstrass's M-test, the givens series is uniformly convergent on $[-\pi, \pi]$ and if f(x) is its sum, then the given series is the Fourier Series f f(x).

Example 15(b):

Find the Fourier Series generated by the periodic function | x | of period 2π.

Also compute the value of the series at 0, 2π, = 3π.

Solution:

The given function is monotonic and continuous on $[-\pi, \pi]$. Since $f(x) = |x|$ is an even function, its Fourier series will consist of cosine terms only.

$$\text{We have } f(x) = \begin{cases} -x, & \text{for } -\pi \le x \le 0, \\ x, & \text{for } \ 0 \le x \le \pi. \end{cases}$$

$$\text{We have } a_0 = \frac{1}{\pi}\int_{-\pi}^{\pi} f(x)\, dx = \frac{2}{\pi}\int_0^{\pi} f(x)\, dx = \frac{2}{\pi}\int_0^{\pi} x\ dx = \pi.$$

$$\text{Now } \ a_n = \frac{2}{\pi}\int_0^{\pi} x \cos nx\, dx = \frac{2}{\pi n^2}(\cos n\pi - 1)$$

$$\therefore \quad a_n = \begin{cases} 0, & \text{for n even,} \\ \dfrac{-4}{\pi n^2}, & \text{for n odd.} \end{cases}$$

Hence the Fourier series of | x | is

$$f(x) = \frac{1}{2}a_0 + \sum_{n=1}^{\infty} a_n \cos nx$$

$$= \frac{\pi}{2} + \frac{4}{\pi}\sum_{n=1}^{\infty} \frac{\cos nx}{n^2} \text{ (n odd)}$$

$$\therefore\ f(x) = \frac{\pi}{2} - \frac{4}{\pi}\left[\frac{\cos x}{1^2} + \frac{\cos 3x}{3^2} + \frac{\cos 5x}{5^2} - \dots\dots\dots\right]$$

We shall now compute the value of the series at 0, 2π, –3π.

The above series converges at all points and so its sum is equal to the given function f(x) = | x |.

The sum of the series at 2π is the same as at 0.

At x = 0

$$|0| = 0 = \frac{\pi}{2} - \frac{4}{\pi}\left[\frac{1}{1^2} + \frac{1}{3^2} + \frac{1}{5^2} + \dots\dots\dots\right]$$

$$\therefore\ 1 + \frac{1}{3^2} + \frac{1}{5^2} + \dots\dots = \frac{\pi^2}{8}.$$

The sum of the series at –3π is the same as at π and so

$$|\pi| = \pi = \frac{\pi}{2} + \frac{4}{\pi}\left[1 + \frac{1}{3^2} + \frac{1}{5^2} + \dots\dots\dots\right]$$

$$\therefore\ 1 + \frac{1}{3^2} + \frac{1}{5^2} + \dots = \frac{\pi^2}{8}.$$

Example 16:

Find (a) the Fourier sine series and (b) the Fourier cosine series which represents f(x) = (π – x) in 0 < x < π.

Solution:

Firstly we find sine series. We have

$$b_n = \frac{2}{\pi}\int_0^{\pi} (\pi - x) \sin nx\, dx$$

$$= \frac{2}{\pi}\left[\left| -(\pi - x)\frac{\cos nx}{n}\right|_0^{\pi} - \int_0^{\pi} \frac{\cos nx}{n} dx\right]$$

$$= \frac{2}{\pi}\left[\frac{\pi}{n}+0\right] = \frac{2}{n}.$$

Since f is continuous in $0 < x < \pi$, we have

$$\Sigma b_n \sin nx = ½\,[f(x+0) + f(x-0)] = [\frac{1}{2} f(x) + f(x)] = f(x)$$

$$\text{or } \pi - x = 2\Sigma \tfrac{1}{n} \sin nx = 2\left(\frac{\sin x}{1} + \frac{\sin 2x}{2} + \frac{\sin 3x}{3} + ...\right)$$

We know that the sum of the series must be 0 for $x = 0$ and $x = \pi$ and this fact can also be verified directly. The representation holds for $x = \pi$ but not for $x = 0$.

Now we find cosine series. We have

$$a_n = \frac{2}{\pi}\int_0^{\pi} (\pi - x)\cos nx\,dx = \frac{2[1-(-1)^n]}{\pi n^2},\ n \neq 0.$$

$$\text{Also, } \quad a_0 = \frac{2}{\pi}\int_0^{\pi} f(x)\,dx = \frac{2}{\pi}\int_0^{\pi}(\pi - x)\,dx = \frac{2}{\pi}.\frac{\pi^2}{2} = \pi.$$

Since f is continuous in $0 < x < \pi$, we have

$$\pi - x = \tfrac{1}{2}\pi + \sum \frac{2[1-(-1)^n]}{\pi n^2}\cos nx$$

$$= \tfrac{1}{2}\pi + \frac{4}{\pi}\left(\frac{\cos x}{1^2} + \frac{\cos 3x}{3^2} + \frac{\cos 5x}{5^2} + ...\right)$$

We know that the sum of the series must be $f(0+) = \pi$ for $x = 0$ and $f(\pi - 0) = 0$ for $x = \pi$. At $x = 0$ or π, we obtain

$$\frac{\pi^2}{8} = 1 + \frac{1}{3^2} + \frac{1}{5^2} + \frac{1}{7^2} + ...$$

Example 17:

Expand in a series of sines and cosines of multiple angles of x, the periodic function f with period in defined as

$$f(x) = \begin{cases} -1, & \text{for } -\pi < x < 0, \\ 1, & \text{for } \ 0 \le x \le \pi. \end{cases}$$

Also calculate the sum f the series at $x = 0$, ½ π, $\pm\, \pi$.

Solution:

The given function is piecewise monotonic, bounded and integrable on $[-\pi, \pi]$.

We have $a_0 = \frac{1}{\pi}\int_{-\pi}^{\pi} f(x)\,dx$

$$= \frac{1}{\pi}\left[\int_{-\pi}^{0}(-1)\,dx + \int_{\pi}^{0} 1.dx\right] = 0. \qquad \therefore a_n = 0$$

Now $a_n = \frac{1}{\pi}\int_{-\pi}^{\pi} f(x)\cos nx\,dx$

$$= \frac{1}{\pi}\left[\int_{-\pi}^{0} -\cos nx\,dx + \int_{0}^{\pi}\cos nx\,dx\right]$$

$$= \frac{1}{\pi}\left[-\left|\frac{\sin nx}{n}\right|_{-\pi}^{0} + \left|\frac{\sin nx}{n}\right|_{0}^{\pi}\right] = \quad \forall n.$$

Now $b_n = \frac{1}{\pi}\int_{-\pi}^{\pi} f(x)\sin nx\,dx$

$$= \frac{1}{\pi}\left[\int_{\pi}^{0} -\sin nx\,dx + \int_{0}^{\pi}\sin nx\,dx\right]$$

$$= \frac{1}{\pi}\left[\left|\frac{\cos nx}{n}\right|_{-\pi}^{0} - \left|\frac{\cos nx}{n}\right|_{0}^{\pi}\right]$$

$$= \frac{1}{\pi}\left[\frac{1}{n}\{1 - \cos n\pi - \cos n\pi + 1\}\right]$$

$$\therefore \qquad b_n = \frac{2}{n\pi}(1 - \cos n\pi) = \begin{cases} 0, & \text{for n even,} \\ \frac{4}{\pi n}, & \text{for n odd.}\end{cases}$$

Since the function f(x) is continuous at all points of $[-\pi, \pi]$ except 0, $\pm\pi$, therefore

$$f(x) = \frac{1}{2}a_0 + \sum(a_n \cos nx + b_n \sin nx)$$

$$= (b_1 \sin x + b_2 \sin 3x + b_2 \sin 5x + ...)$$

$$(\therefore a_0 = 0,\ a_n = 0\ \ \forall n \text{ and } b_n = 0 \text{ for n even})$$

Hence $f(x) = \frac{4}{\pi}\left(\frac{\sin x}{1} + \frac{\sin 3x}{3} + \frac{\sin 5x}{5} + ...\right)$, ...(1)

which is true for all x in $[-\pi, \pi]$ except t 0, $\pm\pi$.

At x = ± π the sum of the series is equal to

$$\frac{1}{2}[f(\pi-) + f(-\pi+)] = \frac{1}{2}(1 - 1) = 0.$$

At x = 0, the sum of the series

$$= \frac{1}{2}[f(0-) + f(0+)] = \frac{1}{2}(-1+1) = 0.$$

At $x = \frac{1}{2}\pi$, the function is continuous and so by (1)

$$f(\pi/2) = \frac{4}{\pi}\left(\frac{\sin \pi/2}{1} + \frac{\sin 3\pi/2}{3} + \frac{\sin 5\pi/2}{5} + \ldots\ldots\right)$$

or $$1 = \frac{4}{\pi}\left(1 - \frac{1}{3} + \frac{1}{5} - \ldots\ldots\right).$$

Hence $1 - \frac{1}{3} + \frac{1}{5} - \ldots = \frac{\pi}{4}.$

Example 18:

Find the Fourier Series for the function | sin x | in the interval [−π, π].

Solution:

Since f(x) = | sin x | is an even function and so the Fourier Series of f(x) will have cosines terms only.

We know $|\sin x| = \begin{cases} -\sin x, & \text{for } -\pi \le x \le 0, \\ \sin x, & \text{for } 0 \le x \le \pi. \end{cases}$

Now $$a_0 = \frac{1}{\pi}\int_{-\pi}^{\pi} f(x)\,dx = \frac{2}{\pi}\int_0^{\pi} f(x)\,dx$$

$$= \frac{2}{\pi}\int_0^{\pi} \sin x\,dx = -\frac{2}{\pi}\Big|\cos x\Big|_0^{\pi} = \frac{4}{\pi}.$$

Also $$a_n = \frac{2}{\pi}\int_0^{\pi} f(x)\cos x\,dx$$

$$= \frac{2}{\pi}\int_0^{\pi} \sin x \cos nx\,dx = \frac{2}{\pi} I_n,$$

where $$I_n = \int_0^{\pi} \sin x \cos nx\,dx$$

$$= \left|\sin x \frac{\sin nx}{n}\right|_0^{\pi} - \int_0^{\pi} \frac{\sin nx}{n} \cos x\,dx$$

$$= 0 - \frac{1}{n}\left[\left|\frac{-\cos nx}{n}\cos x\right|_0^{\pi} + \int_0^{\pi}\frac{\cos nx}{n}(-\sin x)\,dx\right]$$

$$\therefore \qquad I_n = \frac{1}{n^2}(-\cos nx - 1) + \frac{1}{n^2 I_n}$$

$$\text{or } \left(1 - \frac{1}{n^2}\right) i_n = -\frac{1}{n^2}(1 + \cos n\pi) = -\frac{1}{n^2}[1+(-1)^n]$$

$$\therefore \qquad I_n = -\frac{1}{(n^2-1)}[1+(-1)^n]$$

Thus, $$a_n = \frac{2}{-\pi(n^2-1)}[1+(-1)^n]$$

$$\therefore \qquad a_n = \begin{cases} 0, & \text{if n is odd,} \\ \dfrac{4}{-\pi(n^2-1)}, & \text{if n is even} \end{cases}$$

Hence the Fourier Series of | sin x |

$$= \frac{1}{2}a_0 + \sum_{n=1}^{\infty} a_n \cos nx$$

$$= \frac{1}{2}\cdot\frac{4}{\pi} - \sum_{n=1}^{\infty}\frac{4\cos nx}{\pi(n^2-1)}, \text{ for n even}$$

$$= \frac{4}{\pi}\left[\frac{1}{2} - \sum_{k=1}^{\infty}\frac{\cos 2kx}{(4k^2-1)}\right].$$

Example 19:

Expand the function f(x) = 2x in a series of sine and cosines in [0, π].

Solution:

(i) We have

$$b_n = \frac{2}{\pi}\int_0^{\pi} 2x \sin nx\,dx$$

$$= \frac{4}{\pi}\left[\left|-x\frac{\cos nx}{n}\right|_0^{\pi} + \int_0^{\pi}\frac{\cos nx}{n}\,dx\right]$$

$$= \frac{4}{\pi}\left(-\frac{\pi}{n}\cos n\pi + 0\right) = \frac{-4}{n}(-1)^n = \frac{4(-1)^{n+1}}{n}.$$

Since f(x) is continuous in [0, 1], therefore

$$\Sigma b_n \sin nx = \frac{1}{2}[f(x+0) + f(x-0)]$$

$$= \frac{1}{2}[f(x) + f(x)] = f(x).$$

Hence $2x = 4 \sum (-1)^{n+1} \dfrac{\sin nx}{n}$.

(ii) We have

$$a_0 = \frac{2}{\pi}\int_0^{\pi} f(x)\,dx = \frac{2}{\pi}\int_0^{\pi} 2x\,dx = 2\pi.$$

$$\text{Now } a_n = \frac{2}{\pi}\int_0^{\pi} 2x \cos nx\,dx$$

$$= \frac{4}{\pi}\left[\left|\frac{x \sin nx}{n}\right|_0^{\pi} - \int_0^{\pi} \frac{\sin nx}{n}\right]$$

$$= \frac{4}{\pi}\left[0 + \left|\frac{\cos nx}{n^2}\right|_0^{\pi}\right] = \frac{4}{\pi n^2}[\cos n\pi - 1]$$

$$= \frac{4}{\pi n^2}\left[(-1)^n - 1\right].$$

$$\therefore a_n = \begin{cases} 0, & \text{for n even,} \\ -8/\pi n^2, & \text{for n odd.} \end{cases}$$

Since f(x) is continuous in [0, 1], therefore

$$2x = \frac{1}{2}a_0 + \Sigma a_n \cos nx$$

$$= \pi - \frac{8}{\pi}\left[\frac{\cos x}{1^2} + \frac{\cos 3x}{3^2} + \frac{\cos 5x}{5^2} + \ldots\right].$$

Example 20:

If f is bounded and integrable on [–p, p] and if a_n, b_n are its Fourier coefficients, then prove that

$$\sum_{n=1}^{\infty} (a_n^2 + b_n^2) \text{ converges.}$$

Solution:

We have

$$0 \le \int_{-\pi}^{\pi} \left[f - \sum_{n=1}^{m} (a_n \cos nx + b_n \sin nx) \right]^2 dx$$

$$= \int_{-\pi}^{\pi} f^2 dx + \int_{-\pi}^{\pi} \left[\sum_{n=1}^{m} (a_n \cos nx + b_n \sin nx) \right]^2 dx$$

$$- 2 \sum_{n=1}^{m} \left[a_n \int_{-\pi}^{\pi} f \cos nx \, dx \right] - 2 \sum_{n=1}^{m} \left[b_n \int_{-\pi}^{\pi} f \sin nx \, dx \right]$$

$$= \int_{-\pi}^{\pi} f^2 dx + \pi \sum_{n=1}^{m} a_n^2 + \pi \sum_{n=1}^{m} b_n^2 - 2\pi \sum_{n=1}^{m} a_n^2 - 2\pi \sum_{n=1}^{m} b_n^2$$

$$= \int_{-\pi}^{\pi} f^2 dx + \pi \sum_{n=1}^{m} (a_n^2 + b_n^2)$$

Thus, $\sum_{n=1}^{m} (a_n^2 + b_n^2) \le \frac{1}{\pi} \int_{-\pi}^{\pi} f^2 \, dx.$

Taking limit as $m \to \infty$, we obtain

$$\sum_{n=1}^{m} (a_n^2 + b_n^2) \le \frac{1}{\pi} \int_{-\pi}^{\pi} f^2 \, dx < \infty,$$

since f and so f^2 is integrable on $[-\pi, \pi]$

Hence, the series $\Sigma(a_n^2 + b_n^2)$ is convergent.

If, Define the Fourier coefficients a_n and b_n of the bounded and integrable function on $[-\pi, \pi]$. Show that a_n, $b_n \to 0$ as $n \to \infty$.

Hint. Since $\Sigma(a_n^2 + b_n^2)$ converges, therefore

$(a_n^2 + b_n^2) \to 0$ as $n \to \infty$.

Hence $a_n \to 0$, $b_n \to 0$ as $n \to \infty$.

Example 21:

Find the Fourier sine series in $0 < x < \pi$, for

$$f(x) = \begin{cases} \pi/3, & \text{for } 0 < x < \pi/3 \\ 0, & \text{for } \pi/3 < x < 2\pi/3, \\ -\pi/3, & \text{for } 2\pi/3 < x < \pi. \end{cases}$$

Find the sum of the series when $x = \pi$, $2\pi/3$.

Solution:

We have

$$b_n = \frac{2}{\pi} \int_0^{\pi} f(x) \sin nx \, dx$$

$$= \frac{2}{\pi}\left[\int_0^{\pi/3} \frac{\pi}{3} \sin nx\, dx + \int_{\pi/3}^{2n/3} 0.\sin nx\, dx - \int_{2\pi/3}^{\pi} \frac{\pi}{3} \sin nx\, dx\right]$$

$$= \frac{2}{3}\left[\int_0^{\pi/3} \sin nx\, dx + (-1)^n \int_0^{\pi/3} \sin ny\, dy.\right] \qquad y = \pi - x$$

It is clear that $b_n = 0$, for n odd

and $b_n = \frac{4}{3}\int_0^{\pi/3} \sin nx\, dx$, for n even

$$= -\frac{4}{3}\left|\cos nx\right|_0^{\pi/3} = -\frac{4}{3\pi}\left[\cos\frac{n\pi}{3} - 1\right]$$

$$\therefore\ b_n = \frac{4}{3n}\left(2\sin^2\frac{n\pi}{6}\right), \text{ for n even.}$$

Replacing n by 2n,

$$b_{2n} = \frac{4}{3n}\sin^2\left(\frac{n\pi}{3}\right)$$

$$= \begin{cases} 0, & \text{if } n = 3k \\ \frac{4}{3n}\sin^2\left(\frac{\pi}{3}\right), & \text{if } n \neq 3k \end{cases}$$

or $$b_{2n} = \begin{cases} 0, \text{ if } n = 3k, \\ 1/n, \text{if } n \neq 3k. \end{cases}$$

Hence the Fourier Series of f(x) is

$$\sigma(x) = \sum b_n \sin nx = \sum b_{2n} \sin 2nx\ (n \neq 3k)$$

$$= \sum_1^{\infty} \frac{\sin 2nx}{n}\ (n \neq 3k)$$

$$= \sin 2x + \frac{\sin 4x}{2} + \frac{\sin 8x}{4} + \frac{\sin 10x}{5} + \frac{\sin 14}{7} + ...$$

We know that the sum of the series must be 0 for $x = \pi$ and this fact can be verified directly $(\sigma(\pi) = 0)$.

At $x = \frac{2\pi}{3}$, the sum of the series

$$= \frac{1}{2}\left[f\left(\frac{2}{3}\pi + 0\right) f\left(\frac{2}{3}\pi - 0\right)\right]$$

$$= \frac{1}{2}\left[-\frac{\pi}{3} + 0\right] = -\frac{\pi}{6}.$$

It can be shown that the sum of the series is $\pi/6$ at $x = \pi/3$.

Example 22(a):

Show that for all values of x in $[-\pi, \pi]$

$$|\cos x| = \frac{4}{\pi}\left[\frac{1}{2} + \sum_{n=1}^{\infty} \frac{(-1)^{n+1} \cos 2nx}{(4n^2 - 1)}\right].$$

Solution:

The function $f(x) = |\cos x|$ is piecewise monotonic, bounded and integrable in $[-\pi, \pi]$. Since $f(x)$ is an even function of x, so $b_n = 0\ \forall\ n$. Further cos x changes sign in $[0, \pi]$.

Thus,

$$a_0 = \frac{1}{\pi}\int_{-\pi}^{\pi} |\cos x|\, dx = \frac{2}{\pi}\int_0^{\pi} |\cos x|\, dx$$

$$= \frac{2}{\pi}\int_{-\pi/2}^{\pi/2} |\sin y|\, dy,\ x = y + \tfrac{1}{2}\pi$$

$$= \frac{4}{\pi}\int_0^{\pi/2} \sin y\, dy = \frac{4}{\pi}. \qquad \therefore a_0 = \frac{4}{\pi}$$

Now, $$a_n = \frac{1}{\pi}\int_{-\pi}^{\pi} |\cos x| \cos nx\, dx$$

$$= \frac{2}{\pi}\int_0^{\pi} |\cos x| \cos nx\, dx$$

$$= \frac{2}{\pi}\int_{-\pi/2}^{\pi/2} |\sin y| \cos\left[n\left(\tfrac{1}{2}\pi + y\right)\right] dy, \quad \text{where } x = y + \tfrac{1}{2}\pi$$

Thus, $$a_{2n} = \frac{2}{\pi}(-1)^n \int_{-\pi/2}^{\pi/2} |\sin y| \cos 2ny\, dy \qquad \text{(Even integrand)}$$

$$= \frac{2(-1)^n}{\pi}\int_0^{\pi/2} \sin y \cos 2ny\, dy$$

$$= \frac{2(-1)^n}{\pi}\int_0^{\pi/2} [\sin (2n+1)y - \sin(2n-1)\, dy$$

$$= \frac{2(-1)^n}{\pi}\left[\frac{1}{2n+1}-\frac{1}{2n-1}\right] = \frac{4(-1)^{n+1}}{\pi(4n^2-1)}.$$

Again $a_{2n-1} = \dfrac{2(-1)^{n+1}}{\pi}\displaystyle\int_{-\pi/2}^{\pi/2} |\sin y| \sin(2n-1)\, y\, dy = 0$

(Odd integrand)

Since $|\cos x|$ is continuous for all x in $[-\pi, \pi]$

$$|\cos x| = \frac{1}{2}a_0 + \sum_{n=1}^{\infty}(a_n \cos nx + b_n \sin nx)$$

Hence, $|\cos x| = \dfrac{4}{\pi}\left[\dfrac{1}{2} + \displaystyle\sum_{n=1}^{\infty}\frac{(-1)^{n+1}\cos 2nx}{(4n^2-1)}\right]$ $(\because b_n = 0)$

Example 22(b):

Find the Fourier Series of the periodic function f with period in defined as follows:

$$f(x) = \begin{cases} 0, & \text{for } -\pi \le x \le 0, \\ x, & \text{for } 0 \le x \le \pi. \end{cases}$$

What is the sum of the series at $x = 0, \pm\pi$.

Solution:

The function is bounded, piecewise monotonic, and integrable on $[-\pi, \pi]$.

We have

$$a_0 = \frac{1}{\pi}\int_{-\pi}^{\pi} f(x)\, dx$$

$$= \frac{1}{\pi}\left[\int_{-\pi}^{0} 0.\, dx + \int_{\pi}^{0} x.dx\right] = \frac{\pi}{2}. \qquad \therefore a_0 = \frac{1}{2}\pi.$$

Now $a_n = \dfrac{1}{\pi}\displaystyle\int_{-\pi}^{\pi} f(x)\cos nx\, dx = \frac{1}{\pi}\left[0 + \int_0^{\pi} x\cos nx\, dx\right]$

$$= \frac{1}{\pi}\left[\left|\frac{x\sin nx}{n}\right|_0^{\pi} - \frac{1}{n}\int_0^{\pi}\sin nx\, dx\right]$$

$$= \frac{1}{n\pi}\left|\frac{\cos nx}{n}\right|_0^{\pi} = \begin{cases} 0, & \text{if n even,} \\ \dfrac{-2}{\pi n^2}, & \text{for n odd.} \end{cases} \qquad ...(1)$$

Again

$$b_n = \frac{1}{\pi}\int_{-\pi}^{\pi} f(x)\sin nx\,dx = \frac{1}{\pi}\int_{-\pi}^{\pi} x\sin nx\,dx$$

$$= \frac{1}{\pi}\left[\left|\frac{-x\cos nx}{n}\right|_0^{\pi} + \frac{1}{n}\int_0^{\pi}\cos nx\,dx\right]$$

$$= \frac{1}{\pi}\left[\left|\frac{-x\cos nx}{n}\right|_0^{\pi} + \frac{1}{n}\left|\frac{\sin nx}{n}\right|_0^{\pi}\right]$$

$$= \frac{1}{\pi}\left[\frac{-x\cos nx}{n}\right] + 0$$

$$= -\frac{\cos n\pi}{n}\begin{cases}-1/n, & \text{if } n \text{ is even,}\\ 1/n, & \text{if } n \text{ is odd.}\end{cases} \qquad ...(2)$$

In [■π, π], the only points of discontinuity of f are $\pm\pi$.
For all $x \in\ ]-\pi, \pi[$.

$$f(x) = \frac{1}{2}a_0 + \Sigma\,(a_n\cos nx + b_n\sin nx)$$

$$f(x) = \frac{\pi}{4} - \frac{2}{\pi}\left\{\frac{\cos x}{1^2} + \frac{\cos 3x}{3^2} + \frac{\cos 5x}{5^2} + \ldots\ldots\right\}$$

$$+\left\{\frac{\sin x}{1} - \frac{\sin 2x}{2} + \frac{\sin 3x}{3} - \ldots\ldots\right\} \qquad ...(3)$$

Since f is continuous at $x = 0$,

$$f(0) = \frac{\pi}{4} - \frac{2}{\pi}\left[\frac{1}{1^2} + \frac{1}{3^2} + \frac{1}{5^2} + \ldots\right] + 0, \text{ by (3)}$$

$$\therefore\ 1 + \frac{1}{3^2} + \frac{1}{5^2} + \ldots\ldots = \frac{\pi^2}{8} \qquad (\because f(0) = 0)$$

Since $x = \pm\pi$ are the points of discontinuity of f, so the sum of series is $= \frac{1}{2}[\,f(\pi-) + f(-\pi+)]$, where

$$f(n-) = \lim_{x\to\pi} f(x) = \lim_{x\to\pi} x = \pi,$$

$$f(-\pi+) = \lim_{x\to-\pi+} f(x) = 0.$$

Hence the sum of series at $x = \pm\pi$ is $\frac{1}{2}(\pi + 0) = \frac{1}{2}\pi$.

$$\therefore \quad \frac{\pi}{2}=\frac{\pi}{4}+\frac{2}{\pi}\left\{\frac{1}{1^2}+\frac{1}{3^2}+\frac{1}{5^2}+........\right\}$$

$$\text{or } 1+\frac{1}{3^2}+\frac{1}{5^2}+.......=\frac{\pi^2}{8}.$$

Example 22(c):

Expand the periodic function x^2, $0 \le x \le 1$, of period 1 in a series of sines only. Also find the sum of the series at $x = 0, 1$.

Solution:

The substitution $y = \pi x\, l$ transforms the given function with period π on $[0, \pi]$. To get the series of sines only, we have

$$b_n = \frac{2}{l}\int_0^l x^2 \sin(n\pi x/l)\,dx$$

$$= \frac{2}{l}\left[-\frac{l}{n\pi}\left|x^2 \cos\left(\frac{n\pi x}{l}\right)\right|_0^l + \frac{2l}{n\pi}\int_0^l x\cos\left(\frac{n\pi x}{l}\right)dx\right]$$

$$= \frac{2}{l}\left[-\frac{l^3}{n\pi}\cos n\pi + \frac{2l^2}{n^2\pi^2}\left\{\left|x\sin\left(\frac{n\pi x}{l}\right)\right|_0^l - \int_0^l \sin\left(\frac{n\pi x}{l}\right)dx\right\}\right]$$

$$= \frac{2}{l}\left[-\frac{l^3}{n\pi}\cos n\pi + 0 + \frac{2l^3}{n^3\pi^3}\left|\cos\frac{n\pi x}{l}\right|_0^l\right]$$

$$= \frac{2}{l}\left[-\frac{l^3}{n\pi}\cos n\pi + \frac{2l^3}{n^3\pi^3}(\cos n\pi - 1)\right]$$

$$= -\frac{2l^2}{n\pi}\cos n\pi + \frac{4l^2}{n^3\pi^3}(\cos n\pi - 1)$$

$$\therefore \quad = -\frac{2(-l)^{n/2}}{n\pi} - \frac{4l^2}{n^3\pi^3}\{1-(-1)^n\}$$

Since the function x^2 is continuous in $[0, 1]$, therefore

$$\therefore \quad x^2 = \sum_{n=1}^{\infty} b_n \sin\frac{n\pi x}{l} = \sum_{n=1}^{\infty}\left[-\frac{2(-1)^n)l^2}{n\pi} - \frac{4l^2(-1(-1)^n)}{n^2\pi^2}\right]$$

$$\times \sin\frac{n\pi x}{l} \quad \forall\, x \in [0, 1]$$

For $x = 0$ or l, the sum of the series is 0.

Example 22(d):

Show that the Fourier series which converges to f(x) in $-\pi \le x \le \pi$, where f(x) = $x + x^2$ when $-\pi < x < \pi$ and f(x) = π^2 when x = $\pm\pi$ is $\frac{\pi^2}{3}+4$ $\sum(-1)^n\left(\frac{\cos nx}{n^2}-\frac{\sin nx}{2n}\right)$.

Deduce that $\pi^2/6 = 1 + 1/2^2 + 1/3^2 + 1/4^2 + \ldots$.

Solution:

We have

$$a_0 = \frac{1}{\pi}\int_{-\pi}^{\pi}(x^2+x)\,dx$$

$$= \frac{1}{\pi}\left[\int_{-\pi}^{\pi}x^2\,dx + \int_{-\pi}^{\pi}x.dx\right] = \frac{2}{\pi}.\frac{\pi^2}{3} = \frac{2\pi^2}{3}.$$

Now $a_n = \frac{1}{\pi}\int_{-\pi}^{\pi}(x^2+x)\cos nx\,dx$

$$= \frac{1}{\pi}\int_{-\pi}^{\pi}x^2\cos nx\,dx + 0 \quad (\because x \cos nx \text{ is an odd function})$$

$$= \frac{2}{\pi}\int_0^{\pi}x^2\cos nx\,dx \quad (\because x^2 \cos nx \text{ is an even function})$$

$$= \frac{2}{\pi}\left[\left|\frac{x^2\sin nx}{n}\right|_0^{\pi} - \int_0^{\pi}\frac{2x\sin nx}{n}dx\right]$$

$$= \frac{2}{\pi}\left[0 - \frac{2}{\pi}\left\{\left|x\left(-\frac{\cos nx}{n}\right)\right|_0^{\pi} + \int_0^{\pi}\frac{\cos nx}{n}.1dx\right\}\right]$$

$$= \frac{2}{\pi}\left[\frac{2}{\pi^2}\left|x\cos nx\right|_0^{\pi} - \left|\frac{\sin nx}{n}\right|_0^{\pi}\right]$$

$$= \frac{4}{\pi^2\pi}[\pi\cos n\pi] = \frac{4}{n^2}(-1)^n$$

$$\therefore \quad a_n = \frac{4}{\pi^2}(-1)^2.$$

Now $b_n = \frac{1}{\pi}\int_{-\pi}^{\pi}(x^2+x)\sin nx\,dx$

$$= \frac{1}{\pi}\left[0 + 2\int_{-\pi}^{\pi}x\sin nx\,dx\right]$$

$$= \frac{2}{\pi}\left[\left|x\left(-\frac{\cos nx}{n}\right)\right|_0^{\pi} + \int_0^{\pi} 1.\frac{\cos nx}{x}\,dx\right]$$

$$= \frac{2}{\pi}\left[-\frac{\pi(-1)^n}{n}\right] = \frac{2}{n}(-1)^{n+1}$$

$$\therefore \quad b_n = \frac{2}{n}(-1)^{n+1}.$$

Since f is continuous at all points of $[-\pi, \pi]$ except at $\pm \pi$.

$$f(x) = \frac{1}{2}a_0 + \Sigma\,(a_n \cos nx + b_n \sin nx) \text{ in}]-\pi, \pi[.$$

$$\text{or } \left|x^2 + x = \frac{1}{2}\cdot\frac{2}{3}\right|\pi^2 + \sum\left[\frac{4(-1)^n}{n^2}.\cos nx + \frac{2}{n}(-1)^{n+1}\sin nx\right]$$

$$\text{Hence, } x^2 + x = \frac{\pi^2}{3} + 4\,\Sigma(-1)^n\left[\frac{\cos nx}{n^2} - \frac{\sin nx}{2n}\right] \text{ in }]-\pi, \pi[. \quad ...(1)$$

Now, $f(n-) = \underset{h\to 0}{\text{Lt}}\ [(\pi - h)^2 + (\pi - h)] = \pi^2 + \pi.$

and $\qquad f(-\pi+) = \underset{h\to 0}{\text{Lt}}\ [(-\pi + h)^2 + (-\pi + h) = \pi^2 - \pi.$

At $x = \pm\,\pi$, the sum of the series

$$= \frac{1}{2}\ [f(\pi -) + f(-\pi +)] = ½\ (\pi^2 + \pi + \pi^2 - \pi] = \pi^2.$$

From (1), we obtain

$$f(\pi-) \neq \frac{\pi^2}{3} + \lim_{h\to 0}\sum_{n=1}^{\infty} 4(-1)^n\left[\frac{1}{n^2}.\cos n(\pi - h) - \frac{1}{2n}\sin n\,(\pi - h)\right]$$

$$\text{and } f(-\pi+) = \frac{\pi^2}{3} + \lim_{h\to 0}\sum_{n=1}^{\infty} 4(-1)^n\left[\frac{1}{n^2}.\cos n(-\pi + h) - \frac{1}{2n}\sin n(-\pi - h)\right]$$

$$\therefore \qquad \frac{1}{2}[f(\pi-) + f(-\pi+)]$$

$$= \frac{1}{2}\left[\frac{\pi^2}{3} + \sum 4(-1)^n\left\{\frac{\cos n\pi}{n^2} - \frac{\sin n\pi}{2n}\right\} + \frac{\pi^2}{3} + \sum 4(-1)^n\left\{\frac{\cos n\pi}{n^2} + \frac{\sin n\pi}{2n}\right\}\right]$$

$$= \frac{\pi^2}{3} + \sum \frac{4(-1)^n (-1)^n}{n^2} = \frac{\pi^2}{3} + \sum \frac{4(-1)^{2n}}{n^2}$$

$$\therefore \quad \pi^2 = \frac{\pi^2}{3} + \sum_{n=1}^{\infty} \frac{4}{n^2} \Rightarrow \frac{2}{3}\pi^2 = 4\sum \frac{1}{n^2}$$

Hence, $= \dfrac{\pi^2}{6} = \sum \dfrac{1}{n^2}$.

Example 23:

Expand the periodic function, of period 2l > 0.

$f(x) = \left|\cos\left(\frac{\pi x}{l}\right)\right|$, in a Fourier Series.

Solution:

We have

$$f(x) = \begin{cases} -\cos(\pi x/l), & -l \le x \le -l/2, \\ \cos(\pi x/l), & -l/2 \le x \le -l/2, \\ -\cos(\pi x/l), & -l/2 \le x \le -l \end{cases}$$

Since f(x) is an even function, therefore $b_n = 0$, $\forall n$.

Now

$$a_0 = \frac{1}{l}\int_{-l}^{l} |\cos(\pi x/l)| dx$$

$$= -\frac{1}{l}\int_{-l}^{-l/2} \cos(\pi x/l)\, dx + \int_{-l/2}^{l/2} \cos(\pi x/l)\, dx$$

$$= \frac{1}{l}\int_{l/2}^{l} \cos(\pi x/l)\, dx$$

$$= -\frac{2}{l}\int_{l/2}^{l} \cos(\pi x/l)\, dx + \frac{2}{l}\int_{0}^{l/2} \cos(\pi x/l)\, dx = 4/\pi$$

Similarly,

$$a_n = -\frac{1}{l}\int_{-l}^{-l/2} \cos(\pi x/l)\cos(n\pi x/l)\, dx$$

$$+ \int_{-l/2}^{l/2} \cos(\pi x/l)\cos(n\pi x/l)\, dx - \int_{l/2}^{l} \cos(\pi x/l)\cos(n\pi x/l)\, dx$$

$$= -\frac{1}{l}\int_{l/2}^{l} \left[\cos\left\{\frac{(n+1)\,\pi x}{l}\right\} + \cos\left\{\frac{(n-1)\pi x}{l}\right\}\right]$$

$$+ \frac{1}{l}\int_{0}^{l/2} \left[\cos\left\{\frac{(n+1)\,\pi x}{l}\right\} - \cos\left\{\frac{(n-1)\pi x}{l}\right\}\right] dx$$

$$= -\frac{2\cos(n\pi/2)}{(n^2-1)\pi}$$

$$\therefore\ a_{2n} = \frac{2(-1)^{n+1}}{(4n^2-1)\pi},\ \forall n, \text{ and } a_{2n-1} = 0,\ \forall n.$$ **Hence Proved.**

Example 24:

Expand the function f(x) = cos 2x in a series of sines in [0, π].

Solution:

We have

$$b_n = \frac{2}{\pi}\int_0^\pi \cos 2x \sin nx\,dx$$

$$= \frac{1}{\pi}\int_0^\pi [\sin(n+2)x + \sin(n-2)x]\,dx$$

$$= \frac{1}{\pi}\left[\frac{\cos(n+2)x}{n+2} + \frac{\cos(n-2)x}{n-2}\right]_0^\pi$$

$$= -\frac{1}{\pi}\left[\left\{\frac{(-1)^{n+2}}{n+2} + \frac{(-1)^{n-2}}{n-2}\right\} - \left\{\frac{1}{n+2} + \frac{1}{n-2}\right\}\right]$$

$$= -\frac{1}{\pi}\left[(-1)^n\left\{\frac{1}{n+2} + \frac{1}{n-2}\right\} - \left\{\frac{1}{n+2} + \frac{1}{n-2}\right\}\right]$$

$$= -\frac{2n}{\pi(n^2-4)}[(-1)^n - 1].$$

$$\therefore\quad b_n = \begin{cases} 0, & \text{for n even,} \\ \dfrac{4n}{\pi(n^2-4)}, & \text{for n odd.} \end{cases}$$

Equivalently, $b_n = -\dfrac{4}{\pi}\dfrac{(2n-1)}{[2^2-(2n-1)^2]}$.

Since cos 2x is continuous in [0, π], we have

$\cos 2x = \sum b_n \sin nx$

$$= -\frac{4}{\pi}\sum \frac{(2n-1)\sin(2n-1)x}{[2^2-(2n-1)^2]}.$$

Example 25:

Find the Fourier series of the function x^2, periodic with period 2l, on the interval [–l, l].

Solution:

Let us substitute $y = \pi x/l$ so that the given function becomes a periodic function with period 2π on $[-\pi, \pi]$. Since x^3 is an even function, so $b_n = 0$ for all n. We have

$$a_0 = \frac{2}{l}\int_0^l x^2\, dx = \frac{2}{3} l^2.$$

$$\text{Now}\quad a_n = \frac{2}{l}\int_0^l x^2 \cos\frac{n\pi x}{l}\, dx$$

$$= \frac{2}{l}\left[\left|\frac{l}{n\pi} x^2 \sin\frac{n\pi x}{l}\right|_0^l - \frac{2l}{n\pi}\int_0^l x \sin\frac{n\pi x}{l}\, dx\right]$$

$$= \frac{2}{l}\left[0 + \frac{2l^2}{n^2\pi^2}\left\{\left| x\cos\frac{n\pi x}{l}\right|_0^l - \int_0^l \cos\frac{n\pi x}{l}\, dx\right\}\right]$$

$$= \frac{4l^2}{n^2\pi^2}\cos n\pi - 0 = \frac{(-1)^2 4l^2}{n^2\pi^2}$$

Since x^2 is continuous on [–l, l], therefore

$$x^2 = \tfrac{1}{2}a_0 + \sum_{n=1}^{\infty} a_n \cos\frac{n\pi x}{l}.$$

$$\text{Hence}\quad x^2 = \frac{l^2}{3} + \frac{4l^2}{\pi^2}\sum_{n=1}^{\infty}\frac{(-1)^n}{n^2}\cos\left(\frac{n\pi x}{l}\right).$$

Example 26:

Show that for all values f x in $[-\pi, \pi]$, when k is not an integer,

$$\cos kx = \frac{\sin k\pi}{\pi}\left[\frac{1}{k} + \sum_{n=1}^{\infty}\frac{(-1)^n\, 2k\cos nx}{k^2 - n^2}\right]$$

Deduce that $\pi \cot k\pi = \dfrac{1}{k} + \displaystyle\sum_{n=1}^{\infty}\frac{2k}{k^2 - n^2}.$

Solution:

The function f(x) = cos kx is piecewise monotonic, bounded and integrable for all x in $[-\pi, \pi]$. Now

$$a_0 = \frac{1}{\pi}\int_{-\pi}^{\pi} f(x)\,dx = \frac{2}{\pi}\int_0^{\pi}\cos kx\,dx = \frac{2}{k\pi}\sin k\pi.$$

$$a_n = \frac{1}{\pi}\int_{-\pi}^{\pi}\cos kx\cos nx\,dx = \frac{2}{\pi}\int_0^{\pi}\cos kx\cos nx\,dx$$

$$= \frac{1}{\pi}\int_0^{\pi}[\cos(k+n)\cos(k-n)x]\,dx$$

$$= \frac{1}{\pi}\left[\frac{\sin(k+n)x}{k+n} + \frac{\sin(k-n)x}{k-n}\right]_0^{\pi}$$

$$= \frac{1}{\pi}\left[\frac{\sin(k+n)\pi}{k+n} + \frac{\sin(k-n)\pi}{k-n}\right]$$

$$= \frac{\sin k\pi\cos n\pi}{\pi}\left[\frac{1}{k+n} + \frac{1}{k-n}\right] \qquad (\because \sin n\pi = 0)$$

$$= \frac{2(-1)^n k\sin k\pi}{n(k^3-n^2)}$$

Now $b_n = \frac{1}{n}\int_{-\pi}^{\pi}\cos kx\sin nx\,dx = 0$ (integrand being odd)

Since cos kx is continuous in $[-\pi, \pi]$, therefore,

$$\cos kx = \frac{1}{2}a_0 + \sum_{n=1}^{\infty}(a_n\cos nx + b_n\sin nx)$$

$$\cos kx = \frac{\sin k\pi}{\pi}\left[\frac{1}{k} + \sum_{n=1}^{\infty}\frac{(-1)^n 2k\cos nx}{k^2-n^2}\right] \qquad ...(1)$$

Putting $x = \pi$ in (1), we get

$$\pi\cot k\pi = \frac{1}{k} + \sum_{n=1}^{\infty}\frac{(-1)^n 2k\cos n\pi}{k^2-n^2}$$

Hence $\pi\cot k\pi = \frac{1}{k} + \sum_{n=1}^{\infty}\frac{2k}{k^2-n^2}.$

$$[\because (-1)^n\cos n\pi = (-1)^{n+n} = 1]$$

Example 27:

Expand | x | in a Fourier Series on [–1, 1].

Solution:

We have

$|x| = x$ for $0 \le x \le 1$, $|x| = -x$ for $-l \le x \le 0$.

On substituting $y = \pi x/l$, the given function becomes a periodic function with period 2π on $[-\pi, \pi]$. Since the given function is an even function, so $b_n = 0$ for all n. We have

$$a_0 = \frac{2}{l}\int_0^l x\,dx = l$$

Now $$a_n = \frac{2}{l}\int_0^l x\cos\frac{n\pi x}{l}\,dx$$

$$= \frac{2}{l}\frac{l}{n\pi}\left[\left|x\sin\frac{n\pi x}{l}\right|_0^l - \int_0^l \sin\frac{n\pi x}{l}\,dx\right]$$

$$= \frac{2}{n\pi}\left[0 + \frac{l}{n\pi}\left|\cos\frac{n\pi x}{l}\right|_0^l\right]$$

$$= \frac{2l}{n^2\pi^2}(\cos n\pi - 1) = \frac{2l}{n^2\pi^2}[(-1)^n - 1]$$

$$\therefore \quad a_n = \begin{cases} 0, & \text{for n even} \\ \dfrac{-4l}{n^2\pi^2}, & \text{for n odd.} \end{cases}$$

Hence the Fourier Series of $|x|$ is

$$\frac{1}{2}a_0 + \sum_{n=1}^{\infty} a_n \cos\frac{n\pi x}{l}$$

$$= \frac{1}{2}l - \frac{4l}{\pi^2}\sum\frac{1}{n^2}\cos\left(\frac{n\pi x}{1}\right), \text{ for n odd}$$

$$= \frac{1}{2}l - \frac{4l}{\pi^2}\sum_{n=1}^{\infty}\frac{\cos\{(2n-1)nx/l\}}{(2n-1)^2}.$$

Example 28(a):

Find the Fourier Series f the function

$$f(x) = \begin{cases} x - \pi, & \text{when } -\pi < x < 0, \\ \pi - x, & \text{when } 0 < x < \pi. \end{cases}$$

that is the sum of the series for $x = 0, \pm\pi$?

Solution:

The given function is piecewise monotonic, bounded and integrable in $[-\pi, \pi]$. We have

$$a_0 = \frac{1}{\pi}\int_{-\pi}^{0}(x-\pi)\,dx + \frac{1}{\pi}\int_{0}^{\pi}(\pi - x)\,dx$$

$$= \frac{1}{\pi}\left[-\frac{\pi^2}{2} - \pi^2 + \pi^2 - \frac{\pi^2}{2}\right]$$

$$= -\pi. \quad \therefore a_0 = -\pi.$$

$$a_n = \frac{1}{\pi}\int_{-\pi}^{0}(x-\pi)\cos nx\,dx + \frac{1}{\pi}\int_{0}^{\pi}(\pi - x)\cos nx\,dx$$

$$= \frac{1}{\pi}\left[\left|(x-\pi)\frac{\sin nx}{n}\right|_{-\pi}^{0} - \int_{-\pi}^{0}\frac{\sin nx}{n}dx + \left|(\pi - x)\frac{\sin nx}{n}\right|_{0}^{\pi} + \int_{-\pi}^{0}\frac{\sin nx}{n}dx\right]$$

$$= \frac{1}{\pi}\left[0 + \frac{1}{n^2}\left|\cos nx\right|_{-\pi}^{0} + 0 - \frac{1}{n^2}\left|\cos nx\right|_{0}^{\pi}\right]$$

$$= \frac{1}{n^2\pi}\left[(1-\cos n\pi) - (\cos n\pi - 1)\right]$$

$$= \frac{2}{n^2\pi}\left[1-(-1)^n\right] \qquad [\because \cos n\pi = (-1)^n]$$

$$\therefore \quad a_n = \begin{cases} 0, & \text{if } n \text{ is even,} \\ \dfrac{4}{n^2\pi}, & \text{if } n \text{ is odd.} \end{cases}$$

Similarly,

$$b_n = \frac{1}{\pi}\int_{-\pi}^{\pi} f(x)\sin nx\,dx = \frac{2}{\pi}\left[1-(-1)^n\right]$$

$$\therefore \quad b_n = \begin{cases} 0, & \text{if } n \text{ is even,} \\ \dfrac{4}{n}, & \text{if } n \text{ is odd.} \end{cases}$$

In $[-\pi, \pi]$, the points $0, \pm\pi$ are the only points of discontinuity of f. For $x \neq 0, \pm\pi$, we have

$$f(x) = \tfrac{1}{2}a_0 + \sum_{n=1}^{\infty}(a_n \cos nx + b_n \sin nx)$$

$$= -\frac{1}{2}\pi + \frac{4}{\pi}\cdot\frac{\cos x}{1^2} + 4\,\frac{\sin x}{1} + \frac{4}{\pi}\,\frac{\cos 3x}{3^2} + 4\,\frac{\sin 3x}{3^2} + \ldots..$$

$$= -\frac{1}{2}\pi + \frac{4}{\pi}\cdot\left[\frac{\cos x}{1^2} + \frac{\cos 3x}{3^2} + \frac{\cos 5x}{5^2} + \ldots..\right]$$

$$+ 4\left[\frac{\sin x}{1} + \frac{\sin 3x}{3} + ...\right] \quad ...(1)$$

For x = 0, the sum of the series = $\frac{1}{2}$ [f(0+) + f(0–)] = $\frac{1}{2}$ [π – π] =0.

$$\therefore \quad 0 = -\frac{1}{2}\pi + \frac{4}{\pi}\left[1 + \frac{1}{3^2} + \frac{1}{5^2} + ...\right], \text{ by (1)}$$

$$\text{or} \quad \frac{\pi^2}{8} = 1 + \frac{1}{3^2} + \frac{1}{5^2} + ...$$

For x = ± π, the sum of the series

$$= \frac{1}{2}[f(\pi - 0) + f(-\pi + 0] = ½ [0 + (-2\pi)] = -\pi.$$

Putting x = ± π in (1) and using f(± π) = –π, we again obtain

$$\frac{\pi^2}{8} = 1 + \frac{1}{3^2} + \frac{1}{5^2} + ...$$

Example 28(b):

Expand the periodic function x^2, $0 \le x \le l$, of period 1 in a series of cosines only. Also find the sum of the series at x = 0, 1.

Solution:

The substitution y = πx/1 transforms the given function with period π on [0, π]. To get the series of cosines only, we have

$$a_n = \frac{2}{l}\int_0^l x^2 \cos\frac{n\pi x}{l}\,dx \quad \frac{(-1)^n 4}{n^2\pi^2}$$

$$\text{Also} \quad a0 = \frac{2}{l}\int_0^l x^2\,dx = \frac{2}{3}l^2.$$

Since x^2 is continuous on [0, 1], therefore

$$x^2 = \tfrac{1}{2}a_0 + \sum_{n=1}^{\infty} a_n \cos\frac{n\pi x}{l}.$$

$$\text{Hence} \quad x^2 = \frac{l^2}{3} + \frac{4l^2}{\pi^2}\sum_{n=1}^{\infty}\frac{(-1)^n}{\pi^2}\cos\frac{n\pi x}{l} \qquad \forall\, x \in [0, 1]$$

$$\text{At } x = 0,\; 0 = \frac{l^2}{3} + \frac{4l^2}{\pi^2}\left(1 - \frac{1}{2^2} + \frac{1}{3^2} - \frac{1}{4^2} + ...\right)$$

$$\therefore 1 - \frac{1}{2^2} + \frac{1}{3^2} - \frac{1}{4^2} + ... = \frac{\pi^2}{12}.$$

$$\text{At } x = l,\ l^2 = \frac{l^2}{3} + \frac{4l^2}{\pi^2}\sum_{n=1}^{\infty}\frac{(-1)^n}{n^2}\cos n\pi$$

$$= \frac{l^2}{3} + \frac{4l^2}{\pi^2}\sum_{n=1}^{\infty}\frac{1}{n^2} \qquad [\because \cos n\pi = (-1)^n]$$

$$1 + \frac{1}{2^2} + \frac{1}{3^2} + \ldots\ldots = \frac{\pi^2}{6}.$$

Example 29:

Expand f in a Fourier series in $[-\pi, \pi]$

$$f(x) = \begin{cases} 1, & \text{for } -\pi < x \le 0, \\ -2, & \text{for } 0 < x \le \pi. \end{cases}$$

Solution:

The given function is bounded, piecewise monotonic and integrable in $[-\pi, \pi]$.

We have

$$a_0 = \frac{1}{\pi}\int_{-\pi}^{\pi} f(x)\,dx = \frac{1}{\pi}\left[\int_{-\pi}^{0} 1.dx + \int_{0}^{\pi} -2\,dx\right]$$

$$= \frac{1}{\pi}[\pi - 2\pi] = -1, \quad \therefore a_0 = -1.$$

$$\text{Now } a_n = \frac{1}{\pi}\int_{-\pi}^{\pi} f(x)\cos nx\,dx$$

$$= \frac{1}{\pi}\int_{-\pi}^{0}\cos nx\,dx - \frac{2}{\pi}\int_{0}^{\pi}\text{cs } nx\,dx$$

$$= \frac{1}{\pi}\left|\frac{\sin nx}{n}\right|_{-\pi}^{0} - \frac{2}{\pi}\left|\frac{\sin nx}{n}\right|_{0}^{\pi} = 0 \quad \therefore a_n = 0\ \forall\ n$$

$$\text{Now } b_n = \frac{1}{\pi}\int_{-\pi}^{\pi} f(x)\sin nx\,dx$$

$$= \frac{1}{\pi}\int_{-\pi}^{0}\sin nx\,dx - \frac{2}{\pi}\int_{0}^{\pi}\sin nx\,dx$$

$$= \frac{1}{\pi}\left|\frac{\cos nx}{n}\right|_{-\pi}^{0} + \frac{2}{\pi}\left|\frac{\cos nx}{n}\right|_{0}^{\pi}$$

$$= -\frac{1}{n\pi} + \frac{\cos n\pi}{n\pi} + \frac{2\cos n\pi}{n\pi} - \frac{2}{n\pi}$$

$$= \frac{3(-1)^n}{n\pi} - \frac{3}{n\pi}$$

$$\therefore \quad b_n = \begin{cases} 0, \text{ if n is even,} \\ -\dfrac{6}{n\pi}, \text{ if n is odd.} \end{cases}$$

Hence, the Fourier Series of the given function

$$= \frac{1}{2} a_0 + \Sigma(a_n \cos nx + b_n \sin nx)$$

$$= -\frac{1}{2} - \frac{6}{\pi} \sum_{n=1}^{\infty} \frac{\sin(2n-1)x}{2n-1}.$$

Example 30:

In the i interval [0, 2], expand the function

$f(x) = x$, when $0 < x \leq 1$; $f(x) = 2 - x$, when $1 < x \leq 2$

as a (i) series of cosines, (ii) series of sines.

Solution:

(i) We have $l = 2$. To get the Fourier Series of cosine terms only in $[0, l]$, we have to extend the definition of f(x) to that it becomes even (i.e., $f(-x) = f(x)$) and then $b_n = 0 \ \forall$ n. Now

$$a_n = \frac{2}{l} \int_0^1 f(x)\, dx = \int_0^2 f(x)\, dx \qquad (\because l = 2)$$

$$= \int_0^1 x\, dx + \int_1^2 (2-x)\, dx = 1.$$

Again $\quad a_n = \dfrac{2}{l} \displaystyle\int_0^1 f(x) \cos \frac{n\pi x}{l}\, dx = \int_0^2 f(x) \cos\left(\frac{nax}{2}\right) dx$

$$= \int_0^1 x \cos\left(\frac{n\pi x}{2}\right) dx + \int_1^2 (2-x)\cos\left(\frac{n\pi x}{2}\right) dx$$

$$= \frac{2}{n\pi}\left[\left| x \sin\left(\frac{n\pi x}{2}\right)\right|_0^1 - \int_0^1 \sin\left(\frac{n\pi x}{2}\right) dx \right.$$

$$\left. + \left|(2-x) \sin\left(\frac{n\pi x}{2}\right)\right|_1^2 + \int_1^2 \sin\left(\frac{n\pi x}{2}\right) dx\right]$$

$$= \frac{2}{n\pi}\left[\sin\frac{n\pi}{2} + \frac{2}{n\pi}\left|\cos\left(\frac{n\pi x}{2}\right)\right|_0^1 - \sin\frac{n\pi}{2}\right.$$

$$\left. - \frac{2}{n\pi}\left|\cos\left(\frac{n\pi x}{2}\right)\right|_1^2\right]$$

$$= \frac{4}{n^2\pi^2}\left[\left\{\cos\frac{n\pi}{2} - 1\right\} - \left\{\cos n\pi - \cos\frac{n\pi}{2}\right\}\right]$$

$$= \frac{4}{n^2\pi^2}\left[2\cos\frac{n\pi}{2} - (-1)^n - 1\right]$$

It is clear that $a_n = 0$, for n odd

and $\quad a_{2n} = -\frac{2}{n^2\pi 2}\left[1 + (-1)^{n+1}\right]$

Thus, $\quad a_{4n} = 0, a_{4n-2} = -\frac{4}{(2n-1)^2\pi 2}.$

Since the given function is continuous in [0, 2],

$$f(x) = \frac{1}{2}a_0 + \sum_{n-1}^{\infty} a_n \cos\left(\frac{n\pi x}{2}\right)$$

$$= \frac{1}{2} - \frac{4}{\pi^2}\sum \frac{\cos(2n-1)\pi x}{(2n-1)^2}.$$

(ii) To get the Fourier series of sine terms only, we have to extend the definition of f(x) so that it becomes odd and then

$$a_n = 0 \quad \forall n.$$

Now, $\quad b_n = \frac{2}{1}\int_0^1 f(x)\sin\left(\frac{n\pi x}{1}\right)dx$

$$= \int_0^2 f(x)\sin\left(\frac{n\pi x}{1}\right)dx$$

$$= \int_0^1 x\sin\left(\frac{n\pi x}{2}\right)dx + \int_1^2 (2-x)\sin\left(\frac{n\pi x}{2}\right)dx$$

$$= \frac{2}{n\pi}\left[-x\cos\left(\frac{n\pi x}{2}\right)\bigg|_0^1 + \int_0^1 \cos\left(\frac{n\pi x}{2}\right)dx\right.$$

$$\left. - \left|(2-x)\cos\left(\frac{n\pi x}{2}\right)\right|_1^2 - \int_0^1 \cos\left(\frac{n\pi x}{2}\right)dx\right]$$

$$= \frac{2}{n\pi}\left[-\cos\frac{n\pi}{2} + \frac{2}{n\pi}\left|\left(\frac{n\pi x}{2}\right)\right|_0^1 + \cos\frac{n\pi}{2}\right.$$

$$\left. - \frac{2}{n\pi}\left|\sin\left(\frac{n\pi x}{2}\right)\right|_1^2\right]$$

$$= \frac{4}{n^2\pi^2}\left[\sin\frac{n\pi}{2} - \left\{0 - \sin\frac{n\pi}{2}\right\}\right] = \frac{8}{n^2\pi^2}\sin\frac{n\pi}{2}$$

$\therefore$ $b_n = 0$, for n even.

and $b_n = \dfrac{8}{n^2\pi^2}(-1)^{n-1}$, for n odd.

Hence the Fourier sine series is

$$f(x) = \sum_{n=1}^{\infty} b_n \sin\left(\frac{n\pi x}{2}\right) = \frac{8}{\pi^2}\sum_{n=1}^{\infty}(-1)^{n-1}\frac{\sin[(2n-1)\,nx/2]}{(2n-1)^2}.$$

Example 31:

Obtain the Fourier Series in $[-\pi, \pi]$ for the function:

$$f(x) = \begin{cases} x, \text{ if } -\pi < x \le 0, \\ 2x, \text{ if } 0 \le x < \pi. \end{cases}$$

Solution:

The given function is bounded, integrable and continuous in $[-\pi, \pi]$ except at the points $\pm\,\pi$. Now

$$a_0 = \frac{1}{\pi}\int_{-\pi}^{\pi} f(x)\,dx.$$

$$= \frac{1}{\pi}\int_{-\pi}^{0} x\,dx + \frac{1}{\pi}\int_{0}^{\pi} 2x\,dx$$

$$= \frac{1}{\pi}\left|-\frac{\pi^2}{2}\right| + \frac{1}{\pi}(\pi^2) = \frac{\pi}{2}.$$

We have $a_n = \dfrac{1}{\pi}\displaystyle\int_{-\pi}^{\pi} f(x)\cos nx\,dx$

$$= \frac{1}{\pi}\int_{-\pi}^{0} x\cos nx\,dx + \frac{1}{\pi}\int_{0}^{\pi} 2x\cos nx\,dx$$

$$= \frac{1}{\pi}\left|\frac{x\sin nx}{n}\right|_{-\pi}^{0} - \frac{1}{\pi}\int_{0}^{\pi} 2x\cos nx\,dx$$

$$+ \frac{2}{\pi}\left|\frac{x\sin nx}{n}\right|_{0}^{\pi} - \frac{2}{\pi}\int_{0}^{\pi}\frac{\sin nx}{n}\,dx$$

$$= 0 + \frac{1}{n^2\pi}\left|\cos nx\right|_{-\pi}^{0} + 0 + \frac{2}{n^2\pi}\left|\cos nx\right|_{0}^{\pi}$$

$$= \frac{1}{n^2\pi}(1 - \cos n\pi) + \frac{2}{n^2\pi}(\cos n\pi - 1)$$

$$= \frac{\cos n\pi - 1}{n^2\pi}.$$

$$\therefore \quad a_n = \begin{cases} = 0, \text{ when n is even} \\ -\frac{1}{\pi n^2}, \text{ when n is odd} \end{cases}$$

Now, $b_n = \frac{1}{\pi}\int_{-\pi}^{\pi} f(x) \sin nx\, dx$

$$= \frac{1}{\pi}\int_{-\pi}^{0} x \sin nx\, dx + \frac{1}{\pi}\int_{0}^{\pi} 2x \sin nx\, dx$$

$$= \frac{1}{\pi}\left[\left|-x\frac{\cos nx}{n}\right|_{-\pi}^{0} + \int_{-\pi}^{0}\frac{\cos nx}{n}dx\right]$$

$$+ \frac{1}{\pi}\left[\left|\frac{-2x\cos nx}{n}\right|_{0}^{\pi} + 2\int_{0}^{\pi}\frac{\cos nx}{n}dx\right]$$

$$= \frac{1}{\pi}\left[\frac{-\pi\cos n\pi}{n} - \frac{2\pi\cos n\pi}{n}\right] = \frac{3(-1)^{n+1}}{n}.$$

$$[\because \cos n\pi = (-1)^n]$$

Hence the Fourier Series of f(x)

$$= \frac{1}{2}a_0 + \sum_{n=1}^{\infty}(a_n \cos nx + b_n \sin nx)$$

$$= \frac{\pi}{4} - \frac{2}{\pi}\left[\frac{\cos x}{1^2} + \frac{\cos 3x}{3^2} + \frac{\cos 5x}{5^2} = \ldots\ldots\right]$$

$$+ \left[\frac{\sin x}{1} + \frac{\sin 2x}{2} + \frac{\sin 3x}{3} = \ldots\ldots\right]$$

Example 32(a):

If $f(x) = cx$, when $-\pi < x < \pi$,

$f(-\pi) = f(\pi) = 0$ and $f(x + 2\pi) = f(x)$ $\forall$ x; show that

$f(x) = 2c\ [\sin x - \frac{1}{2} 2x + \frac{1}{2} \sin 3x - .....]$ $\forall$ x.

Solution:

The given function is bounded, integrable and piecewise monotonic in $[-\pi, \pi]$.

$$\text{Now} \quad a_0 = \frac{1}{\pi}\int_{-\pi}^{\pi} f(x)\,dx = \frac{1}{\pi}\int_{-\pi}^{\pi} cx\,dx = 0$$

$$a_n = \frac{1}{\pi}\int_{-\pi}^{\pi} f(x)\cos nx\,dx = \frac{c}{\pi}\int_{-\pi}^{\pi} x\cos nx\,dx$$

$$= \frac{c}{\pi}\left[\left|\frac{x\sin nx}{n}\right|_{-\pi}^{\pi} - \frac{1}{n}\int_{-\pi}^{\pi}\sin nx\,dx\right]$$

$$= \frac{c}{\pi}\left[0 + \frac{1}{n^2}|\cos nx|_{-\pi}^{\pi}\right] = 0. \qquad \therefore a_n = 0\ \forall\ n.$$

$$b_n = \frac{1}{\pi}\int_{-\pi}^{\pi} f(x)\sin nx\,dx = \frac{c}{\pi}\int_{-\pi}^{\pi} x\sin nx\,dx$$

$$= \frac{c}{\pi}\left[\left|-x\frac{\cos nx}{n}\right|_{-\pi}^{\pi} + \frac{1}{\pi}\int_{-\pi}^{\pi}\cos nx\,dx\right]$$

$$= -\frac{c}{n\pi}\times 2\pi\cos n\pi = -\frac{2c}{\pi}(-1)^n = \frac{2c}{n}(-1)^{n+1}.$$

Since f is continuous for all x, therefore

$$f(x) = \frac{1}{2}a_0 + \sum_{n=1}^{\infty}(a_n\cos nx + b_n\sin nx)$$

$$\text{or} \quad f(x) = \sum_{n=1}^{\infty} b_n \sin nx = b_1\sin n + b_2\sin 2x + ...$$

Hence $f(x) = 2c\ [\sin x - \frac{1}{2}\sin 2x + \frac{1}{3}\sin 3x - .... 1.$

Example 32(b):

Let $\frac{1}{2}a_0 + \Sigma(a_n\cos nx + b_n\sin nx)$ be a Fourier Series which converges uniformly to f(x) on $[-\pi, \pi]$. Show that

$$\frac{1}{2}a_0^2 + \sum_{n=1}^{\infty}(a_n^2 + b_n^2) = \frac{1}{\pi}\int_{-\pi}^{\pi}\{f(x)\}^2\,dx.$$

Solution:

We have $f(x) = \frac{1}{2} a_0 + \Sigma (a_n \cos nx + b_n \sin nx)$.

Multiplying both sides by f(x) an integrating term by term from $-\pi$ to π (which is justified since the series is uniformly convergent), we obtain

$$\int_{-\pi}^{\pi} \{f(x)\}^2 \, dx = \tfrac{1}{2} a_0 \int_{-\pi}^{\pi} f(x) \, dx +$$

$$\sum_{n=1}^{\infty} \left\{ a_n \int_{-\pi}^{\pi} f(x) \cos nx \, dx + b_n \int_{-\pi}^{\pi} f(x) \sin nx \, dx \right\}$$

$$= \tfrac{1}{2} a_0^2 \, \pi + \pi \sum_{n=1}^{\infty} (a_n^2 + b_n^2), \quad \text{of a0, an, } b_n.$$

using the definitions

Hence the result is proved.

Ex. If $\phi(x) = \frac{1}{2} x$ for $0 < x < \alpha$, $\phi(x)$

$= \frac{1}{2} \alpha$ for $0 < x < \pi - \alpha$ and $\phi(x)$

$= \frac{1}{2} (\pi - \alpha)$ for $\pi - \alpha < x < \pi$; prove that

$$\phi(x) = \tfrac{2}{\pi} \left(\sin \alpha \sin x + \tfrac{1}{9} \sin 3\alpha \sin 3x + \ldots\right).$$

Example 32(c):

Let $f(x) = \begin{cases} -\cos x, \text{ for } -\pi < x < 0 \\ \cos x, \text{ for } \quad 0 < x < \pi. \end{cases}$

Show that the Fourier Series which converges to f(x) is

$$\frac{\pi}{4}\left(\frac{2}{1.3} \sin 2x + \frac{4}{3.5} \sin 4x + \frac{6}{5.7} \sin 6x + \ldots\ldots\ldots\right).$$

Solution:

We have

$$c_0 = \frac{1}{\pi} \int_{-\pi}^{\pi} f(x) \, dx$$

$$= \frac{1}{\pi} \int_{-\pi}^{0} -\cos x \, dx + \frac{1}{\pi} \int_{0}^{\pi} \cos x \, dx = 0.$$

Now $a_n = \frac{1}{\pi} \int_{-\pi}^{\pi} f(x) \cos nx \, dx$

$$= \frac{1}{\pi}\int_{-\pi}^{0} -\cos x \cos nx \, dx + \frac{1}{\pi}\int_{0}^{\pi} \cos x \cos nx \, dx$$

$$= -\frac{1}{\pi}\int_{-\pi}^{0} \cos x \cos nx \, dx + \frac{1}{\pi}\int_{0}^{\pi} \cos x \cos nx \, dx$$

$$\therefore \quad a_n = 0 \quad \forall n.$$

$$\text{Now,} \quad b_n = \frac{1}{\pi}\int_{-\pi}^{0} \cos x \sin nx \, dx + \frac{1}{\pi}\int_{0}^{\pi} \cos x \sin nx \, dx$$

$$= \frac{2}{\pi}\int_{0}^{\pi} \cos x \sin nx \, dx$$

$$= \frac{1}{\pi}\int_{0}^{\pi} \{\sin (n+1)x + \sin (n-1)x\} \, dx$$

$$= \frac{-1}{\pi}\left[\frac{\cos (n+1)x}{n+1} + \frac{\cos (n-1)x}{n-1}\right]_0^{\pi}$$

$$= -\frac{1}{\pi}\left[\frac{1}{(n+1)}\{(-1)^{n+1} - 1\} + \frac{1}{(n-1)}\{(-1)^{n-1} - 1\}\right]$$

$$= -\frac{1}{\pi}\left[(-1)^{n+1}\left\{\frac{1}{n+1} + \frac{1}{n-1}\right\} - \left\{\frac{1}{n+1} - \frac{1}{n-1}\right\}\right]$$

$$= -\frac{2n}{\pi(n^2-1)}[(-1)^{n+1} - 1]$$

$$\therefore \quad b_n = -\begin{cases} 0, & \text{when n is odd,} \\ \dfrac{4n}{\pi(n^2-1)}, & \text{when n is even.} \end{cases}$$

Hence the Fourier Series of | cos x | is

$$\frac{1}{2}a_0 + \sum_{n=1}^{\infty}(a_n \cos nx + b_n \sin nx)$$

$$= 0 + \sum \frac{4n}{n(n^2-1)} \sin nx, \text{ (n even)} \quad [_n = 0 \ \forall \ n]$$

$$= \frac{4}{\pi}\left[\frac{2}{1.3}\sin 2x + \frac{4}{3.5}\sin 4x + \frac{6}{5.7}\sin 6x + \ldots\ldots\right]$$

SERIES OF ARBITRARY TERMS

In this chapter we will discuss the following terms:

(i) Series of arbitrary terms, (ii) Abel's and Dirichlet's Tests, (iii) Rearrangement of Series, (iv) Multiplication of Series and (v) Double Series.

The reader is advised to go through the chapter *'Infinite Series'* from the Author's Book entitled *Real Analysis (Brilliant Series)*. The results of Infinite Series will be frequently used in this chapter.

Throughout this chapter we shall use the fact that the series

$1-\frac{1}{2}+\frac{1}{3}-\frac{1}{4}+\frac{1}{5}=....$ is convergent and that its sum of log 2.

Theorem 1: Abel's Lemma

If the sequence $< u_n >$ be bounded and if the sequence $< u_n >$ be positive and monotonically decreasing, then the series $\Sigma u_n (v_n - v_{n+1})$ is absolutely convergent.

Proof:

Since the sequence $< u_n >$ is bounded, therefore, there exists a positive number K such that $| u_n | < K$, for all n.

Since $< v_n >$ is monotonically decreasing $v_{n+1} \le v_n \ \forall$ n.

Now $\sum_{n=1}^{m} |u_n (v_n - v_{n+1})| = \sum_{n=1}^{m} |u_n (v_n - v_{n+1})|$, since $v_n - v_{n+1} \ge 0$.

$$< K \sum_{n=1}^{m} (v_n - v_{n+1})| = K (v_1 - v_{m+1})$$

$$< K v_1 \qquad (\because v_1 = v_{m+1} < v_1, v_{m+1} > 0)$$

Thus the sequence of partial sums of the positive term series $\Sigma | u_n (v_n - v_{n+1}) |$ is bounded above by Kv_1 and so $\Sigma | v_n (V_n - v_{n+1}) |$ converges. Hence the series $\Sigma v_n (v_n - v_{n+1})$ is absolutely convergent.

Theorem 2:

Abel's Test

If the series Σu_n converges and if $< u_n <$ is a positive monotonically creasing sequence, then the series $\Sigma u_n v_n$ converges.

Proof:

Let $< T_n >$ be the sequence of partial series of $\Sigma u_n v_n$,

$T_n = u_1 v_1 + u_2 v_2 + ... + u_n v_n$.

We know that

$$T_n = \sum_{k=1}^{n-1} S_k\,(v_k - v_{k+1}) + S_n v_n. \qquad ...(1)$$

Since $\Sigma\, v_n$ is convergent, so $< S_n >$ is convergent and therefore bounded.

Since $< v_n >$ is positive and monotnically decreasing, therefore by Abel's Lemma.

$$= \sum_{k=1}^{n-1} S_k\,(v_k - v_{k+1}) \text{ tends to a finite limits as } n \to \infty. \qquad ...(2)$$

It is clear that the sequence $< v_n >$ converges, since it is monotonically decreasing and bounded below by zero. Also $< S_n >$ is convegent, so

$S_n v_n$ tends to a finite limit as $n \to \infty$. ...(3)

Using (2) and (3) in (1), we see that T_n tends to a finite limit as $n \to \infty$. Hence, the series $\Sigma\, u_n\, v_n$ converges.

***Theorem 3:* Dirichlet's Test**

If for a series $\Sigma\, u_n$, the sequence $< S_n >$ of partial sums of bounded, and $< v_n >$ is a positive monotonically decreasing sequence with lim $v_n = 0$, then the series å $u_n\, v_n$ converges.

Proof:

Let $S_n = u_1 + u_2 + ... u_n$ and

$$\begin{aligned} T_n &= u_1 v_1 + u_2 v_2 + ... + u_n v_n \\ &= S_1 v_1 + (S_2 - S_1)\, v_2 + ... + (S_n - S_{n-1})\, v_n \\ &= S_1(v_1 - v_2) + S_2(v_2 - v_3) + ... + S_{n-1}\,(v_{n-1} - v_n) + S_n v_n \end{aligned}$$

$$\therefore\ T_n = \sum_{k=1}^{n-1} S_k\,(v_k - v_{k+1}) + S_n v_n. \qquad ...(1)$$

Since the sequence $< S_n >$ is bounded, and the sequence $< v_n >$ is positive and monotonically decreasing, therefore by Abel's Lemma.

$$= \sum_{k=1}^{n-1} S_k\,(v_k - v_{k+1}) \text{ tends to a finite limit (say } l) \text{ as } n \to \infty. \qquad ...(2)$$

Also, since $v_n \to 0$ as $n \to \infty$ and since $< S_n >$ is bounded, therefore

$S_n v_n \to 0$ as $n \to \infty$. ...(3)

Using (2) and (3) in (1). $\lim_{n\to\infty} T_n = l$.

Thus, the sequence $< T_n >$ of partial sums of the series $\Sigma\, u_n v_n$ converges. Hence, the series $\Sigma u_n v_n$ converges.

Corollary:

Leibnitz's test is a particular case of Dirichlet's test.

Proof:

The sequence of partial sums of the series $\Sigma(-1)^{n-1}$ is bounded (in fact, $S_n = 0$ if n is even, and $S_n = 1$ if n is odd). Let us take $u_n = (-1)^{n-1}$ in Dirichlet's test.

$\therefore \Sigma u_n v_n = v_1 - v_2 + v_3 - ... + (-1)^{n-1} v_n + ...$

Thus Dirichlet's test can be stated as:

If $< v_n >$ is a positive monotonically decreasing sequence with that $v_n = 0$, then the sereis $v_1 = v_2 + v_3 - ... + (-1)^{n-1} v_n + ...$ converges.

This is Leibnitz's test.

REARRANGEMENT OF SERIES

A series $\Sigma\, v_n$ is said to be a *rearrangement* of a series $\Sigma\, u_n$ if there exists a one-to-one correspondence between the terms of $\Sigma\, u_n$ and those of $\Sigma\, v_n$ such that the corresponding terms are equal.

For example, the series $1 + \frac{1}{3} - \frac{1}{2} + \frac{1}{5} + \frac{1}{7} - \frac{1}{4} + ...$

is a rearrangement of the series

$$1 - \frac{1}{2} + \frac{1}{3} - \frac{1}{4} + \frac{1}{5} + ...$$

Theorem 4: Dirichlet's Theorem

Any series obtained from an absolutely convergent series by a rearrangement of terms is absolutely convergent, and has the same sum as the original series.

Proof:

Case 1:

We shall first prove the theorem for a series of positive temrs. Let Σu_n be a convergent series of positive terms with sum s and let Σv_n be any rearrangement of Σu_n. Let $< S_n >$ and $< T_n >$ be the sequences of partial sums Σu_n of and Σv_n respectively. If T_m be any member of $< T_n >$, then the terms of T_m consist of a finite number of terms of Σu_n, and therefore form a part of some partial sum S_n of Σu_n. Since the series are a positive temrs, therefore $T_m \leq S_n \leq s$. Thus the sequence $< T_n >$ is bounded above by s and so Σv_n also converges, say to t and $l \leq s$. Similarly Σu_n can also be though of as a rearrangement of Σv_n and so $s \leq l$. Hence $l = s$.

Case 2:

Suppose that Σu_n is any absolutely convergent series and that Σv_n is a rearrangement of Σu_n. We define the sequence

$< P_n >, < q_n >, < r_n >, < S_n >$ as follows:

$$p_n = \begin{cases} u_n, \text{ if } u_n > 0 \\ 0, \text{ if } u_n \le 0 \end{cases}, \quad q_n = \begin{cases} -u_n, \text{ if } u_n \le 0 \\ 0, \text{ if } u_n > 0 \end{cases}$$

$$r_n = \begin{cases} v_n, \text{ if } v_n > 0 \\ 0, \text{ if } v_n \le 0 \end{cases}, \quad q_n = \begin{cases} -v_n, \text{ if } v_n \le 0 \\ 0, \text{ if } v_n > 0 \end{cases}$$

Then $u_n = p_n - q_n \mid u_n \mid = p_n + q_n$, $v_n = r_n - s_n$, $\mid v_n \mid = r_n + s_n$. ...(1)

From (1), we have

$$\left.\begin{aligned} p_n &= \tfrac{1}{2}(u_n + |u_n|, \\ q_n &= \tfrac{1}{2}(|u_n| - u_n \end{aligned}\right\} \qquad ...(2)$$

$$\left.\begin{aligned} r_n &= \tfrac{1}{2}(v_n + |v_n|, \\ s_n &= \tfrac{1}{2}(|v_n| - v_n \end{aligned}\right\} \qquad ...(3)$$

Since Σu_n, $\Sigma \mid u_n \mid$ are both convergent, therefore by (2), Σp_n and Σq_n are also convergent.

Now Σp_n and Σq_n are convergent series of positive terms and Σr_n, Σs_n are rearrangements of these series respectively, therefore, by case 1 Σr_n and Σs_n converge. If p, q, r, s denote the sum of the series Σp_n, Σq_n, Σr_n, Σs_n respectively, then

$p = r,\ q = s.$ ...(4)

From (1) and (4), we see that Σv_n and $\Sigma \mid v_n \mid$ converge and

$$\sum_{n=1}^{\infty} v_n = r - s = p - q = \sum_{n=1}^{\infty} u_n,$$

$$\sum_{n=1}^{\infty} |v_n| = r + s = p + q = \sum_{n=1}^{\infty} |u_n|.$$

which shows that Σv_n converges absolutely and has the same sum as the series Σu_n.

Remark:

The conclusion of the Dirichtel's theorem may not hold if a series is not absolutely convergent.

Consider the series $S : \sum \frac{(-1)^{n-1}}{n} = 1 - \frac{1}{2} + \frac{1}{3} - \frac{1}{4} + ...$

which is not absolutely convergent. However, by Leibnitz's test the above series is convergent.

Consider the following rearrangement of the given series:

$$S' = 1 - \frac{1}{2} - \frac{1}{4} + \frac{1}{3} - \frac{1}{6} - \frac{1}{8} + \frac{1}{5} - \frac{1}{10} - \frac{1}{12} + ...$$

Suppose we write it as

$$S' = \left(1 - \frac{1}{2}\right) - \frac{1}{4} + \left(\frac{1}{3} - \frac{1}{6}\right) - \frac{1}{8} + \left(\frac{1}{5} - \frac{1}{10}\right) - \frac{1}{12} + ...$$

$$= \frac{1}{2} - \frac{1}{4} + \frac{1}{6} - \frac{1}{8} + \frac{1}{10} - \frac{1}{12} + ...$$

$$S' = \frac{1}{2}\left(1 - \frac{1}{2} + \frac{1}{3} - \frac{1}{4} + \frac{1}{5} - .....\right) = \frac{1}{2} S.$$

Thus $S' \neq S$.

It may be noted that the series $\sum \frac{(-1)^{n-1}}{n}$ is conditionally convergent.

Ex: Show that the series $\sum \frac{(-1)^{n-1}}{n}$ *is conditionally convergent. use this series to show that rearrangement of terms and bracketing of terms can lad to a series with a different sum.*

Theorem 5: (Riemann's Rearrangement Theorem)

If $\sum u_n$ *be a conditionally convergent series, then the terms of this series can be rearranged to give either a divergent series or an oscillating series or a convergent series whose sum is an arbitrary preassigned number.*

The proof is omitted.

DOUBLE SEQUENCE

A double sequence is a rectangular array of numbers whose rows and columns continue indefinitely to the right and downward respectively.

$$\begin{matrix} a_{11} & a_{12} & a_{13} & \cdots\cdots & a_{1n} & \cdots\cdots \\ a_{21} & a_{22} & a_{23} & \cdots\cdots & a_{2n} & \cdots\cdots \\ \vdots & \vdots & \vdots & \cdots\cdots & \vdots & \cdots\cdots \end{matrix}$$

a_{m1} a_{m2} a_{m3} .... a_{mn} ....
: : : .... : ....

This double sequence is denoted as $< a_{mn} >$.

Definition:

A double sequence $< a_{mn} >$ converges to a number t as (m, n) $\to \infty$ if for each $\varepsilon > 0$, there exists a positive integer N such that

$| a_{mn} - 1 | < \varepsilon$ whenever $m > N$ and $n > N$.

We write this limit as

$$\lim_{m,n\to\infty} a_{mn} = 1$$

It can be shown that if such a number l exists, then it must be unique.

DOUBLE SERIES

If a_{mn} > is a double sequence, then an expression of the form

$a_{11} + a_{12} + \ldots a_{21} + a_{22} + \ldots a_{m1} + a_{m2} + \ldots$

is called a double series and is denoted by $\sum_{m,\,n=1}^{\infty} a_{mn}$ or simply $\sum a_{mn}$

Let $S_{mn} \sum_{r=1}^{m} \sum_{s=1}^{n} a_{rs}$

In other words S_{mn} is the sum of all the elements of the $m \times n$ matrix obtained by selecting the *m* rows and *n* coloumns in the upper left-hand corner of the double sequence $< a_{mn} >$. The sequence $< S_{mn} >$ is called the *sequence of partial sums of the series* $\sum a_{mn}$.

If $\lim_{(m,n)\to\infty} S_{mn}$ exists, we say that the double series $\sum a_{mn}$ *converges*; otherwise we say it *diverges.* The convergence of the double series as defined above is sometimes called *convergence by rectangles.* Many of the theorems on double series are direct extensions of those for single series. We state a few theorems on double series.

Theorem 6:

A necessary condition for the convergence of a double series $\sum a_{mn}$ *is that* $\lim_{(m,n)\to\infty} a_{mn} = 0$. *The converse, however, is not true*

Proof:

For $m, n \geq 1$, let $S_{m,n} = \sum_{r=1}^{m} \sum_{s=1}^{n} a_{rs}$; $S_{m,0} = S_{0,n} = S_{0,0} = 0$

Then $a_{mn} = S_{m,n} - S_{m-1,\,n} - S_{m,n-1} + S_{m-1,\,n-1}$

Since $\sum a_{mn}$ converges, so the sequence $< S_{mn} >$ converges.

Suppose $< S_{mn} >$ converges to S. Therefore.

$$\lim_{m,n\to\infty} a_{nm} = \lim_{m,n\to\infty} (S_{m,n} - S_{m-1,n} - S_{m,n-1} + S_{m-1,n-1})$$

Remark:

If $\lim_{m,n\to\infty} a_{mn} \neq 0$ *then the double series* $\sum a_{mn}$ *does not converge.*

Theorem 7:

Let $\sum a_{mn}$ *and* $\sum b_{mn}$ *be convergent double series and that c and d are constants. Then* $\sum (ca_{nm} + db_{mn}) = c \sum a_{mn} + d \sum b_{mn}$ *is convergent and.*

$\Sigma(ca_{mn} + db_{mn}) = c\, \Sigma a_{mn} + d\Sigma b_{mn}$

Theorem 8:

Suppose that $a_{mn} \geq 0$ *for all and n.*

(i) *If there is a number M such that* $S_{mn} \leq M$ *for all m and n, then* $\sum a_{mn}$ *is convergent.*

(ii) (Comparison Test)

If $0 \leq a_{nm} \leq bmn$ and $\sum b_{mn}$ converges, then $\sum a_{mn}$ also converges.

Theorem 9: (Cauchy's Principle of Convergence)

A necessary and sufficient condition for a double series Σa_{mn} *to be convergent is that, for each* $\varepsilon > 0$*, there exists a positive integer N such that*

$| S_{m+p,\, n+q} - S_{m,n} | < \varepsilon \ \forall\ m, n \geq N$ and $\forall\ p, q \geq 1$

A FUNDAMENTAL THEOREM FOR POSITIVE TERM DOUBLE SERIES

Theorem 10:

Let $a_{mn} \geq 0$ *for all m, n. The* following statements are equivalent:

1. $\sum a_{mn}$ *converges by rectangles.*

2. $\sum a_{mn}$ *converges by rows.*

3. $\sum a_{mn}$ *converges by columns.*

4. $\sum a_{mn}$ *converges by squares.*

5. $\sum a_{mn}$ *converges by triangles.*

Further, $\sum a_{mn}$ *has the* same sum under all the five modes of convergence.

Proof:

We show that (1) $\Rightarrow$ (2).

Suppose $\sum a_{mn}$ converges by rectangles to σ. Then for each m, n $\geq$ 1, $S_{mn} \leq \sigma$. Since $a_{mn} \geq 0$ for each m, n $\geq$ 1, it follows that for m $\geq$ 1.

$a_{m1} + a_{m2} + ... + a_{mn} \leq S_{mn} \leq \sigma$ for each n $\geq$ 1.

Thus, for m $\leq$ 1, the sequence of partial sums of the series $\sum_{n=1}^{\infty} a_{mn}$ is bounded above by σ and so $\sum_{n=1}^{\infty} a_{mn}$ converges for each m $\geq$ 1. We denote the sum of $\sum_{n=1}^{\infty} a_{mn}$ by R_m. Again

$S_{mn} \leq \sigma \Rightarrow \lim_{n\to\infty} S_{mn} \leq \sigma$ for m $\geq$ 1. But $\lim_{n\to\infty} S_{mn} = R_1 + R_2 + ... + R_m$.

Thus, $R_1 + R_2 + ... + R_m \leq \sigma$ for each m $\geq$ 1. This shows that the sequence of partial sums of $\sum_{m=1}^{\infty} R_m$ is bounded above by σ. Since $R_m \geq 0$ for each m $\geq$ 1, the series $\sum_{m=1}^{\infty} R_m$ converges. If $\sum_{m=1}^{\infty} R_m$ converges to R, then R $\leq \sigma$.

Thus, Σa_{mn} converges by rows to R, R $\leq \sigma$.

Now, we show that (2) $\Rightarrow$ (3).

Suppose, Σa_{mn} converges by rows to R. For by R_m. For m $\geq$ 1, we denote the sum of the series $\sum_{n=1}^{\infty} a_{mn}$ a given j $\geq$ 1, $a_{mj} \leq R_m$ so that

$$\sum_{m=1}^{k} a_{mj} \leq \sum_{m=1}^{k} R_m \leq R.$$

Thus, $\sum_{m=1}^{\infty} a_{mj}$ converges for each $j \geq 1$.

We denote the sum of $\sum_{m=1}^{\infty} a_{mj}$ by C_j. For $n \geq 1$,

$$C_1 + C_2 + ... + C_n = \sum_{m=1}^{\infty} a_{m1} + \sum_{m=1}^{\infty} a_{m2} + ... + \sum_{m=1}^{\infty} a_{mn}$$

$$= \sum_{m=1}^{\infty} (a_{m1} + a_{m2} + ... + a_{mn})$$

$$= \sum_{m=1}^{\infty} (\sum_{k=1}^{m} a_{mj})$$

$$\leq \sum_{m=1}^{\infty} (\sum_{j=1}^{m} a_{mj}) = \sum_{m=1}^{\infty} R_m \leq R.$$

This shows that the sequence of partial sums of $\sum_{n=1}^{\infty} C_n$ is bounded above by R. Thus the series $\sum_{n=1}^{\infty} C_n$ converges, that is, Σa_{mn} converges by columns. If Σa_{mn} converges by columns to C, then $C \leq R$.

Next we show that (3) ⇒ (4).

Suppose Σa_{mn} converges by Columns to C.

$$\text{Let } S_n = = \sum_{r=1}^{n} \sum_{s=1}^{n} a_{rs} = \sum_{s=1}^{n} (\sum_{r=1}^{n} a_{rs}) \leq \sum_{s=1}^{n} (\sum_{r=1}^{\infty} a_{rs}).$$

$\therefore S_n \leq \sum_{s=1}^{n} C_s \leq \sum_{s=1}^{\infty} C_s = C$ for each $n \geq 1$. This shows that the sequence $< S_n >$ of partial sums of $\sum_{n=1}^{\infty} a_{mn}$ is bouned above by C and so $< S_n >$ converges.

Thus $\Sigma\, a_{mn}$ converges by squares. If $\Sigma\, a_{mn}$ converges by squares to S, then $S \leq C$.

We now prove that (4) ⇒ (5).

Suppose $\Sigma\, a_{mn}$ converges by squares to S. For $n \geq 2$, let $T_n = \sum_{k=2}^{n} (\sum_{r+s=k}^{\infty} a_{rs})$. Then $T_n \leq S_n = \sum_{r=1}^{n} \sum_{s=1}^{\infty} a_{rs} \leq S$ for each $n \geq 1$. Thus $< T_n >$ is bouned above by S. As $< T_n >$ is a monotonically increasing sequence, which is bounded above by S, it follows that $< T_n >$ converges. Hence Σa_{mn} converges by triangles. If Σa_{mn} converges by triangles to T, then $T' \leq S$. Lastly we show that (5) $\Rightarrow$ (1).

Suppose $\Sigma\, a_{mn}$ converges by triangles to T. The $T_n \leq T$ for all $n \geq 1$. Since S_{mn} contains lesser number of terms of the array (a_{ij}) than T_{m+n}. So $S_{mn} \leq T_{m+n} \leq T$ for each $m, n \geq 1$. Thus Σa_{mn} converges by rectangles. If it converges by rectangles to σ, then $\sigma \leq T$. From the above discussion, we have

$$\sigma \geq R\ C \geq S \geq T \geq \sigma.$$

Hence $\sigma = R = C = S = T$.

ABSOLUTE CONVERGENCE OF A DOUBLE SERIES

Definition:

A double series $\Sigma\, a_{mn}$ as said to converge absolutely if the series $\Sigma\, |\, a_{mn}\, |$ converges.

It may be noted that the mode of convergence in both the sries must be the same.

Theorem 11:

If $\Sigma\, a_{mn}$ *converges absolutely, then* $\Sigma\, p_{mn}$, $\Sigma\, q_{mn}$ *converges.*

Proof:

Do your self.

Theorem 12:

If $\Sigma\, a_{mn}$ *converges absolutely, then* $\Sigma\, a_{mn}$ *converges.*

Proof:

Since $-1\, |\, a_{mn}\, | \leq a_{mn} \leq |\, a_{mn}\, |$ for all $m, n \geq 1$. ...(1)

$\therefore\ 0 \leq a_{mn} + |\, a_{mn}\, | \leq 2\, |\, a_{mn}\, |$ for all $m, n \geq 1$

Since $\Sigma\, |\, a_{mn}\, |$ converges, so by the comparison test $\Sigma\, (a_{mn} + |\, a_{mn}\, |)$ converges.

Since $a_{mn} = (a_{mn} + |\, a_{mn}\, |) - |\, a_{mn}\, |$, the result follows by Theorem 6.

Definition:

We define two special sequences $< p_n >$ and $< q_n >$ as follow:

$$p_{mn} = \frac{1}{2}(|\, a_{mn}\,| + a_{mn}),\ q_{mn} = \frac{1}{2}(|a_{mn}| - a_{mn})$$

From (1), we obtain

$p_{mn} \geq 0$ and $q_{mn} \geq 0$ for all m, n ≥ 1.

Also $\Sigma a_{mn} = \Sigma (p_{mn} - q_{mn})$ and $\Sigma [a_{mn}] = [p_{mn} + q_{mn}]$

Theorem 13:

If Σa_{mn} be a double series, then the following statements are equivalent:

1. Σa_{mn} *converges absolutely be rectangles*
2. Σa_{mn} *converges absolutely by rows.*
3. Σa_{mn} *converges absolutely by columns.*
4. Σa_{mn} *converges absolutely by squares.*
5. Σa_{mn} *converges absolutely by triangles.*

Further Σa_{mn} has the same sum under all the five modes of convergence.

Proof:

Let Let $p_{mn} = \frac{1}{2}(a_{mn} + |a_{mn}|)$, $q_{mn} = \frac{1}{2}(|a_{mn}| - a_{mn})$

We show that (1) ⇒ (2).

Suppose Σa_{mn} converges absolutely by rectangles. then $\Sigma | a_{mn} |$ and Σa_{mn} converge by rectangles and consequently and Σp_{mn} and Σq_{mn} converge by rectangles.

We know $p_{mn} \geq q_{mn} \geq 0$, m, n ≥ 1

By Theorem 11, Σp_{mn} and Σq_{mn} converge by rows and

$$\Sigma p_{mn} = R(\Sigma p_{mn}),\ \Sigma q_{mn} = R(\Sigma q_{mn}).$$

By Theorem 6, $\Sigma | a_{mn} | = \Sigma (p_{mn} + q_{mn})$ converges by rows and so Σa_{mn} converges absolutely by rows.

$$\begin{aligned}\text{Now } R(\Sigma a_{mn}) &= R\left(\sum (p_{mn} - q_{mn})\right)\\ &= R\left(\sum p_{mn}\right) - R\left(\sum q_{mn}\right)\\ &= \sum p_{mn} - \sum q_{mn}\\ &= \sum (p_{mn} - q_{mn}) = \sum a_{mn}\end{aligned}$$

Similarly, we can prove the other parts

Theorem 14: (Cauchy's Theorem)

Let $\sum a_n$ and $\sum b_n$ converge absolutely. *Then the Cauchy product $\sum c_n$ of $\sum a_n$ and $\sum b_n$ converges absolutely and*

$$\sum c_n = (\sum a_n)\,(\sum b_n)$$

In other words, if $\sum a_n = A$ *and* $\sum b_n = B$, *then* $\sum c_n = AB$.

Proof:

Consider the double series

$$\begin{array}{llll} a_0b_0 + a_0b_1 + a_0b_0 + a_0b_3 + \ldots\ldots & & & \\ + 0 \quad + a_1b_0 + a_1b_1 + a_1b_2 + \ldots\ldots & & & \ldots(1) \\ + 0 \quad + 0 \quad + a_2b_0 + a_2b_1 + \ldots\ldots & & & \\ + \ldots\ldots\ldots\ldots\ldots\ldots & & & \end{array}$$

Since $\sum a_n$ and $\sum b_n$ converge absolutely, $\sum |a_n|$ and $\sum |b_n|$ converge. Let $\sum |a_n| = a$ and $\sum |b_n| = b$. We have

$$\sum_{m=0}^{\infty}\left(\sum_{n=0}^{\infty} |a_m b_n|\right) = \sum_{m=0}^{\infty} |a_m| \left(\sum_{n=0}^{\infty} |b_n|\right) = \sum_{m=0}^{\infty} b\,|a_m| = ba$$

Thus, the double series in (1) converges absolutely by rows and hence converges by columns and the two sums by rows and columns are equal. Hence the column series given by (1)

$$a_0b_0 + (a_0b_1 + a_1b_0) + (a_0b_2 + a_1b_0 + a_2b_0) + \ldots = \sum c_n \text{ converges}$$

absolutely and $\sum c_n = AB = (\sum a_n)\,(\sum b_n)$.

Corollary:

Let $\sum_{n=0}^{\infty} a_n$ *and* $\sum_{n=0}^{\infty} b_n$ *be two positive term convergent series.*

If $C_n = \sum_{k=0}^{n} a_n b_{n-k}$, then $\sum_{n=0}^{\infty} c_n$ *is convergent and its sum is*

$$\left(\sum_{n=0}^{\infty} a_n\right)\left(\sum_{n=0}^{\infty} b_n\right)$$

Proof:

Do your self.

Theorem 15: (Abel's Theorem)

Let $\sum c_n$ *be the Cauchy rpoduct of two convergent series* $\sum a_n$ *and* $\sum b_n$ *with sums A and B respectively. If* $\sum c_n$ *is convergent, then its sum* is AB.

Proof:

Let $A_n = a_0 + a_1 + ... + a_n$, $B_n = B_0 + b_1 + ... + b_n$. $C_n = c_0 + c_1 + ... + c_n$. As done in the above therem, we have

$C_n = c_0 B_n + a B_{n-1} + ... + a_n B_0$.

$\therefore C_0 + C_1 + ... + C_n$

$= a_0 B_0 + (a_0 B_1 + a_1 B_0) + ... + (a_0 B_n + a_1 B_{n-1} + ... + a_n B_0)$

$= (a_0 + a_1 + ... + a_n) B_0 + (a_0 + a_1 + ... + a_{n-1}) B_1 + ... + a_0 B_n$

$= A_n B_0 + A_{n-1} B_1 + ... + A_0 B_n$

If $C_n \rightarrow$ (by the convergence of $\sum c_n$), then by Cauchy's First Theorem on limits.

$$\lim_{n\to\infty} \frac{C_0 + C_1 + ... + C_n}{n-1} = C$$

or $$C = \lim_{n\to\infty} \frac{A_n B_0 + A_{n-1} + ... + A_0 B_n}{n+1} \text{ using (1)}$$

$$= \frac{\{AB + (AB) .... + AB\} \text{ (n+1 tir}}{n+1}$$

$(\because A_n \rightarrow A, B_n \rightarrow B)$

Hence $C = AB$

or $\sum c_a = (\sum a_n)(\sum b_n)$

MULTIPLICATION OF SERIES

Definition:

If $\sum_{n=0}^{\infty} a_n$ and $\sum_{n=0}^{\infty} b_n$ are two infinite series, then their product called the Cauchy product, is defined as $\sum_{n=0}^{\infty} c_n$, where

$c_0 = a_0 b_0$,

$c_1 = a_0 b_1 + a_1 b_0$,

$c_2 = a_0 b_2 + a_1 b_1 + a_2 b_0$

.

$c_n = a_0 b_n + a_1 b_{n-1} + ... + a_n b_0 = \sum_{n=0}^{\infty} a_k b_{n-k}$

It may be noted that

$$c_n = \sum_{k=0}^{\infty} a_k b_{n-k} = \sum_{k=0}^{n} a_{n-k} b_k$$

Theorem 16: (Merten's Theorem)

If a series $\sum_{n=0}^{\infty} a_n$ *is absolutely convergent and a series* $\sum_{n=0}^{\infty} b_n$ *is convergent, then the Cauchy product* $\sum_{n=0}^{\infty} c_n$ *is convergent and*

$$\Sigma C_n = (\Sigma a_n)(\Sigma b_n). \quad ...(1)$$

Equivalently, $\Sigma C_n = AB$, *where* $A = \Sigma a_n$, $B = \Sigma b_n$.

Proof:

Let $A_n = a_0 + a_1 + ... + a_n$, $B_n = b_0 + b_1 + ... + b_n$,

and $C_n = c_0 + c_1 + ... + c_n$

or $C_n = a_0b_0 (a_0b_1 + a_1b_0) + ... + (a_nb_0 + a_{n-1} b_1 + ... + a_0b_n)$

$= a_0 (b_0 + b_1 + ... + b_n) + a_1 (b_0 + b_1 + ... + b_{n-1}) +....+ a_nb_0$

$\therefore \quad C_n = a_0 B_n + a_1 B_{n-1} + ... + a_nB_0.$

Put $B_n = B + t_n$, where $t_n \to 0$ as $n \to \infty$. (Here we use fact that Σb_n is convergent and $< B_n >$ is the sequence of its partial sums.)

$\therefore \quad C_n = a_0 (B + t_n) + a_1 (B + t_{n-1}) + ... + a_n (B_0 + t_0)$

$= B (a_0 + a_1 + ... + a_n) + (a_0 t_n + a_1 t_{n-1} + ... a_n t_0)$

$= B A_n + (a_n t_1 + a_1 t_{n-1} + ... + a_{n-m} t_m) + (a_{1-m} t_{m-1} + ... + a_n t_0)$

or $C_n = BA_n + P + Q$. say. ...(2)

Since Σa_n converges

find some number $k > 0$ such that

$$| a_0 | + | a_1 | + ... | a_n | < k \ \forall \ n. \quad ...(3)$$

Since $t_n \to 0$, so for any $\varepsilon > 0$, there exists a positive integer m such that

$$| t_n | < \frac{\varepsilon}{2k} \ \forall \ n \geq m. \quad ...(4)$$

Now $p \leq | a_0 | | t_n | | a_1 | | t_{n-1} | + ... + | a_{n-m} | | t_m |$, by (2)

$< \frac{\varepsilon}{2k} [| a_0 | + | a_1 | + ... + a_{n-m} |]$, using (4)

Using (3), we obtain

$$P < \frac{\varepsilon}{2k} . k = \frac{\varepsilon}{2} \quad ...(5)$$

Since every absolutely convergent series is convergent, so $\Sigma\, a_n$ is convergent. Thus each of a_n, ... , $a_{n-m+1} \to 0$ as $n \to \infty$ and so $Q = a_{n-m+1}\, t_{m-1} + \ldots + a_n\, t_0 \to 0$

$$\Rightarrow \qquad Q < \frac{1}{2}\,\varepsilon \;\forall\; n \geq N > m \qquad \ldots(6)$$

Using (5) and (6) in (2) we obtain

$$C_n = B\, A_n + \varepsilon \;\forall\; n \geq N.$$

Hence, $\lim C_n = \lim B\, A_n = B \lim A_n = B\, A = AB$.

SOME MODES OF CONVERGENCE

We defined convergence by rectangles. Now we define some other modes of convergence in the following.

Definition:

We say that a double series $\sum a_{mn}$ converges

1. *By rectangles* if the sequence $< S_{mn} >$, where $S_{mn} = \sum_{r=1}^{m} \sum_{r=1}^{n} a_{rs}$, converges to some number, say S.S is called the *sum by rectangles* or simply the *sum* of the double series $\sum a_{mn}$. Thus

$$S = \lim_{m,n\to\infty} S_{mn} = \sum_{r,s=1}^{\infty} a_{rs} \text{ (simultaneous limit exists)}$$

2. *By squares* if the sequence $< S_n >$, where $S_n = \sum_{r=1}^{n} \sum_{s=1}^{n} a_{rs}$ converges
3. *By triangles (or diagonals)* if the sequence $< T_n >$, where $T_n = a_{11} + (a_{21} + a_{12}) + (a_{31} + a_{22} + a_{13}) + \ldots.$

or $\quad T_n = \sum_{k=2}^{n} \left(\sum_{p+q=k} a_{pq} \right) (n \geq 2)$ converges

4. *By rows* if for *m*, the sum R_m of the series $\sum_{n=1}^{\infty} a_{mn}$ exists and the series $\sum_{m=1}^{\infty} R_m$ converges
5. *By columns* if for each n, the sum C_n of the series $\sum_{m=1}^{\infty} a_{mn}$ exists and the series $\sum_{n=1}^{\infty} C_n$ converges.

We shall denote the sum by squares by S $(\sum a_{mn})$, the sum by diagonals by D $(\sum a_{mn})$, the sum by rows by R $(\sum a_{mn})$ or by $\sum_{m=1}^{\infty}\left(\sum_{n=1}^{\infty} a_{mn}\right)$ and the sum by columns by C $(\sum a_{mn})$ or by $\sum_{n=1}^{\infty}\left(\sum_{m=1}^{\infty} a_{mn}\right)$.

Remark:

The sum of a double series depends upon the mode of convergence as seen in the following examples :

Theorem 17: (Pringsheim's Theorem)

First we prove the following

Lemma

If the sum R_m of the series $\sum_{n=1}^{\infty} a_{mn}$ exists for all positive integers m, then $\lim_{n\to\infty} S_{mn} = R_1 + R_2 + \ldots + R_m$

Proof:

we know

$$S_{mn} = \sum_{r=1}^{m}\sum_{s=1}^{n} a_{rs}$$

$$= \sum_{s=1}^{n} a_{1s} \sum_{s=1}^{n} a_{2s} + \ldots\ldots + \sum_{s=1}^{n} a_{ms}$$

Hence $\lim_{n\to\infty} S_{mn} = R_1 + R_2 + \ldots\ldots + R_m$.

Theorem 18: (Pringsheim's Theorem)

Let $\sum a_{mn}$ be a double series which converges by rows, columns and rectangles; then their sums are equal i.e.,

$$R(\sum a_{mn}) = C(\sum a_{mn}) = \sum a_{mn}$$

Proof:

Suppose $\sum a_{mn}$ converges by rectangles to S. Since $\sum a_{mn}$ converges by rows, it follows that the sum R_m of the series $\sum_{n=1}^{\infty} a_{mn}$ exist for each positive integer m. Convergence by rectangles implies that for any $\varepsilon > 0$, there exists a positive integer N such that

$$| S_{mn} - S | < \varepsilon \text{ for } m \geq N, n \geq N;$$

Where $S_{mn} = \sum_{r=1}^{m} \sum_{s=1}^{n} a_{mn}$

Thus for m ≥ N, we obtain

$$\lim_{n\to\infty} | S_{mn} \; S | < \varepsilon \quad ...(1)$$

By above lemma, $\lim_{n\to\infty} | S_{mn} \; R_1 + R_2 + ... + R_m$ ...(2)

$\therefore \quad | R_1 + R_2 + ... + R_m - S | < \varepsilon$ for all m ≥ N, by (1) and (2).

This shows that $\sum_{m=1}^{\infty} R_m$ converges to S, i.e.,

$$R\left(\sum a_{mn}\right) = S$$

Similarly we can show that

$$C\left(\sum a_{mn}\right) = S.$$

MISCELLANEOUS EXAMPLES

Example 1:

Test the convergence of the series

$$\sum \frac{\{(n^3 + 1)^{1/3} - n\}}{\log n}$$

Solution:

Let $u_n = (n^3 + 1)^{1/3} - n$ and $v_n = 1/\log n$.

Then the given series can be written us $\sum u_n v_n$

$$\text{Now } u_n = n\left(1 + \frac{1}{n^3}\right)^{1/3} - n = n\left[\left(1 + \frac{1}{n^3}\right)^{1/3} - 1\right]$$

$$= n\left[\left(1 + \frac{1}{3n^3}\right) - 1\right] = \frac{1}{3n^2} + ...\left(-\frac{1}{n^2}\right)$$

Thus, $\sum u_n$ converges.

Now $< v_n >$ is a positive monotonically decreasing sequence, therefore by Abel's test, $\sum u_n v_n$ converges.

Example 2:

Test for convergence : $\sum \frac{(-1)^{n+1}}{n}\left(1+\frac{1}{n}\right)^{-n}$

Solution:

Do your self.

Hint. $\sum u_n = \sum \frac{(-1)^{n+1}}{n}$ converges, by Leibnitz's Test. $< v_n >$ $= <\left(1+\frac{1}{n}\right)^{-n}> = <\frac{1}{2}.\frac{4}{9}.\frac{27}{64},.....>$ is a positive and monotonically decreasing sequence. Apply Abel's Test.

Example 3:

Show that the series $\sum_{n=2}^{\infty} \frac{\sin nx}{\log n}$ *converges* $\forall x \in R$

Solution:

Do yourself.

Example 12:

Examine the convergence of the series

$$\sum\left\{\frac{1}{(n+1)^2}+\frac{1}{(n+2)^2}+...+\frac{1}{(2n)^2}\right\}\cos n\,\theta.$$

Solution:

Let $\Sigma\, u_n = \Sigma \cos n\,\theta$ and

$$v_n = \frac{1}{(n+1)^2}+\frac{1}{(n+2)^2}+...+\frac{1}{(2n)^2}.$$

Then the given series can be written as $\Sigma\, u_n\, v_n$

Let $< S_n >$ be the sequence of partial sums of $\Sigma\, u_n$. Then

$$S_n = \sum_{r=1}^{n}\cos r\,\theta = \frac{[\cos (n-1)\,\theta/2].\sin (n\theta/2)}{\sin(\theta/2)}, \qquad \theta \neq 0,\ 2\,k\pi$$

$$\Rightarrow |S_n| \le \frac{1}{|\sin(\theta/2)|}\ \forall \qquad (\theta \neq 0,\ 2\,k\pi)$$

Thus $< S_n >$ is bounded. $(\theta \neq 0,\ 2\,k\pi)$

Let $a_n = \frac{n}{(2n)^2} = \frac{n}{(n+n)^2} = \frac{1}{4n}$ so that $\lim_{n\to\infty} a_n = 0.$

By Cauchy's First Theorem on Limits. We have

$$\lim_{n\to\infty} \frac{a_1 + a_2 + ... + a_n}{n} = 0$$

$$\Rightarrow \lim_{n\to\infty}\left[\frac{1}{(n+1)^2} + \frac{1}{(n+2)^2} + ... + \frac{1}{(2n)^2}\right] = 0 \Rightarrow \lim v_n = 0.$$

Now we have

$$v_{n+1} - v_n = \frac{1}{(2n+2)^2} + \frac{1}{(2n+1)^2} - \frac{1}{(n+1)^2} = \frac{-8n^2 - 4n + 1}{4(n+1)^2 (2n+1)^2} < 0 \ \forall\ n$$

$\therefore v_{n+1} \le v_n \ \forall\ n$ and so $< v_n >$ is a monotonically decreasing sequence which tends to zero. By Dirichlet's test, it follows that $\Sigma\, u_n v_n$ i.e., the given series converges for $\theta \neq 0, 2\,k\pi$.

Example 4:

Show that the series

$$0 - \frac{1}{2} + \frac{1}{2^2} - \frac{1}{3} + \frac{2}{3^2} - \frac{1}{4} + \frac{3}{4^2} + \ ... + \textit{ is convergem.}$$

Solution:

Let $\Sigma\, u_n = 1 - 1 + \frac{1}{2} - \frac{1}{2} + \frac{1}{3} - \frac{1}{3} + ...$

and $< v_n > = < 0, \frac{1}{2}, \frac{1}{2}, \frac{2}{3}, \frac{2}{3}, \frac{2}{3}, \frac{3}{4}, \frac{3}{4}, ...>$

The the given series can be written as $\Sigma\, u_n v_n$.

Let $< S_n >$ be the sequence of partial sums of the series $\Sigma\, u_n$. Then

$$\lim_{n\to\infty} S_{2n} = 0, \quad \lim_{n\to\infty} S_{2n+1} = \lim_{n\to\infty} \frac{1}{n+1} = 0.$$

$\therefore \lim\limits_{n\to\infty} S_n = 0$ and so the series $\Sigma\, n_u$ converges.

It is clear that the sequence $< v_n >$ is monotonic and bounded. By Abel's test, $\Sigma\, u_n v_n$ i.e., the given series is convergent.

Example 5:

Examine the converges of the series:

(i) $\Sigma \frac{\cos n\theta}{n^\alpha}$ (ii) $\Sigma \frac{\sin n\theta}{n^\alpha}$.

Solution:

(i) The sequence $< S_n >$ of partial sums of the series $\Sigma \cos n\theta$ is bounded for $\theta \neq 0, 2\,k\,\pi$.

For a > 0, $\sum \frac{1}{n^\alpha}$ is a monotonically decreasing sequence tending to zero. By Dirichlet's test, the given series converges for a > 0 ($\theta \neq 0, 2\,k\,\pi$).

For $\alpha \leq 0$, the series diverges since the nth term does not tend to zero as $n \to \infty$.

For $\theta = 0$ or $2\,k\pi$, the given series becomes $\sum \theta/n^\alpha$ which converges for $\alpha > 1$.

(ii) The sequence $< S_n >$ of partial sums of the series $\sum \sin n\,\theta$ is bounded for $\theta \neq 0, 2\,k\,\pi$.

For a > 0, $\sum 1/n^\alpha$ is a monotonically decreasing sequence tending to zero. By Dirichlet's test, the given series converges for $\alpha > 0$ ($\theta \neq 0, 2\,k\,\pi$).

For $\alpha \leq 0$, the series diverges since the nth term does not tend to zero as $n \to \infty$.

For $\theta = 0, 2k\,\pi$ the given series become a series of zeros and is therefore conversent for all values of α.

Example 6:

If a series $\sum u_n$ is convergent, then show that the series $\sum \left(\frac{n+1}{n}\right) u_n$ is also convergent.

Solution:

Let $v_n = \frac{n+1}{n} = 1 + \frac{1}{n}$.

It is clear that $< v_n >$ is a positive monotonically decreasing sequence.

By Abel's test, $\sum u_n\, v_n$ is convergent. Hence the series $\sum \left(\frac{n+1}{n}\right) u_n$ is convergent.

Example 7:

Show that the series $\sum \frac{\sin nx}{n}$ converges for all real values of x.

Solution:

Let $< S_n >$ be the sequence of partial sum of the series $\sum \sin nx$. Then

$$S_n = \sin x + \sin 2x + ... + \sin nx = \frac{\cos(x/2) - \cos|(2n+1)x/2|}{2\sin(x/2)}$$

$u \neq 0.2\,k\pi$

$$\therefore |S_n| < \frac{1+1}{2|\sin(x/2)|} = \frac{1}{|\sin(x/2)|} \ \forall u.$$

Thus the sequence $< S_n >$ of partial sums of the series $\Sigma \sin nx$ is bounded. Also $< 1/n >$ is a monotonically decreasing sequence tending to zero. By Dirichlet's test, it follows that $\Sigma \frac{\sin nx}{n}$ converges for $x \neq 0.2 k\pi$.

For $x = 0$ or $2 k \pi$, the given sereis is obviously convergent.

Hence the result follows.

Example 8:

Show that the series

$$1 - \frac{1}{4.3} + \frac{1}{4^2 5} - \frac{1}{4^3.7} + ... \text{ converges.}$$

Solution:

Let $u_n = \left(-\frac{1}{4}\right)^{n=1} . v_n = \frac{1}{2n-1}.$

Then the given series can be written as $\Sigma u_n v_n$.

Since $|u_n| = \left(\frac{1}{4}\right)^{n=1}$, therefore $\Sigma |u_n|$ being a geometric series with common ratio $\frac{1}{4}$ is convergent.

Thus Σu_n is absolutely convergent and so Σu_n is convergent. Since $< v_n >$ is a positive, monotonically decreasing sequence, therefore by Abel's test, the series $\Sigma u_n v_n$ converges.

Ex.: Show that if a series Σu_n is convergent then

(i) $\Sigma \frac{u_n}{n}$ (ii) $\sum_{n=2}^{\infty} \frac{u_n}{\log n}$ are convergent.

Example 9:

Show that the series

$$1 - \frac{1}{3.2^2} + \frac{1}{5.3^2} - \frac{1}{7.4^2} + ...$$

is convergent.

Solution:

Let $\Sigma u_n = \frac{1}{1^2} - \frac{1}{2^2} + \frac{1}{3^2} - \frac{1}{4^2} + ... v_n = \frac{1}{2n-1}.$

The the given series can be written as $\Sigma u_n v_n$. By Leibnitz's Test. Σu_n converges.

Since $< v_n >$ is a positive and monotonically decreasing requence, there.

Example 10:

Examine the convergence of the series

$$\sum_{n=1}^{\infty}\left(1+\frac{1}{2}+\frac{1}{3}+\ldots+\frac{1}{n}\right)\frac{\sin n\theta}{n}$$

Solution:

The sequence $< S_n >$ of partial sums of Σu_n is bounded ($\theta \neq 0, 2k\pi$).

Let $v_n = \frac{1}{n}\left(1+\frac{1}{2}+\frac{1}{3}+\ldots+\frac{1}{n}\right)$.

Then the given series can be written as $\Sigma u_n v_n$.

Let $a_n = \frac{1}{n}$ so that $\lim_{n\to\infty} a_n = 0$.

By Cauchy's First Theorem on limits.

$$\lim_{n\to\infty}\frac{a_1+a_2+\ldots. a_n}{n} = \lim_{n\to\infty}\frac{1}{n}\left(1+\frac{1}{2}+\ldots.+\frac{1}{n}\right) = 0.$$

Thus $\lim_{n\to\infty} v_n = 0$, where $< v_n >$ is a positive monotonically decreasing sequence. Hence by Dirichlet's test, $\Sigma u_n v_n$ is convergent for $\theta \neq 0, 2k\pi$.

For $\theta = 0$ or $2k\pi$, the given series reduces to the series of zero which is obviously convergent.

Hence the given series is convergent for all values of θ.

Ex: Show that the series

$$\sum_{n=1}^{\infty}\left(1+\frac{1}{2}+\frac{1}{3}+\ldots+\frac{1}{n}\right)\frac{\sin n\theta}{n}$$

converges, absolutely for $\theta = k\pi$, k is any integer, and conditionally for all other real values of θ.

Example 11:

Determine the values of x for which the series

$1+\frac{1}{2}(5x-3x^2)+\frac{1}{2^2}(5x-3x^2)^2+\frac{1}{2^3}(5x-3x^2)^3+$. *converges and can be rearranged in ascending power of x.* ...(1)

Solution:

The given series being a geometric series with common ratio

$$r = \frac{1}{2}(5x - 3x^2) \text{ is convergent if } \left|\frac{1}{2}(5x - 3x^2)\right| < 1$$

$\Rightarrow -2 < 5x - 3x^2 < 2$

$\Rightarrow \quad 3x^2 - 5x - 2 < 0$ and $3x^2 - 5x + 2 > 0$

$\Rightarrow (3x - 1)(x - 2) < 0$ and $3x - 2(x - 1) > 0$

Thus the given series converges if

$$\left(-\frac{1}{3} < x < 2\right) \text{ and } \left(x < \frac{2}{3} \text{ or } x > 1\right)$$

i.e. if $x \in \left]-\frac{1}{3}, \frac{2}{3}\right[$ or ε [1, 2[.

The given series can be written as

$$1 + \frac{1}{2}(5x - 3x^2) + \frac{1}{4}(25x^2 - 30x^3 + 9x^4) + \ldots.$$

rearranging in ascending powers of x as follows :

$$1 + \frac{5}{2}x - \frac{3}{2}x^2 + \frac{25}{4}x^2 - \frac{30}{4}x^3 + \frac{125}{8}x^3 + \frac{9}{4}x^4 + \ldots.$$

if the series (1) is absolutely convergent. The series

$$1 + \frac{1}{2}(5|x| + 3|x|^2) + \frac{1}{2^2}(5|x| + 3|x|^2)^2 + \ldots.$$

converges if $\frac{1}{2}(5|x| + 3|x|^2) < 1 \Rightarrow 5|x| + 3|x|^2 - 2 < 0$

$$\Rightarrow (3|x| - 1)(|x| + 2) < 0$$

This is possible if $|x| < \frac{1}{3}$

Hence the given can be rearranged in ascending powers of x if $|x| < \frac{1}{3}$.

Example 12:

Explain the fallacy:

$$1 - \frac{1}{2} + \frac{1}{3} - \frac{1}{4} + \frac{1}{5} - \frac{1}{6} + \ldots..$$

$$= 1 + \frac{1}{2} + \frac{1}{3} + \frac{1}{4} + \frac{1}{5} + \frac{1}{6} + \ldots. - 2\left(\frac{1}{2} + \frac{1}{4} + \frac{1}{6} + \ldots.\right)$$

$$= 1+\frac{1}{2}+\frac{1}{3}+....-\left(1+\frac{1}{2}+\frac{1}{3}+....\right)=0$$

Hence the series $1-\frac{1}{2}+\frac{1}{3}+\frac{1}{4}+....$ *converges to zero!*

Explanation

The given series is conditionally convergent and so, by Riemann's Rearrangement Theorem, can be made to converge to any limit by a rearrangement of the terms.

Example 13(a):

Explain the fallacy of the following:

$$1-\frac{1}{2}+\frac{1}{3}-\frac{1}{4}+\frac{1}{5}-\frac{1}{6}+....$$

$$=(2-1)-\frac{1}{2}+\left(\frac{2}{3}-\frac{1}{3}\right)-\frac{1}{4}+\left(\frac{2}{5}-\frac{1}{5}\right)-\frac{1}{6}+....$$

$$=2-1-\frac{1}{2}+\frac{2}{3}-\frac{1}{3}-\frac{1}{4}+\frac{2}{5}-\frac{1}{5}-\frac{1}{6}+....$$

$$=2-1+\frac{2}{3}-\frac{1}{2}+\frac{2}{5}-\frac{1}{3}+\frac{2}{7}-\frac{1}{4}+\frac{2}{9}....$$

$$=2\left[1-\frac{1}{2}+\frac{1}{3}-\frac{1}{4}+\frac{1}{5}-\frac{1}{6}+\frac{1}{7}-....\right]$$

$\therefore$ S = 2S $\Rightarrow$ 1 = 2 ?

Example 13(b):

If the terms of the series $1-\frac{1}{2}+\frac{1}{3}-\frac{1}{4}+......$ *be rearranged, so that the ratio of the number of positive terms to the number of negative terms in succession is k > 0, then the sum of the rearranged series is 1/2 log 4K.*

Solution:

For k = 1, the result is evident.

Let k > 1. Let T_n > be the sequence of partial sums of the rearranged series, where n = m + p such that lim p/m = k gives n $\to \infty$ $\Rightarrow$ m $\to \infty$ and

$$T_n = T_{m+p} = \left(1-\frac{1}{2}+....+\frac{1}{2m-1}-\frac{1}{2m+1}\right)+\left(\frac{1}{2m+1}+.....+\frac{1}{2p-1}\right) \quad ...(1)$$

As $m \to \infty, \left(1-\frac{1}{2}+\frac{1}{3}-\frac{1}{4}+....-\frac{1}{2m}\right) \to \log 2$ ...(2)

Now $\lim_{m\to\infty}, \left[\frac{1}{2m+1}+....+\frac{1}{2p-1}\right] \lim_{m\to\infty}, \frac{1}{m}\sum_{r=1}^{p-m}\frac{1}{2+\left(\frac{2r-1}{m}\right)}$

$$= \int_0^{k-1}\frac{dx}{2+2x}dx = \frac{1}{2}\,|\log(1+x)|_0^{k-1} = \frac{1}{2}\log k \quad ...(3)$$

Hence $\lim_{n\to\infty} T_n = \log 2 + \frac{1}{2}\log k$, using (2) and (3) in (1)

$$= \frac{1}{2}\log\ 4k.$$

Similarly the result holds for $1 > k > 0$.

Example 14:

Find the sum of the series

$$1-\frac{1}{2}-\frac{1}{4}+\frac{1}{3}-\frac{1}{6}-\frac{1}{8}+\frac{1}{5}-......$$

Solution:

Let T_{3n} denote the sum of the first 3n terms of the series. Then

$$T_{3n} = \left(1-\frac{1}{2}-\frac{1}{4}\right)+\left(\frac{1}{3}-\frac{1}{6}-\frac{1}{8}\right)+......\left(\frac{1}{2n-1}-\frac{1}{4n-2}-\frac{1}{4n}\right)$$

$$= \left(\frac{1}{2}-\frac{1}{4}\right)+\left(\frac{1}{6}-\frac{1}{8}\right)+...\left(\frac{1}{4n-2}-\frac{1}{4n}\right)$$

$$= \frac{1}{2}\left(1-\frac{1}{2}+\frac{1}{3}-\frac{1}{4}+....+\frac{1}{2n}\right) = \frac{1}{2}S_{2n}$$

$$\therefore \lim_{n\to\infty} T_{3n} = \frac{1}{2}\lim_{n\to\infty} S_{2n} = \frac{1}{2}\log 2$$

Also $\lim_{n\to\infty} T_{(3n-i)} = \frac{1}{2}\log 2$ for $i = 1, 2$.

Hence $\lim_{n\to\infty} T_n = \frac{1}{2}\log 2$

Example 15:

Find the sum of the series

$$1+\frac{1}{3}+\frac{1}{5}-\frac{1}{2}+\frac{1}{7}+\frac{1}{9}+\frac{1}{11}-\frac{1}{4}+......$$

Solution:

Let T_{4n} denote the sum of the first 4n terms of the series. Then

$$T_{4n} = \left(1-\frac{1}{3}+\frac{1}{5}-\frac{1}{2}\right)+\left(\frac{1}{7}+\frac{1}{9}+\frac{1}{11}-\frac{1}{4}\right)+....$$

$$+\left(\frac{1}{6n-5}+\frac{1}{6n-3}+\frac{1}{6n-1}-\frac{1}{2n}\right)$$

$$=\left(-\frac{1}{2}+\frac{1}{3}+\frac{1}{4}+.....+\frac{1}{2n-1}-\frac{1}{2n}\right)$$

$$+\left(\frac{1}{2n+1}+\frac{1}{2n+3}+.....+\frac{1}{2n+2r-1}+....+\frac{1}{6n-1}\right)$$

$$= S_{2n} + \sum_{r=1}^{2n} \frac{1}{2n+2r-1}$$

$$\lim_{n\to\infty} T_{4n} = \lim_{n\to\infty} S_{2n} + \lim_{n\to\infty} \sum_{r=1}^{2n} \frac{1}{2n+2r-1}$$

$$= \log 2 + \lim \sum_{r=1}^{2n} \frac{1/n}{\left(2+\frac{2r}{n}\right)}$$

$$= \log 2 + \int_0^2 \frac{dx}{2+2x}$$

$$= \log 2 + \frac{1}{2} \mid \log(1+x)|_0^2 = \log 2 + \frac{1}{2}\log 3$$

$$= \frac{1}{2}[2\log 2 + \log 3]\frac{1}{2}\log(2^2 \times 3) = \frac{1}{2}\log 12$$

Also $\lim_{n\to\infty} T_{(4n-i)} = \frac{1}{2}\log 12$ for i = 1,2,3

Hence $\lim_{n\to\infty} T_n = \frac{1}{2}\log 12$.

Example 16:

Examine for convergence the following series :

$$1-\frac{1}{2}-\frac{1}{4}-\frac{1}{6}-\frac{1}{8}+\frac{1}{3}-\frac{1}{10}-\frac{1}{12}-\frac{1}{14}-\frac{1}{16}+\frac{1}{5}+... .$$

Solution:

Let $< T_n >$ denote trhe sequence of partial sums of the given series. Then

$$T_{5n} = \left(1-\frac{1}{2}-\frac{1}{4}-\frac{1}{6}-\frac{1}{8}\right)+\left(\frac{1}{3}-\frac{1}{10}-\frac{1}{12}-\frac{1}{14}-\frac{1}{16}\right)+....$$

$$+\left(\frac{1}{2n-1}-\frac{1}{8n-6}-\frac{1}{8n-4}-\frac{1}{8n-2}-\frac{1}{8n}\right)$$

$$=\left(1-\frac{1}{2}+\frac{1}{3}-\frac{1}{4}+\frac{1}{5}-.....+\frac{1}{2n-1}-\frac{1}{2n}\right)$$

$$-\left(\frac{1}{2n+2}+\frac{1}{2n+4}+.....+\frac{1}{2n+2r}+....+\frac{1}{8n}\right)$$

$$= S_{2n} - \sum_{r=1}^{3n} \frac{1}{2n+2rn}$$

$$\therefore \lim_{n\to\infty} T_{5n} \lim_{n\to\infty} S_{2n} - \lim_{n\to\infty} \sum_{r=1}^{3n} \frac{1}{2n+2r}$$

$$= \log 2-\frac{1}{2} \lim_{n\to\infty} \sum_{r=1}^{3n} \frac{1/n}{\left(1+\frac{r}{n}\right)}$$

$$= \log 2-\frac{1}{2}\int_0^3 \frac{dx}{1+x}$$

$$= \log 2-\frac{1}{2}|\log(1+x)\ |_0^3 = \log 2-\frac{1}{2}\log 4$$

$$= \frac{1}{2}(2\log 2 - \log 4) = \frac{1}{2}(\log 4 - \log 4) = 0$$

Also $\lim_{n\to\infty}$ (5n–i) = 0 of i = 1, 2, 3, 4

Hence $\lim_{n\to\infty} T_n = 0$.

Example 17:

Show that the sum of the series

$$1+\frac{1}{3}-\frac{1}{2}+\frac{1}{5}+\frac{1}{7}-\frac{1}{4}+\frac{1}{9}+\frac{1}{11}-\frac{1}{6}+ ...... \text{ is } \frac{3}{2}\log 2$$

Solution:

The sum of series is $S = \frac{1}{2}\log(4 \times 2) = \frac{3}{2}\log 2.$

Now we prove it independently. Let $< T_n >$ denote the sequence of partial sums of the series. Then

$$T_{5n} = \left(1+\frac{1}{3}-\frac{1}{2}\right)+\left(\frac{1}{5}+\frac{1}{7}-\frac{1}{4}\right)+....+\left(\frac{1}{4n-3}+\frac{1}{4n-1}-\frac{1}{2n}\right)$$

$$=\left(1-\frac{1}{2}+\frac{1}{3}-\frac{1}{4}+.....+\frac{1}{2n-1}-\frac{1}{2n}\right)$$

$$+\left(\frac{1}{2n+1}+\frac{1}{2n+3}....+\frac{1}{2n+2r-1}+....+\frac{1}{4n-1}\right)$$

$$= S_{2n} + \sum_{r=1}^{n} \frac{1}{2n+2r-1}$$

$$\therefore \lim_{n\to\infty} T_{3n} \lim_{n\to\infty} S_{2n} + \lim_{n\to\infty} \sum_{r=1}^{n} \frac{1}{2n+2r-1}$$

$$= \log 2 + \lim \sum_{r=1}^{n} \frac{1/n}{\left(2+\frac{2r}{n}\right)}$$

$$= \log 2 + \int_0^1 \frac{dx}{2+2x} = \log 2 + \frac{1}{2}|\log(1+x)|_0^1$$

$$= \log 2 + \frac{1}{2}\log 2 = \frac{3}{2}\log 2$$

Also lim T(3n–e) $=\frac{3}{2}$ log 2 for i = 1, 2

Hence $\lim_{n\to\infty} T_n = \frac{3}{2}\log 2.$

Example 18:

Examine for convergence the following series :

$$1+\frac{1}{2}-\frac{1}{3}+\frac{1}{4}+\frac{1}{5}-\frac{1}{6}+.....$$

Solution:

Here the number of positive terms is m = 2 and those of negative terms n = 1 (in each succession)

$$\therefore \quad k = \frac{m}{n} = \frac{2}{1} = 2$$

The sum of the series is

$$S = \frac{1}{2}\log 4k = \frac{1}{2}\log(4\times 2) = \frac{1}{2}\log 2^3 = \frac{3}{2}\log 2 \ .$$

Example 19:

Define rearrangement of a series. Show that any rearrangement of a convergent series of positive terms converges to the same sum. Show by an example that the result may not be true if the series is not that of positive terms.

Example 20(a):

Show that the series

$$1-1+\frac{1}{2}-\frac{1}{2}+\frac{1}{3}-\frac{1}{3}+\frac{1}{4}-\frac{1}{4}+.... \quad ...(A)$$

is convergent. Show that the following rearrangement of this series

$$1+\frac{1}{2}-1+\frac{1}{3}+\frac{1}{4}-\frac{1}{2}+\frac{1}{5}+\frac{1}{6}-\frac{1}{3}+.... \quad ...(B)$$

$$1+\frac{1}{2}+\frac{1}{3}-1+\frac{1}{4}+\frac{1}{5}+\frac{1}{6}-\frac{1}{2}+.... \quad ...(C)$$

converge respective to log 2 and log 3.

Solution:

For the series (A), we have

$$\lim_{n\to\infty} S_{2n} = 0, \quad \lim_{n\to\infty} S_{2n-1} = \lim_{n\to\infty} \frac{1}{n} = 0.$$

Hence $\lim_{n\to\infty} S_n = 0$**, which shows that the series (A) is convergent.**

Let $< T_n >$ **denote the sequence of partial sums of the series (B). Then** T_{3n} **is the sum of the firt 3n terms of the series so that**

$$T_{3n} = \left(1+\frac{1}{2}-1\right)+\left(\frac{1}{3}+\frac{1}{4}-\frac{1}{2}\right)+....+\left(\frac{1}{2n-1}+\frac{1}{2n}-\frac{1}{n}\right)$$

$$=\left(1-1+\frac{1}{2}-\frac{1}{2}+....+\frac{1}{n}-\frac{1}{n}\right)+\left(\frac{1}{n+1}+\frac{1}{n+2}+....+\frac{1}{2n-1}+\frac{1}{2n}\right)$$

$$= 0 + \sum_{r=1}^{n} \frac{1}{n+r}$$

$$\therefore \quad \lim_{n\to\infty} T_{3n} = \lim_{n\to\infty} \sum_{r=1}^{n} \frac{1}{n+r} = \lim_{n\to\infty} \sum_{r=1}^{n} \frac{1/n}{1+r/n}$$

$$= \int_0^1 \frac{dx}{1+x} = |\log(1+x)|_0^1 \log 2.$$

Also $\lim_{n\to\infty} T_{(3n-i)} = \log 2$ **for** $i = 1, 2.$

Hence $\lim_{n\to\infty} T_n = \log 2$ and so the series (B) converges to log 2.

Let T_{4n} denote the sume of the first 4n terms of the series (C). Then.

$$T_{4n} = \left(1-1+\frac{1}{2}-\frac{1}{2}+....+\frac{1}{n}-\frac{1}{n}\right)+\left(\frac{1}{n+1}+\frac{1}{n+2}+....+\frac{1}{3n}\right)$$

$$= 0 + \sum_{r=1}^{2n} \frac{1}{n+r}$$

$$\therefore \quad \lim_{n\to\infty} T_{4n} \int_0^2 \frac{dx}{1+x} = \log 3.$$

Also $\lim_{n\to\infty} T_{(4n-i)} = \log 3$ for i = 1, 2, 3.

Hence $\lim_{n\to\infty} T_n = \log 3$.

Example 20(b):

Find the values of x for which the series

$1 + (x - x^2) + (x - x^2)^2 + (x - x^2)^3 + ...$

(i) converges and (ii) can be rearranged in ascending powers of x.

[Hint. (i) converges if $|x - x^2| < 1$ i.e., $- x - x^2 < x - x^2 < 1$ i.e., $x^2 - x\ 1 < 0$ and , $x^2 - x + 1 > 0$]

[Ans. : $\left]\frac{1}{2}(1-\sqrt{5}),\frac{1}{2}(1+\sqrt{5})\right[$, (ii) $\left]\frac{1}{2}(1-\sqrt{5}),\frac{1}{2}(\sqrt{5}-1)\right[$.]

Example 20(c):

Show that

(i) $1+\frac{1}{3}+\frac{1}{5}-\frac{1}{2}-\frac{1}{4}+\frac{1}{7}+\frac{1}{9}+\frac{1}{11}-\frac{1}{6}-\frac{1}{8}+....=\frac{1}{2}\log 6$.

$$\left[\text{Hint. } T_{5n} = S_{4n} + \sum_{r=1}^{n} \frac{1}{4n+2r-1}\right]$$

(ii) $1+\frac{1}{3}-\frac{1}{2}-\frac{1}{4}-\frac{1}{6}+\frac{1}{5}+\frac{1}{7}-\frac{1}{8}-\frac{1}{10}-\frac{1}{12}+....=\frac{1}{2}\log\frac{8}{3}$.

(iii) $1-\frac{1}{2}-\frac{1}{4}+\frac{1}{5}+\frac{1}{7}-\frac{1}{8}-\frac{1}{10}+\frac{1}{11}+\frac{1}{13}+....=\frac{2}{3}\log 2$.

Example 21:

Find the values of x for which the series

$$1+\left(\frac{5x}{2}-x^2\right)+\left(\frac{5x}{2}-x^2\right)^2+.....$$

can be arranged as a convergent series in ascending powers of x.

$$\left[\textbf{Ans. } -\frac{1}{2}(\sqrt{41}-5)<x<\frac{1}{2}(\sqrt{41}-5).\right]$$

[**Hint**. For absolute convergence, $\frac{5}{2}|x|+|x|^2<1$ or $2|x|^2+5|x|-2<0.1$]

(iv) $1-\frac{1}{2}-\frac{1}{4}-\frac{1}{6}+\frac{1}{3}-\frac{1}{8}-\frac{1}{10}-\frac{1}{12}+\frac{1}{5}-\frac{1}{14}-\frac{1}{16}-\frac{1}{18}+\frac{1}{7}-....=\frac{1}{2}\log\frac{4}{3}$

(v) Examine the convergence of $1-\frac{1}{2}-\frac{1}{3}+\frac{1}{4}+\frac{1}{5}-\frac{1}{6}-\frac{1}{7}+....$

Example 22:

Show that the double series

$$\sum_{m,n=1}^{\infty}\frac{1}{m^4+n^4} \textit{ is convergent.}$$

Solution:

We have $(m^2-n^2)^2=m^4+n^4-2m^2n^2\geq 0$

$\Rightarrow m^4+n^4\geq 2m^2n^2$

$\Rightarrow \frac{1}{m^4+n^4}\leq\frac{1}{2}\frac{1}{m^2n^2}$ for all m, n.

By Eample 20, $\sum\frac{1}{m^2n^2}$ is convergent. Hence by the comparison test, the given double series is convergent.

Example 23:

Show that $\sum_{m=1}^{\infty}\sum_{n=1}^{\infty}\frac{1}{m^a+n^a}$*, $\alpha>2$, is convergent.*

Solution:

We have

$$0\leq\left(m^{\alpha/2}-n^{\alpha/2}\right)^2=m^{\alpha}+n^{\alpha}-2\,m^{\alpha/2}n^{\alpha/2}$$

$$\Rightarrow m^\alpha + n^\alpha \geq 2m^{\alpha/2}n^{\alpha/2} \Rightarrow \frac{1}{m^\alpha + m^\alpha} \leq \frac{1}{2}.\frac{1}{m^{\alpha/2}}.\frac{1}{n^{\alpha/2}}$$

$$\therefore\ S_{mn} = \sum_{p,q=1}^{m,n} \frac{1}{p^\alpha + q^\alpha} \leq \frac{1}{2}\left(\sum_{p=1}^{m} \frac{1}{p^{\alpha/2}}\right)\left(\sum_{q=1}^{n} \frac{1}{q^{\alpha/2}}\right)$$

We know $\sum_{p=1}^{\infty} \frac{1}{p^{\alpha/2}}$ is convergent if $\frac{\alpha}{2} >$ i.e., $\alpha > 2$.

Let $\sum_{p=1}^{\infty} \frac{1}{p^{\alpha/2}}\ M < \infty$ if $\alpha > 2$.

If follows that, $S_{mn} < \frac{1}{2}M^2$ for all values of *m* and *n* if $\alpha > 2$.

Hence the given double series is convergent for $\alpha > 2$.

Ex: *Show that the double series* $\sum \frac{1}{m^2(1+n^{5/2})}$ *is convergent.*

[Hint. $m^2(1 + n^{5/2}) \geq m^2n^{5/2}$**]**

Example 24:

Show that the double series $\sum_{m,n=1}^{\infty} \frac{1}{m^2n^2}$ *is convergent.*

Solution:

We have

$$S_{mn} = \sum_{p,q=1}^{m,n} \frac{1}{p^2q^2} = \left(\sum_{p=1}^{m} \frac{1}{p^2}\right)\left(\sum_{q=1}^{n} \frac{1}{q^2}\right)$$

Since $\sum \frac{1}{n^2}$ is convergent, so $\sum_{n=1}^{\infty} \frac{1}{2} = M < \infty$.

If follows that

$S_{mn} < M^2$ for all m, n.

Hence the given double series is convergent.

Example 25:

For the double series

$$\begin{array}{l} 2 + 2 + 2 + 2 + \ldots\ldots \\ -2 - 2 - 2 - 2 - \ldots\ldots \\ +2 - 2 + 0 + 0 + \ldots\ldots \\ +2 - 2 + 0 + 0 + \ldots\ldots \\ +2 - 2 + 0 + 0 + \ldots\ldots \\ + \ldots\ldots\ldots\ldots\ldots\ldots\ldots\ldots \end{array}$$

Show that the sums by rows and columns do not exist; the sum by rectangle and squares is each zero and the sum by diagonals is 4.

Solution:

We see that the first two rows R_1 and R_2 are not convergent, so the sum by rows does not exist. Similarly the first two columns C_1 and C_2 are not convergent so the sum by columns does not exist. The sums of successive diagonals are

$$2, 0, 2, 0, 0, 0, \ldots..$$

Thus the sum by diagonals is 4. We have $S_{mn} = 0$ if m, n > 1 and so the sum by rectangles is 0. Also the sum by squares is 0.

Example 26:

For the double series

$$\begin{aligned}&0 + 2 + 0 + 0 + 0 + \ldots\ldots.\\ &-2 + 0 + 2 + 0 + 0 + \ldots\ldots.\\ &+0 - 2 + 0 + 2 + 0 + \ldots\ldots.\\ &+0 + 0 - 2 + 0 + 2 + \ldots\ldots.\\ &+0 + 0 + 0 - 2 + 0 + \ldots\ldots.\end{aligned}$$

show that the sums by rows is 2, that by columns –2, that by diagonals 0, and that by squares is 0. Also show that the sum by rectangles does not exist.

Solution:

We see that the sums of successive rows are 2, 0, 0, 0, ... Consequently, the sum by rows is 2.

Again the sums of successive columns are –2, 0, 0, 0, ...

Consequently, the sum by columns is –2.

The sum of successive diagonals are 0, 0, 0, ... and so the sum by liagonals is 0. Similarly the sum by squares is 0. Now

$$\begin{aligned}S_{mn} &= (a_{11} + a_{12}) + (a_{21} + a_{22} + a_{23} + a_{13}) + \ldots. + (a_{m1} + a_{mn} + \ldots. + a_{1n})\\ &= (0 + 2) + (-2 + 0 + 2 + 0) + 0 + 0 + \ldots + 0 = 2.\end{aligned}$$

$$\therefore \lim_{(m,n)\to\infty} S_{mn} = 2 \neq 0$$

Hence the sum by rectangles does not exist.

Example 27:

Show that the convergence by rows of a double series does not imply convergence by columnus, but if the sum by rows, columns and rectangles all exist, then all three must be equal. Show also that the result may not be true if the convergence by rectangles is not assumed.

Solution:

(i) Consider the double series

$$\frac{1}{2}-\frac{1}{5}+\frac{1}{5}-\frac{1}{10}+\frac{1}{10}-\frac{1}{15}+\frac{1}{15}-\ldots\ldots$$
$$+\frac{1}{2^2}-\frac{4}{5^2}+\frac{4}{5^2}-\frac{9}{10^2}+\frac{9}{10^2}-\frac{14}{15^2}+\frac{14}{15^2}-\ldots\ldots$$
$$+\frac{1}{2^3}-\frac{4^2}{5^3}+\frac{4^2}{5^3}-\frac{9^2}{10^3}+\frac{9^2}{10^3}-\frac{14^2}{15^3}+\frac{14^2}{15^3}-\ldots\ldots$$
$$+\frac{1}{2^4}-\frac{4^3}{5^4}+\frac{4^3}{5^4}-\frac{9^3}{10^4}+\frac{9^3}{10^4}-\frac{14^3}{15^4}+\frac{14^3}{15^4}+\ldots\ldots$$
$$+\ \ldots\ldots\ldots\ldots\ldots\ldots\ldots\ldots\ldots\ldots$$

Thus sums of the successive rows are $\frac{1}{2},\frac{1}{2^2},\frac{1}{2^3}\ldots.$

The sum by rows $=\frac{\frac{1}{2}}{1-\frac{1}{2}}=1$. Hence the given double series converges by rows. The sums of successive columns are

$$1, -1, 1, -1, 1, -1, \ldots .$$

Thus the sum by columns does not exist and so the doubles series does not converge by columns. For part

(ii) See Theorem 10.

(iii) Let S_{mn} denote the sum of the terms in the rectangle bounded by the first m rows and n columns. Then

$$S_{m,\,2n-1}=1-\frac{1}{2^m}$$

and $$S_{m,2n}=\left(1-\frac{1}{2^m}\right)+\left(1-\frac{1}{5^n}\right)^m-1$$

$$=\left(1-\frac{1}{5^n}\right)^m-\frac{1}{2^m}$$

It shows that sums by rectangles and squares do not exist. Thus the sum by rows, columns and rectangle, may not exist (or be equal) if the convergence by rectangles is not assumed.

Note : The double series of Example 23 is not convergent by rectangles, the sum by rows is 2 and the by columns is – 2.

Ex: *Give an example of a double series which is convergent by rectangles bu is not convergent by rows and columns.*

Example 28:

Convert the series

$$\frac{x}{1+x}-\frac{2x^2}{1+x^2}+\frac{3x^3}{1+x^3}-\frac{4x^2}{1+x^4}+\ldots\ldots,\quad |x|<1 \qquad \ldots(1)$$

into a double series and show it is equal to

$$\frac{x}{(1+x)^2}-\frac{x^2}{(1+x^2)^2}+\frac{x^3}{(1+x^3)^2}-\ldots\ldots \qquad \ldots(2)$$

Solution:

For $|x| < 1$, the series in (1) is the sum by rows of the double series:

$$\begin{array}{l} x - x^2 + x^3 - x^4 + x^5 - x^6 + \ldots\ldots \\ -2x^2 + 2x^4 - 2x^6 + 2x^8 - 2x^{10} + 2x^{12} - \ldots\ldots \\ +3x^3 - 3x^6 + 3x^9 - 3x^{12} + 3x^{15} - 3x^{18} + \ldots\ldots \\ - \ldots\ldots\ldots\ldots\ldots\ldots\ldots\ldots \end{array} \qquad \ldots(3)$$

We shall show that the double series in (3) converges absolutely by rows. Replacing x by $|x| = y$ and all negative signs by positive signs in (3), we obtain

$$\begin{array}{l} y + y^2 + y^3 + y^4 + y^5 + y^6 + \ldots\ldots \\ +2y^2 + 2y^4 + 2y^6 + 2y^8 + 2y^{10} + 2y^{12} + \ldots\ldots \\ +3y^3 + 3y^6 + 3y^9 + 3y^{12} + 3y^{15} + 3y^{18} + \ldots\ldots \\ + \ldots\ldots\ldots\ldots\ldots\ldots\ldots\ldots \end{array} \qquad \ldots(4) \quad (0 \le y < 1)$$

For $0 \le y < 1$, the sum by rows of the double series in (4) is

$$\frac{y}{1-y}+\frac{2y^2}{1-y^2}+\frac{3y^2}{1-y^3}+\ldots\ldots \qquad \ldots(5)$$

For $y = 0$, the series in (5) converges.

For $0 < y < 1$, we have

$$\frac{a_n}{a_{n+1}}=\frac{ny^n}{1-y^n}\cdot\frac{1-y^{n+1}}{(n+1)\,y^{n+1}}$$

$$\therefore \lim_{n\to\infty}\frac{a_n}{a_{n+1}}=\lim_{n\to\infty}\frac{n}{n+1}\cdot\frac{1}{y}\cdot\frac{1-y^{n+1}}{1-y^n}=\frac{1}{y}>1 \qquad (\because 0<y<1)$$

By D' Alembert's Ratio Test, the series in (5) converges. Thus the double series in (3) converges absolutely by rows and hence columns and the two sums are equal. For $|x| < 1$, the sum by columns of the double series in (3) is given by

$$\frac{x}{(1+x)^2} - \frac{x^2}{(1+x^2)^2} + \frac{x^3}{(1+y^3)^2} - \ldots.$$

Example 29:

Convert the series on L.H.S. into a double series and show it equals on R.H.S. for $|x| < 1$.

$$\frac{x}{1+x^2} + \frac{x^3}{1+x^6} + \frac{x^5}{1+x^{10}} + \ldots. = \frac{x}{1-x^2} - \frac{x^3}{1-x^6} + \frac{x^5}{1-x^{10}} - \ldots.. \qquad \ldots(1)$$

Solution:

For $|x| < 1$, the series on the L.H.S. of (1) is the sum by rows of the double series:

$$\begin{aligned} & x - x^3 + x^5 - x^7 + x^9 - \ldots.. \\ & +x^3 - x^9 + x^{15} - x^{21} + x^{27} - \ldots.. \qquad \ldots(2) \\ & +x^5 - x^{15} + x^{25} - x^{35} + x^{45} - \ldots.. \\ & \ldots\ldots\ldots\ldots\ldots\ldots\ldots\ldots\ldots \end{aligned}$$

We shall show that the double series in (2) converges absolutely by rows. Replacing x by $|x| = y$ and all negative signs by positive signs in (2), the double series becomes

$$\begin{aligned} & y + y^3 + y^5 + y^7 + y^9 + \ldots\ldots \\ & +y^3 + y^9 + y^{15} + y^{21} + y^{27} + \ldots.. \qquad \ldots(3) \\ & +y^5 + y^{15} + y^{25} + y^{35} + y^{45} + \ldots.. \\ & + \ldots\ldots\ldots\ldots\ldots\ldots\ldots\ldots\ldots \qquad (0 \le y < 1) \end{aligned}$$

For $0 \le y < 1$, the sum by rows of the double series in (3) is

$$\frac{y}{1-y^2} + \frac{y^3}{1-y^6} + \frac{y^5}{1-y^{10}} + \ldots. \qquad \ldots(4)$$

For $y = 0$, the series in (4) converges.

For $0 < y < 1$, we have

$$\frac{a_n}{a_{n+1}} = \frac{y^{2n-1}}{1-y^{4n}-2} \times \frac{1-y^{4n+2}}{y^{2n+1}}$$

$$\therefore \lim_{n\to\infty} \frac{a_n}{a_{n+1}} = \frac{1}{y^2} > 1. \qquad (\because 0 < y < 1)$$

By D' Alembert's Ratio Test, the series in (4) converges for $0 < y < 1$. Ths the double series in (2) converges absolutely by rows and hence by columns and the two sums are equal. for $|x| < 1$, the sum by columns of the double series in (2) is given by

$$\frac{x}{1+x^2} - \frac{x^3}{1-x^6} + \frac{x^5}{1-y^{10}} - \ldots$$

Hence the two sides in (1) are equal.

Example 30:

Show that for $|x| < 1$,

$$\frac{x}{1+x} + \frac{x^2}{1-x^2} + \frac{x^3}{1-x^3} + \ldots = \frac{1+x}{1-x} + x^4 \frac{1+x^2}{1-x^2} + x^9 \frac{1+x^3}{1-x^3} + \ldots \qquad \ldots(1)$$

Solution:

For $|x| < 1$, the L.H.S. of (1) is the sum by rows of the double series

$$\begin{array}{l} x + x^2 + x^3 + x^4 + x^5 + \ldots \\ + x^2 + x^4 + x^6 + x^8 + x^{10} + \ldots \\ + x^3 + x^6 + x^9 + x^{12} + x^{15} + \ldots \\ + x^4 + x^8 + x^{12} + x^{16} + x^{20} + \ldots \\ + \ldots\ldots\ldots \end{array} \qquad \ldots(2)$$

We shall show that the double series in (2) converges absolutely by rows. Replacing x by $|x| = y$ in (2), we obtain the double series.

$$\begin{array}{l} y + y^2 + y^3 + y^4 + y^5 + \ldots \\ + y^2 + y^4 + y^6 + y^8 + y^{10} + \ldots \\ + y^3 + y^6 + y^9 + y^{12} + y^{15} + \ldots \\ + y^4 + y^8 + y^{12} + y^{16} + y^{20} + \ldots \\ + \ldots\ldots\ldots \qquad (0 \le y < 1) \end{array} \qquad \ldots(3)$$

For $0 \le y < 1$, the sum by rows of the double series in (3) is

$$\frac{y}{1-y} + \frac{y^2}{1-y^2} + \frac{y^3}{1-y^3} + \frac{y^4}{1-y^4} + \ldots \qquad \ldots(4)$$

For $y = 0$, the series in (4) converges.

For $0 < y < 1$, $\dfrac{a_n}{a_{n+1}} = \dfrac{y^n}{1-y^n} \cdot \dfrac{1-y^{n+1}}{y^{n+1}} = \dfrac{1-y^{n+1}}{y(1-y^n)}$

Therefore $\lim\limits_{n\to\infty} \dfrac{a_n}{a_{n+1}} = \dfrac{1}{y} > 1 \quad (\because 0 < y < 1)$

By D'Alembert's Ratio Test, the series in (4) converges. Hence the double series in (2) converges absolutely by rows for $|x| < 1$.

Absolute convergence of the double series implies that it can be summed up in several ways, all having the same sum. We sum up the double series in (2) according to P-shaped blocks as shown in (2).

Thus, the L.H.S. of (1) is

$$= (x + 2x^2 + 2x^3 + 2x^4 + ...) + (x^4 + 2x^6 + 2x^8 + 2x^{10} + ...)$$
$$+ (x^9 + 2x^{12} + 2x^{15} + 2x^{18} + ...) + ...$$
$$= \left[x+2x^2\left(\frac{1}{1-x}\right)\right] + \left[x^4+2x^6\left(\frac{1}{1-x^2}\right)\right] + \left[x^9+2x^{12}\left(\frac{1}{1-x^3}\right)\right] + ...$$
$$= x\left(\frac{1+x}{1-x}\right) + x^4\left(\frac{1+x^2}{1-x^2}\right) + \left(\frac{1+x^3}{1-x^3}\right) + ...$$

Example 31:

For $|x| < 1$, prove that

$$1+\frac{8x}{1-x}+\frac{16x^2}{1+x^2}+\frac{24x^3}{1-x^3}+....... \qquad ...(1)$$

$$= 1 + \frac{8x}{(1-x)^2} + \frac{8x^2}{(1+x^2)^2} + \frac{8x^3}{(1-x^3)^2} + ... \qquad ...(2)$$

Solution:

For $|x| < 1$, the series in (1) is the sum by rows of the double series:

$$+ 1 + 8x + 8x^2 + 8x^3 + 8x^4 + ... \qquad ...(3)$$
$$+ 0 + 16x^2 - 16x^4 + 16x^6 - 16x^8 + ...$$
$$+ 0 + 24x^3 + 24x^6 + 24x^9 - 24x^{12} + ...$$
$$+ ....$$

We shall show that this double series converges absolutely by rows. Replacing x by $|x| = y$ and all negative sings by positive signs in (3), the double series in (3) becomes

$$1 + 8y + 8y^2 + 8y^3 + 8y^4 + ...$$

$$+ 0 + 16y^2 + 16x^4 + 16x^6 + 16y^8 + ... \quad ...(4)$$

$$+ 0 + 24y^3 + 24y^6 + 24y^9 + 24y^{12} + ...$$

$$+ .... \quad (0 \le y < 1)$$

For $0 \le y < 1$, the sum by rows of the double series in (4) is

$$1 + \frac{8y}{1-y} + \frac{16y^2}{1-y^2} + \frac{24y^3}{1-y^3} + ... \quad ...(5)$$

For y = 0, the series in (5) converges. For $0 < y < 1$, we have

$$\frac{a_n}{a_{n+1}} + \frac{8(ny^n)}{1-y^n} + \frac{1-y^{n+1}}{8(n+1)\,y^{n+1}} \quad \text{(neglecting the firs term).}$$

$\lim_{n\to\infty} \frac{a_n}{a_{n+1}} = \frac{1}{y} > 1$ (since 0 < y , 1). By D' Alembert's Ratio Test, the series in (5) converges. Thus the series in (3) converges absolutely by rows for $|x| < 1$ and hence by columns and the two sums by rows and columns of the double series in (3) are equal.

For | x | , 1, the sum by columns of the double series in (3) is

$$1 + \frac{8x}{(1-x)^2} + \frac{8x^2}{(1+x^2)^2} + \frac{8x^3}{(1-x^3)^2} + ...$$

Example 32:

If $\sum_{n=1}^{\infty} a_n$ converges absolutely, then

$$\frac{a_1}{1.2} + \frac{a_1 + 2a_2}{2.3} + \frac{a_1 + 2a_2 + 3a_3}{3.4} + ... = \sum_{n=1}^{\infty} a_n \quad ...(1)$$

Solution:

Consider the double series

$$\frac{a_1}{1.2} + \frac{a_1}{2.3} + \frac{a_1}{3.4} + \frac{a_1}{4.5} + ...$$

$$+ 0 + \frac{2a_2}{2.3} + \frac{2a_2}{3.4} + \frac{2a_2}{4.5} + ... \quad ...(2)$$

$$+ 0 + 0 + \frac{a_3}{3.4} + \frac{3a_3}{4.5} + ...$$

$$+ ....$$

The sum of the term in the nth row is a_n. The sum by rows of (2) is $\sum_{n=1}^{\infty} a_n$.

Since $\sum_{n=1}^{\infty} a_n$ converges absolutely, the double series in (2) converges absolutely by rows and hence by columns and the two sums are equal. The sum by columns of (2) is the left-hand side of (1). Hence the result.

Example 33:

Show

$$\sum_{m=1}^{\infty}\sum_{n=1}^{\infty}\frac{(-1)^{n+1}}{(m+n^2)(m+n^2-1)}=\frac{\pi^2}{12}$$

Solution:

Let the given series be denoted as S.

Let us consider the series

$$T=\sum_{n=1}^{\infty}(-1)^{n+1}\sum_{m=1}^{\infty}\frac{(-1)^{n+1}}{(m+n^2)(m+n^2-1)} \qquad ...(1)$$

We have

$$\frac{1}{(m+n^2)(m+n^2-1)}=\frac{1}{m+n^2-1}-\frac{1}{m+n^2}$$

$$\therefore \sum_{m=1}^{k}\frac{1}{(m+n^2)(m+n^2-1)}=\left(\frac{1}{n^2}-\frac{1}{n^2+1}\right)+\left(\frac{1}{n^2+1}-\frac{1}{n^2+2}\right)+......+\left(\frac{1}{k+n^2-1}-\frac{1}{k+n^2}\right)$$

$$=\left(\frac{1}{n^2}-\frac{1}{k+n^2}\right)$$

Now, $\sum_{m=1}^{\infty}\frac{1}{(m+n^2)(m+n^2-1)}=\frac{1}{n^2}$. from (1), we get

$T=\sum\frac{(-1)^{n+1}}{n^2}$ which converges absolutely be Leibnitz Test.

Consequently,

$$S=T=\sum\frac{(-1)^{n+1}}{n^2} \quad \text{[By Dirichlet's Theorem]}$$

$$=1-\frac{1}{2^2}+\frac{1}{3^2}-\frac{1}{4^2}+\frac{1}{5^2}-\frac{1}{6^2}+....$$

$$= \left(1+\frac{1}{2^2}+\frac{1}{3^2}-\frac{1}{4^2}+.....\right)-2\left(\frac{1}{2^2}+\frac{1}{4^2}+\frac{1}{6^2}+....\right)$$

$$= \frac{\pi^2}{6}-\frac{2}{2^2}\left(\frac{\pi^2}{6}\right)=\frac{\pi^2}{12}.$$

Example 34:

Suppose $\sum_{n=0}^{\infty} a_n$ *and* $\sum_{n=0}^{\infty} b_n$ *be two convergent series. A necessary and sufficient condition for the convergence of the Cauchy product series* $\sum_{n=0}^{\infty} c_n$ *of* $\sum_{n=0}^{\infty} a_n$ *and* $\sum_{n=0}^{\infty} b_n$ *is that the sequence* $< T_n >$, *where* $T_n = \sum_{k=1}^{\infty} a_k\,(b_n+b_{n-1} + ... + b_{n-k+1})$ *converges to zero.*

Solution:

Let A_n, B_n, C_n denote the (n + 1)th partial sums of $\Sigma\, a_n$, $\Sigma\, b_n$ and $\Sigma\, c_n$ respectively. Since $\Sigma\, a_n$, $\Sigma\, b_n$ converge, the sequences $< A_n >$ and $< B_n >$ converge. Suppose $A_n \to A$ and $B_n \to B$ as $n \to \infty$.

For $n \le 0$, we have

$C_n = c_1 + c_2 + ... + c_n = a_0b_0 + (a_0b_1 + a_1b_0) + ... + (a_0b_n + ... + a_nb_0)$

$= a_0\,(b_0 + b_1 +....+ b_n) + a_1\,(b_0 +...+ b_{n-1}) + ... + a_{n-1}\,(b_0+b_1) + a_nb_0$

$= a_0\, B_n + a_1\, B_{n-1} + ... + a_n\, B_0$. Now

$T_n = a_1\, b_n + a_2\,(b_n + b_{n-1}) + ... + a_n\,(b_1 + b_{n-1} + ... + b_1)$.

$\therefore\; C_n + T_n = a_0\, B_n + a_1\, B_n + ... + a_n\, B_n = (a_0 + ... + a_n)$

$B_n = A_n\, B_n$ ...(1)

Let $T_n \to 0$ as $n \to \infty$. From (1).

$\lim C_n + \lim T_n = \lim A_n\, B_n \Rightarrow \lim C_n = \lim A_n \lim B_n = AB.$

Hence the series $\Sigma\, c_n$ converges.

Conversely, suppose $\sum_{n=1}^{\infty} c_n$ converges.

By Abel's Theorem, since all the three series $\Sigma\, a_n$, $\Sigma\, b_n$ and $\Sigma\, c_n$ are convergent, so

$$\lim C_n = AB. \qquad ...(2)$$

From (1), $\qquad T_n = A_n\, B_n - C_n$

$\Rightarrow \lim T_n = \lim A_n \lim B_n - AB = AB - AB = 0$, by (2).

Hence the sequence $< T_n >$ converges to zero.

Example 35:

Prove that for certain values of x and θ which are to be stated, that

$$\sum_{n=1}^{\infty} \frac{x^n}{n} \cos n\theta \sum_{n=1}^{\infty} \frac{x^n}{n} \sin n\theta = \sum_{n=2}^{\infty} \frac{x^n}{n}\left(1+\frac{1}{2}+\ldots\ldots+\frac{1}{n-1}\right) \sin n\theta$$

Solution:

Let $a_n = \dfrac{x^{n+1}}{n+1} \cos(n+1)\theta,\ b_n = \dfrac{x^{n+1}}{n+1} \sin(n+1)\theta$

$\therefore |a_n| \le |x|^{n+1}/(n+1) < |x|^{n+1}$; $|b_n| \le |x|^{n+1}/(n+1) < |x|^{n+1}$.

Thus the series $\Sigma |a_n|$ and $\Sigma |b_n|$ converge if $|x| < 1$ i.e., Σa_n and Σb_n are absolutely convergent if $|x| < 1\ \forall\ \theta$. Hence the Cauchy product Σc_n is absolutely convergent for $|x| < 1$ and $\forall\ \theta$. Now

$$c_n = \sum_{k=0}^{n} a_k b_{n-k} = \sum_{k=0}^{n} \frac{x^{k+1} \cos(k+1)\theta}{k+1} \cdot \frac{x^{n-k+1} \sin(n-k+1)\theta}{n-k+1}$$

$$= x^{n+2} \sum_{k=0}^{n} \frac{\cos(k+1)\theta \sin(n-k+1)\theta}{(k+1)(n-k+1)}$$

or $$\frac{2(n+2)C_n}{x^{n+2}} = \sum_{k=0}^{n} [\sin(n+2)\theta - \sin(n-2k)\theta]\left[\frac{1}{k+1}+\frac{1}{n=k+1}\right]$$

$$= \sin(n+2)\theta \sum_{k=0}^{n}\left(\frac{1}{k+1}+\frac{1}{n-k+1}\right)$$

$$- \sum_{k=0}^{n} \frac{\sin(n-2k)\theta}{k+1} + \frac{\sin(n-2k)\theta}{n-k+1}$$

We have

$$\sum_{k=0}^{n} \frac{1}{n-k+1} = \sum_{k=0}^{n} \frac{1}{k+1};\ \Sigma \frac{\sin(n-2k)\theta}{n-k+1} = -\Sigma \frac{\sin(n-2k)\theta}{k+1}$$

$$\therefore \frac{2(n+2)c_n}{x^{n+2}} = 2\sin(n+2)\theta \sum_{k=0}^{n} \frac{1}{k+1}$$

$$\Rightarrow c_n = \frac{x^{n+2}\sin(n+2)\theta}{n+2}\left(1+\frac{1}{2}+\ldots\ldots+\frac{1}{n+1}\right)$$

$$\therefore \sum_{n=0}^{\infty} c_n = \sum_{n=2}^{\infty} c_{n-2} = \sum_{n=2}^{\infty} \frac{x^n \sin n\theta}{n}\left(1+\frac{1}{2}+\ldots\ldots+\frac{1}{n+1}\right)$$

Hence $(\Sigma\, b_n) = \Sigma\, c_n$ proves the desired result.

Example 36:

Give an example to show that the Cauchy product of two convergent series may be divergent.

Solution:

Let $a_n = b_n = \dfrac{(-1)^n}{\sqrt{n+1}}$

By Leibnitz's Rule, $\Sigma\, a_n$ and $\Sigma\, b_n$ are convergent series.

$$\text{Now } |c_n| = \left|\sum_{k=0}^{n} a_k\, b_{n-k}\right| = \sum_{k=0}^{n} \frac{1}{\sqrt{k+1}\,\sqrt{n-k+1}}$$

For $0 \le k \le n$, $\sqrt{k+1} \le \sqrt{n+1}, \le \sqrt{n-k+1} \le \sqrt{n+1}$

$\Rightarrow \sqrt{k+1}\,\sqrt{n-k+1} \le \sqrt{n+1}\,\sqrt{n+1} = (n+1)$

$$\Rightarrow \quad |c_n| \ge \sum_{k=0}^{n} \frac{1}{(n+1)} \ge 1$$

$\Rightarrow \lim\limits_{n\to\infty} |c_n| \ne 0$ and so $\Sigma\, c_n$ is divergent.

Example 37:

Assuming that log $(1 + x) = x = x - \frac{1}{2}x^2 + \frac{1}{3} - \ldots$ for $|x|\ 1$, prove that $\dfrac{1}{2}[\log(1-x)]^2 = \displaystyle\sum_{n=2}^{\infty} (-1)^2 \left[1 + \frac{1}{2} + \frac{1}{3} + \ldots\ldots + \frac{1}{n-1}\right]\frac{x^n}{n}$.

Solution:

Do your self.

Example 38(a):

Show that

$$\sum_{n=0}^{\infty} (-1)^n \left[\frac{1}{(n+1)} + \frac{1}{n.2} + \ldots\ldots + \frac{1}{1.(n+1)}\right] (\log 2)^2 .$$

Solution:

We know $\sum a_n = -\frac{1}{2} + \frac{1}{3} - \frac{1}{4} + \frac{1}{5} - \ldots\ldots = \log 2$

Thus $(\log 2)^2 = (\Sigma\, a_n)\,(\Sigma\, a_n)$. ...(1)

For the Cauchy product series $\sum_{n=0}^{\infty} c_n$, $c_0 = 1$ and for $n \geq 1$.

$$c_n = (-1)^n \frac{1}{(n+1).1} + \frac{1}{n.2} + \ldots.. + \frac{1}{1.(n+1)} \qquad \ldots(2)$$

$$= \frac{(-1)^n}{n+2}\left[\left(1+\frac{1}{(n+1)}\right) + \left(\frac{1}{2}+\frac{1}{n}\right) + \ldots.. + \left(\frac{1}{(n+1)}+1\right)\right] \qquad \ldots(3)$$

The result will follow from Abel's Theorem, if we can show that Σc_n converges. From (3), we have

$$|c_n| = \frac{2}{n+2}\left(1+\frac{1}{2}+\ldots..+\frac{1}{n+1}\right) \to 0 \text{ as } n \to \infty.$$

Now

$$|c_{n+1}| - |c_n| = \left(\frac{2}{n+3}-\frac{2}{n+2}\right)\left(1+\frac{1}{2}+\ldots.+\frac{1}{n+1}\right)+\frac{2}{(n+2)\,(n+3)}$$

$$= \frac{-2}{(n+2)\,(n+3)}\left[\frac{1}{2}+\frac{1}{3}+\ \ldots\ldots + \frac{1}{n+1}\right] < 0$$

$\therefore\ |c_{n+1}| < |c_n|\ \forall\ n$ and $|c_n| \to 0$ as $n \to \infty$.

Thus $\Sigma\,|c_n|$ is convergent and so $\Sigma\, c_n$ is also convergent.

By Abel's theorem, $(\Sigma\, a_n)\,(\Sigma\, a_n) = \Sigma\, c_n$. Hence, by (1) and (2).

$$(\log 2)^2 = \sum_{n=0}^{\infty} (-1)^n \left[\frac{1}{(n+1).\,1} + \frac{1}{n.\,1} + \ldots. + \frac{1}{1.\,(n+1)}\right]$$

Example 38(b):

Give an example to show that the Cauchy product of two divergent series may be convergent.

Solution:

We'll show that if

$$\sum a_n = 1 - \left(-\frac{3}{2}\right)^2 - \left(\frac{3}{2}\right)^3 - \ldots\ldots$$

and $$\sum b_n = 1 + \left[2 + \left(\frac{1}{2}\right)^2\right] + \frac{3}{2}\left[2^2 + \left(\frac{1}{2}\right)^3\right] + \ldots.$$

then the series $\Sigma\, a_n$ and $\Sigma\, b_n$ both diverge but the Cauchy product $\Sigma\, c_n$ is convergent.

Since $\Sigma\, a_n$ is a geometric series having the common ratio $3/2 > 1$, it is divergent. In $\Sigma\, b_n$ each term is greater than a fixed number, say 1, and hence its sum tends to infinity as $n \to \infty$. Therefore Σb_n is divergent. We have

$$a_0 = 1, a_n = -(3/2)^n \ ; \ b_0 = 1, b_n = (3/2)^{n-1}[2^n + \left(\frac{1}{2}\right)^{n+1}], \ n = 1,2,......$$

$$c_n = a_0 b_n + a_n b_{0+} \sum_{k=1}^{n-1} a_{n-k} \ b_k$$

$$= r^{n-1}\left[2^n + \left(\frac{1}{2}\right)^{n+1}\right] - r^n - r^{n-1}\sum_{k=1}^{n-1}\left[2^k + \left(\frac{1}{2}\right)^{k+1}\right], \ \left(r = \frac{3}{2}\right)$$

$$= r^{n-1}\left[2^n + \left(\frac{1}{2}\right)^{n+1}\right] - r^2 - r^{n-1}\left\{2\left(\frac{2^{n-1}-1}{2-1}\right) + \frac{1}{4}\left(\frac{1-(1/2)^{n-1}}{1-1/2}\right)\right\}$$

$$= 2^n r^{n-1} + r^{n-1}\left(\frac{1}{2}\right)^{n+1} - r^n - r^{n-1}\left\{2^n - \left(2 - \frac{1}{2}\right) - \frac{1}{2^n}\right\}$$

$$= (r^{n-1}/2^{n+1}) + (r^{n-1}/2^n) \ = \ (3/2)\ (r^{n-1}/2^n) = (r/2)^n = (3/4)^n.$$

Hence, $\Sigma C_n = \Sigma (3/4)^n$ being a geometric series with common ratio $3/4 < 1$ is convergent.

Example 38(c):

Prove that for certain values of x and θ which are to be stated, that

$$2\sum_{n=0}^{\infty} x^n \cos n\theta \sum_{n=0}^{\infty} x^n \sin n\theta \sum_{n=0}^{\infty}(n+1)x^n \sin n\theta.$$

Solution:

Let $a_n = x^n \cos n\,\theta$, $b_n = x^n \sin n\,\theta$. Then

$| a_n | = | x |^n \, | \cos n\,\theta | \le | x |^n \ ; \ | b_n | = | x |^n \, | \sin n\,\theta | \le | x |^n \ \forall\ \theta.$

Thus the series $\Sigma | a_n |$ and $\Sigma |b_n|$ converge if $| x | < 1$. Now

$$C_n = \sum_{k=0}^{n} a_{n-k}\ b_n = \sum_{k=0}^{n} x^{n-k} \ \cos (n-k) \,.\, x^k \ \sin k\theta.$$

$$= \frac{1}{2}\, x^n \sum_{k=0}^{n} 2\cos (n-k)\,\theta \sin k\theta$$

$$= \frac{1}{2}\, x^n \sum_{k=0}^{n} [\sin n\,\theta - \sin (n-2k)\,\theta]$$

Now, $\displaystyle\sum_{k=0}^{n} \sin n\,\theta = (n+1)\sin n\theta,$

and $\displaystyle\sum_{k=0}^{n} \sin (n-2k) = \sum_{r=n}^{-n} \sin r\theta = 0.$ (Take $n - 2k = r$)

Thus $C_n = \frac{1}{2} x^n (n + 1) \sin n\theta$ for $|x| < 1, \forall \theta$.

From $(\Sigma a_n)(\Sigma b_n) = (\Sigma c_n)$, the result follows.

Example 39:

Show that for $-1 < x \leq 1$,

$$\frac{1}{2}\left(x - \frac{x^3}{3} + \frac{x^5}{5} - \frac{x^7}{7} + ...\right)^2 = \sum_{n=1}^{\infty}\left[1 + \frac{1}{3} + \frac{1}{5} + ... + \frac{1}{2n+1}\right]\frac{x^{2n+2}}{2n+n}$$

Solution:

Let $a_n = b_n = \dfrac{(-1)^n x^{2n+1}}{2n+1}$.

For $|x| < 1$, the series Σa_n and Σb_n are absolutely convergent. Hence, by Cauchy Theorem, the series $\Sigma c_n = (\Sigma a_n)(\Sigma(b_n)$ is absolutely convergent. Now,

$$c_n = \sum_{k=0}^{n} a_k b_{n-k} = \sum_{k=0}^{n}(-1)^k \frac{x^{2k+1}}{2k+1}(-1)^{n-k}\frac{x^{2(n-k)+1}}{2(n-k)+1}$$

$$= (-1)^n x^{2n+2}\sum_{k=0}^{n}\frac{1}{(2k+1)(2n-1=2k)}$$

$$\therefore \quad c_n = (-1)^n x^{2n+2} a_n \qquad ...(1)$$

where $\qquad a_n = \displaystyle\sum_{k=0}^{n}\frac{1}{(2k+1)(2n+2k-1)}$

$$= \frac{1}{(2x-1)} + \frac{1}{3(2x-1)} + \frac{1}{5(2x-3)} + ... + \frac{1}{(2n+1)}$$

$$= \frac{1}{(2n-1)}\left[\left(1 + \frac{1}{2n+1}\right) + \left(\frac{1}{3} + \frac{1}{2n-1}\right)\left(\frac{1}{5} + \frac{1}{2n-3}\right) + ... + \left(\frac{1}{(2n+1)} + 1\right)\right]$$

$$= \frac{2}{2n+2}\left[1 + \frac{1}{3} + \frac{1}{5} .... + \frac{1}{2n+1}\right] \qquad ...(2)$$

From (1) and (2), we get

$$\Sigma c_n = \sum_{n=0}^{\infty} \frac{2(-1)^n x^{2n+2}}{(2n+2)}\left[1+\frac{1}{3}+\frac{1}{5}+....+\frac{1}{2n+1}\right]$$

Hence $(\Sigma a_n)(\Sigma b_n) = (\Sigma c_n)$ gives the required result.

At $x = 1$, Σc_n is convergent as $<\alpha_n>$ is monotonically decreasing and tends to zero as $n \to \infty$.

Example 40:

Show that for all x, y ∈ R

$$\left(\sum_{n=0}^{\infty} \frac{x^n}{n!}\right)\left(\sum_{n=0}^{\infty} \frac{y^n}{n!}\right) = \sum_{n=0}^{\infty} \frac{(x+y)^n}{n!}.$$

Solution:

Let $a_n = x^n/n!$ so that $a_{n+1} = x^{n+1}/(n+1)!$ and

$$\lim_{n\to\infty}\left|\frac{a_n}{a_{n+1}}\right| = \lim_{n\to\infty} \frac{n+1}{|x|} = \infty \quad \text{for all } x \neq 0.$$

By ratio test, $\Sigma\, a_n$ converges absolutely for all $x \neq 0$.

But for $x = 0$, $\Sigma\, a_n$ is obviously convergent.

Thus $\Sigma a_n = \Sigma x^n/n!$ and $\Sigma b_n = \Sigma y^n/n\,!$ are absolutely convergent for all $x, y \in R$. By Cauchy's Theorem, the Cauchy product Σc_n of Σa_n and Σb_n also converges absolutely for all $x, y \in R$.

$$\text{Further } \left(\sum_{n=0}^{\infty} \frac{x^n}{n!}\right)\left(\sum_{n=0}^{\infty} \frac{y^n}{n!}\right) = \sum_{n=0}^{\infty} c_n, \qquad ...(1)$$

where $c_n = a_n b_0 + a_{n-1} b_1 + ... + a_1 b_{n-1} + a_0 b_n$

$$= \frac{x^n}{n!} + \frac{x^{n-1}\, y}{(n+1)!} + \frac{x^{n-2}}{(n-2)!}\cdot\frac{y^2}{2!} + ... + x.\frac{y^{n-1}}{(n-1)!} + \frac{y^n}{n!}$$

$$= \frac{1}{n!}\left[x^n + nc_1\, x^{n-1}\, y + nc_1\, x^{n-2}\, y^2 + ... + y^n\right] = \frac{(x+y)^n}{n!} \qquad ...(2)$$

Hence the result follows by (1) and (2).

Example 41:

Assuming $\tan^{-1} x = x - \frac{x^3}{3} + \frac{x^5}{5} - \frac{x^7}{7} + ...$, *prove that*

$$\frac{1}{2}(\tan^{-1} x)^2 = \sum_{n=0}^{\infty} \frac{(-1)^n\, x^{2n+2}}{(2n+2)}\left[1+\frac{1}{3}+\frac{1}{5}+\ldots+\frac{1}{2n+1}\right].$$

Also prove that the series converges absolutely for | x | < 1 and converges when x = 1.

Solution:

Similar to Example 39.

Example 42:

For | x | < 1, prove that

$$\left[x-\frac{x^2}{2}+\frac{x^3}{3}-\ldots\right]^2 = 2\sum_{n=0}^{\infty}(-1)^n\left[1+\frac{1}{2}+\frac{1}{3}+\ldots+\frac{1}{n-1}\right]\frac{x^n}{n}.$$

Solution:

Let $a_n = (-1)^n x^{n+1}/(n+1) = b_n$.

The series Σa_n, Σb_n are absolutely convergent for | x | < 1. Hence Cauchy product of Σa_n and Σb_n will converge absolutely to Σc_n, where

$$c_n = \sum_{k=0}^{n} a_k\, b_{n-k} = \sum_{k=0}^{n} \frac{(-1)^k\, x^{k+1}}{k+1}\cdot\frac{(-1)^{n-k}\, x^{n-k+1}}{n-k+1}$$

$$= (-1)^n\, x^{n+2}\sum_{k=0}^{n}\frac{1}{(k+1)\,(n-k-1)}$$

Replacing n + 2 by n, we obtain

$$c_{n-2} = (-1)^{n-2}\, x^n \sum_{k=0}^{n-2}\frac{1}{(k+1)\,(n-k-1)} = (-1)^n\, x^n\, a_n, \qquad \ldots(1)$$

$$a_n = \sum_{k=0}^{n-2}\frac{1}{(k+1)(n-k-1)} = \frac{1}{(n-1)}+\frac{1}{2.(n-2)}+\ldots+\frac{1}{(n-1)}$$

$$= \frac{1}{n}\left[\left(1+\frac{1}{n-1}\right)+\left(\frac{1}{2}+\frac{1}{n-2}\right)+\left(\frac{1}{3}+\frac{1}{n-3}\right)+\ldots+\left(\frac{1}{n-1}+1\right)\right]$$

$$= \frac{2}{n}\left[1+\frac{1}{2}+\frac{1}{3}+\ldots+\frac{1}{n-1}\right]. \qquad \ldots(2)$$

From (1) and (2), $c_{n-2} = 2\,(-1)^n\,\dfrac{x^n}{n}\left[1+\dfrac{1}{2}+\dfrac{1}{3}+\ldots+\dfrac{1}{n-1}\right].$

The Cauchy product of the series

$$\left(\sum_{n=0}^{\infty} a_n\right)\left(\sum_{n=0}^{\infty} b_n\right) = \sum_{n=2}^{\infty} c_{n-2} \text{ prove the required result.}$$

Example 43:

For | x | < 1, show that

$$\frac{1}{1-x}\log\left(\frac{1}{1-x}\right) = \sum_{n=1}^{\infty}\left(1+\frac{1}{2}+\frac{1}{3}+....+\frac{1}{n}\right)x^n.$$

Solution:

We know

$$(1-x)^{-1} = \sum_{k=0}^{\infty} x^n, -\log(1-x) = \sum_{n=0}^{\infty}\frac{x^{n+1}}{n+1}$$

Both these series are absolutely convergent for | x | < 1.

The Cauchy product is

$$c_n = \sum_{k=0}^{n} a_{n-k}b_k = \sum_{k=0}^{n} x^{n-k}\left(\frac{x^{k+1}}{k+1}\right) = x^{n+1}\sum_{k=0}^{n}\frac{1}{k+1}$$

Hence $(\Sigma a_n)(\Sigma b_n) = \Sigma c_n$ gives

$$(1-x)^{-1}\log(1-x)^{-1} = \sum_{n=0}^{\infty} x^{n+1}\left(1+\frac{1}{2}+....+\frac{1}{n+1}\right)$$

$$= \sum_{n=1}^{\infty} x^n\left(1+\frac{1}{2}+\frac{1}{3}+....+\frac{1}{n}\right)$$

Example 44:

Suppose $\sum_{n=1}^{\infty} a_n X^n$ *is absolutely convergent for | x | < 1, and its sum is A(x). Show that for | x | < 1,*

$$(1-x)^{-1}A(x) = \sum_{n=0}^{\infty} S_n X^n, \textit{ where } S_n = a_0 + a_1 + ... + a_n.$$

Solution:

The series $\sum_{n=0}^{\infty} x^n$ converges absolutely for | x | < 1 and its sum is $(1-x)^{-1}$. By Cauchy's theorem, the Cauchy product $\sum_{n=0}^{\infty} c_n$ of $\sum_{n=0}^{\infty} x^n$ and $\sum_{n=0}^{\infty} a_n X^n$ converges absolutely for | x | < 1 and

$$\sum_{n=0}^{\infty} c_n = (1-x)^{-1} A(x), |x| < 1. \quad ...(1)$$

We have $c_0 = a_0 = 1$. For $n \geq 1$, we have

$$c_n = 1 . a_n X^n + x(a_{n-1} X^{n-1}) + x^2 (a_{n-2} x^{n-2}) + ... + a_0 x^n$$

or $c_n = (a_n + a_{n-1} + ... + a_0) x^n = S_n x^n$. ...(2)

From (1) and (2), $(1-x)^{-1} A(x) = \sum_{n=0}^{\infty} S_n X^n$.

EXERCISES

1. Prove that the Cauchy product of the series $\sum_{n=2}^{\infty} \frac{(-1)^n}{\log n}$ with itself cannot converge.
2. Let $a_0 = 3$, $a_n = 3^n$ $(n \geq 1)$; $b_0 = -2$, $b_n = 2^n$ $(n \geq 1)$. Show that the Cauchy product of the series Σa_n and Σb_n is convergent although Σa_n and Σb_n are divergent.
3. State at least two conditions ensuring the convergence of the product.
4. If $S(x) = \sum_{n=0}^{\infty} (-1)^n \frac{x^{2n+1}}{(2n+1)!}$, $C(x) = \sum_{n=0}^{\infty} (-1)^n \frac{x^{2n}}{(2n)!}$

   then for all x, y ∈ R, show that

   (i) $S(x) C(y) + S(y) C(x) = S(x+y)$,

   (ii) $C(x) C(y) - S(x) S(y) = C(x+y)$,

   (iii) $2 S(x) C(x) = S(2x)$,

   (iv) $\{C(x)\}^2 - \{S(x)\}^2 = C(2x)$.
5. If $0 < b < a$, prove that

   $$\sum_{n=1}^{\infty} \frac{nb^n}{a^n - b^n} = = \sum_{m=1}^{\infty} \frac{a^m b^m}{(a^m - b^m)^2}$$
6. Examine for convergence

   $$\sum_{m=0}^{\infty} \sum_{n=0}^{\infty} \frac{(-1)^n}{(m+4n^2)(m+4n^2+1)}.$$
7. Prove that

$$\left(\sum_{n=0}^{\infty}\frac{x^n}{(n!)^2}\right)\left(\sum_{n=0}^{\infty}(-1)^n\frac{x^n}{(n!)^2}\right)=\sum_{n=0}^{\infty}\frac{(-1)^n\, x^{2n}}{(n!)^2\,(2n)!},\ x\in R.$$

8. Check the validity of $(\Sigma a_n)(\Sigma b_n) = \Sigma c_n$.

$c_n = \sum_{k=1}^{n} a_k\, b_{n-k}$, when

(i) $a_n = b_n = \dfrac{1}{n^2}$.

(ii) $a_n = b_n = \dfrac{(-1)}{\sqrt{n}}$,

(iii) $a_n = b_n = \dfrac{(-1)}{n}$.

9. For $|x| < 1$, show that

$$\frac{x}{1+x^2}-\frac{x^2}{1+x^4}+\frac{x^3}{1+x^5}-\ldots=\frac{x}{1+x}-\frac{x^3}{1+x^3}+\frac{x^5}{1+x^5}-\ldots$$

10. For $|x| < 1$, show that

$$\frac{x}{(1+x)^2}+\frac{x^2}{(1+x^2)^2}+\frac{x^3}{(1+x^3)^2}+\ldots=\frac{x}{1-x}-\frac{2x^2}{1-x^2}+\frac{3x^3}{1-x^3}-\ldots$$

11. Using double series, prove that for $|x| < 1$

$$\log(1+x) = x-\frac{1}{2}x^2+\frac{1}{3}x^2-\frac{1}{4}x^4+\ldots$$

and $\dfrac{1}{2}[\log(1+x)]^2 = \dfrac{1}{2}x^2-\dfrac{1}{3}x^3\left(1+\dfrac{1}{2}\right)+\dfrac{1}{4}x^4\left(1+\dfrac{1}{2}+\dfrac{1}{3}\right)-\ldots.$

Find the sum of the series when $x = \pi,\ 2\pi/3$.

12. Show that for all values of x in $[-\pi, \pi]$, when k is not an integer, cos

$$k\,x = \frac{\sin k\pi}{\pi}\left[\frac{1}{k}+\sum_{n=1}^{\infty}\frac{(-1)^n\,2k\cos nx}{k^2-n^2}\right]$$

Deduce that

$$\pi\cot k\pi = \frac{1}{k}+\sum_{n=1}^{\infty}\frac{2k}{k^2-n^2},$$

$$\frac{\pi}{\sin k\pi} = \sum_{n=0}^{\infty}(-1)^n\left(\frac{1}{n+k} + \frac{1}{n+1-k}\right)$$

13. Show that the Fourier Series

$$\frac{\pi^2}{3} = 4\sum_{n=1}^{\infty}(-1)^{n-1}\left(\frac{\sin nx}{2n} - \frac{\cos nx}{n^2}\right)$$

converges to the periodic function f in $]-\pi, \pi[$, where $f(x) = x^2 + x$, for $-\pi < x < \pi$, and $f(x) = \pi^2$, for $x = \pm\pi$.

What is the value of the series at

$x = \pm\pi, 0, 9\pi, 19\pi/2, 10\pi$?

deduce that

$$\frac{\pi^2}{6} = 1 + \frac{1}{2^2} + \frac{1}{3^2} + \ldots\ldots$$

14. Find the Fourier Series in $[0, \pi]$ for the function

$$f(x) = \begin{cases} \pi/3, & \text{for } 0 < x < \pi/3 \\ 0, & \text{for } \pi/3 < x < 2\pi/3 \\ -\pi/3, & \text{for } 2\pi/3 < x < \pi \end{cases}$$